T0206340

Bird Conservation and Agriculture

Populations of many species of farmland birds in Britain collapsed during the 20th century, creating one of the biggest conservation problems of the day and sparking a wave of research to find out why this happened, and devise practical solutions. This book summarises this story, exploring relationships between bird populations and agricultural land management. The first part of the book sets the historical context of change in agriculture and bird communities since the 18th century, and introduces the bird communities of agricultural land today. The second part provides an overview of this very active area of applied conservation science, including in-depth case studies of 16 species that have been the subject of detailed research effort and that, taken together, illustrate the many ways that agricultural intensification has affected bird populations. The last part shows how this evidence base, coupled with recent greening of agriculture policy, has provided opportunities to manage agricultural land to better integrate the needs of food production and bird conservation. The book concludes with a look forward to challenges that the conservation of bird populations on agricultural land is likely to face in the near future.

JEREMY D. WILSON has worked on the ecology of birds in agricultural systems since 1991, first at the British Trust for Ornithology and then at Oxford University. He joined RSPB in 1996 and is now Head of Research for RSPB Scotland. He is an Editor for the British Ornithologists' Union journal *Ibis*, and an Associate Editor for the British Ecological Society's *Journal of Applied Ecology*.

ANDREW D. EVANS joined the RSPB as a field biologist in 1988, studying the reasons behind the near extinction of the cirl bunting in the UK. This put him at the forefront of the unfolding story of bird population declines as a result of agricultural intensification. He is currently Head of Species Recovery at the RSPB.

PHILIP V. GRICE has worked professionally as an ornithologist in the government's nature conservation bodies since 1991, and is now a Senior Specialist in ornithology at Natural England. He has worked on a broad range of bird conservation issues but has become increasingly concerned with addressing the declines in farmland birds since the mid-1990s.

ECOLOGY, BIODIVERSITY AND CONSERVATION

The world's biological diversity faces unprecedented threats. The urgent challenge facing the concerned biologist is to understand ecological processes well enough to maintain their functioning in the face of the pressures resulting from human population growth. Those concerned with the conservation of biodiversity and with restoration also need to be acquainted with the political, social, historical, economic and legal frameworks within which ecological and conservation practice must be developed. The new *Ecology, Biodiversity, and Conservation* series will present balanced, comprehensive, up-to-date, and critical reviews of selected topics within the sciences of ecology and conservation biology, both botanical and zoological, and both 'pure' and 'applied'. It is aimed at advanced final-year undergraduates, graduate students, researchers, and university teachers, as well as ecologists and conservationists in industry, government and the voluntary sectors. The series encompasses a wide range of approaches and scales (spatial, temporal, and taxonomic), including quantitative, theoretical, population, community, ecosystem, landscape, historical, experimental, behavioral and evolutionary studies. The emphasis is on science related to the real world of plants and animals rather than on purely theoretical abstractions and mathematical models. Books in this series will, wherever possible, consider issues from a broad perspective. Some books will challenge existing paradigms and present new ecological concepts, empirical or theoretical models, and testable hypotheses. Other books will explore new approaches and present syntheses on topics of ecological importance.

The Ecology of Phytoplankton
C. S. Reynolds

Invertebrate Conservation and Agricultural Ecosystems
T. R. New

Risks and Decisions for Conservation and Environmental Management
Mark Burgman

Nonequilibrium Ecology
Klaus Rohde

Ecology of Populations
Esa Ranta, Veijo Kaitala and Per Lundberg

Ecology and Control of Introduced Plants
Judith H. Myers, Dawn R. Bazely

Systematic Conservation Planning
Christopher R. Margules, Sahotra Sarkar

Assessing the Conservation Value of Fresh Waters
Philip J. Boon, Catherine M. Pringle

Bird Conservation and Agriculture

JEREMY D. WILSON
RSPB

ANDREW D. EVANS
RSPB

PHILIP V. GRICE
Natural England

With line illustrations by

COLIN WILKINSON

CAMBRIDGE
UNIVERSITY PRESS

University Printing House, Cambridge CB2 8BS, United Kingdom

Cambridge University Press is part of the University of Cambridge.

It furthers the University's mission by disseminating knowledge in the pursuit of education, learning and research at the highest international levels of excellence.

www.cambridge.org
Information on this title: www.cambridge.org/9780521734721

First published 2009

A catalogue record for this publication is available from the British Library

Library of Congress Cataloguing in Publication data
Wilson, Jeremy D.
Bird conservation and agriculture : the bird life of farmland, grassland, and heathland / Jeremy D. Wilson, Andrew D. Evans, Philip V. Grice.
p. cm.
Includes bibliographical references and index.
ISBN 978-0-521-57181-4
1. Birds – Ecology – Great Britain. 2. Agricultural ecology – Great Britain. 3. Bird populations – Great Britain. I. Evans, Andrew D. II. Grice, P. V. (Phil V.) III. Title.
QL690.G7W5265 2009
598.175′50941 – dc22 2008053941

ISBN 978-0-521-57181-4 Hardback
ISBN 978-0-521-73472-1 Paperback

Contents

Contents

Preface

Recent estimates suggest that over a third of the world's land surface, over a third of terrestrial primary production and over a half of freshwater primary production are appropriated by human food chains (Gerard 1995; Tilman *et al.* 2001), a proportion that continues to increase as the global human population rises. Conversion of land to agriculture and the subsequent intensification of agricultural management are concerned with maximising the proportion of primary production that is channelled to human consumption, and to the extent that this is achieved, the rest of wild nature is bound to suffer (Krebs *et al.* 1999).

In many parts of the world, conservation is a backs-to-the-wall battle to protect remaining areas of pristine habitat and the richest of global biodiversity from conversion to agricultural use. Over much of Europe, and most of Britain, however, 6000 years of creation of cultural landscapes since the Neolithic means those pristine habitats have long been a thing of the past. As Bill Bryson (2000) wrote in reference to the English countryside, 'it is one of the busiest, most picked over, most meticulously groomed, most conspicuously used, most sumptuously and relentlessly improved landscapes on the planet'. It has been so for centuries and, as a consequence, the wildlife of Britain, including its bird populations, has long coexisted and coevolved with agriculture. The enduring fascination of its people with natural history, and indeed ornithology (the British Ornithologists' Union celebrated its 150th anniversary in 2008) has ensured that Britain's avifauna and its associations with agriculture are well documented back at least to the eighteenth century (Shrubb 2003). During the twentieth century the rise in support for wildlife conservation and for organisations able to coordinate long-term survey and recording effort, for example the British Trust for Ornithology, the Royal Society for the Protection of Birds and the Wildfowl and Wetlands Trust, means that knowledge of more recent change in bird populations is probably unrivalled anywhere in the world. This rich treasury of data was instrumental in enabling conservation scientists to detect declines in bird populations associated with rapid, recent agricultural intensification (Newton 1986; O'Connor & Shrubb 1986; Potts 1986). Two decades of intensive research effort have followed, directed towards understanding the mechanisms through which agricultural change has caused bird population declines and other biodiversity losses in agricultural systems, and what management solutions might be found. The gradual 'greening' of agricultural policy both in Britain and across Europe as the damaging environmental effects of intensive agriculture and overproduction were recognised, redirected subsidy support to management for environmental goods on farmland. This in turn allowed management measures designed to restore biodiversity, including bird populations, to be tested in agri-environment schemes, on set-aside land and through the growth of organic farming. As a consequence, our understanding of the ecology and conservation of bird populations in modern agricultural systems has increased immeasurably.

This history makes a renewed case study of the relationships between birds and agriculture in Britain timely. Our aim is in part to synthesise the very large and scattered literature that has proliferated since O'Connor & Shrubb's seminal *Farming and Birds* (1986) reviewed the subject just over 20 years ago. We focus on Britain, but draw extensively on the wider literature, especially from similar temperate agricultural systems and bird communities across Europe, to provide context. Such a synthesis, however, is merely a pause to draw breath. The relationship between birds, wider biodiversity and agriculture in Europe is at a crossroads. The last 20 years of research and development of agri-environment policy and practice have undoubtedly given us an increasingly effective tool kit to help integrate wildlife conservation and productive agriculture in our fine-grained agricultural landscapes. However, globalisation of world markets, continuing growth of human populations and economic aspirations, and the need to both mitigate and adapt to changing climate pose growing challenges. We face a future in which our tool kit will need to grow and change to help us manage agricultural landscapes for a wider range of so-called ecosystem services – for example carbon sequestration and storage, water quality, flood prevention, wildlife conservation and energy generation, as well as food production – and all in a changing climate. To illustrate just how fast things change, as we go to press the European Union announced that the set-aside rate for 2008 should be set to zero, a change that will result in the removal of a huge area of beneficial fallow from the agricultural landscape. At the same time rising commodity prices and renewed concerns for food security pose new challenges to environmental management in intensive agricultural systems.

If our work has value, it is perhaps mainly to the extent that it draws attention to the gaps in knowledge that we still need to fill to ensure that diverse and thriving bird populations are maintained as part of agricultural landscapes in meeting these challenges.

Acknowledgements

This work could not have been completed without the collaboration, inspiration and insight of friends and colleagues across many institutions and over many years, but especially those at the Royal Society for the Protection of Birds, Natural England, British Trust for Ornithology and University of Oxford. We are especially grateful to Richard Bradbury, Rob Fuller, James Pearce-Higgins, Michael Usher, Juliet Vickery and Ellen Wilson who have each read all or parts of earlier drafts. Their insightful comments and criticism have improved the final manuscript immeasurably, and errors and inelegancies that remain are of our own making.

Blackwell Publishing (Oxford, UK), Elsevier Science (Oxford, UK), A. & C. Black Publishers Ltd (London), the British Trust for Ornithology and the Royal Society of London all kindly gave permission for reproduction of technical diagrams from journals to which they hold copyright. RSPB, Natural England, Guy Anderson, Jamie Boyle and Peter Dennis also kindly allowed us to reproduce photographs from their collections. We owe a particular debt of gratitude to Colin Wilkinson who produced the line drawings that head each chapter. We also greatly appreciate the long-term patience of Cambridge University Press in publishing this work, and the enthusiastic support of Michael Usher and Dominic Lewis for its inclusion in the *Ecology, Biodiversity and Conservation* series.

JDW thanks Peter Lack, Chris Perrins and Lord Krebs for the initial inspiration to embark on this project, and both Chris Perrins and Tracey Sanderson for patience and encouragement as successive deadlines passed. Lynn Giddings and Ian Dawson of the RSPB library, and Alix Middleton of RSPB's Scotland Headquarters have, as always, been of huge help in securing references. My co-authors Andy Evans and Phil Grice have shown great commitment in seeing the work to completion, and the support and encouragement of my wife, Ellen, has been unstinting throughout.

The authors thank the RSPB and Natural England for the time afforded to them in preparation of this book. The views expressed are entirely our own.

Introduction

Bird conservation and agriculture

The scope of this volume is the bird life of all the 'farmed' lands of Britain. To many this will bring to mind a landscape dominated by a patchwork of enclosed pastures and cultivated fields, bounded by hedgerows, stone walls, ditches or other features, and with scattered trees and patches of woodland. Throughout this book, 'farmland' is our shorthand for such enclosed agricultural landscapes. Yet a variety of unenclosed, 'semi-natural' habitats such as moorland and lowland heath also owe their existence in large part to the grazing of livestock. Taken together, these farmlands and grazed, semi-natural grasslands and heathlands make up all our agricultural land. Within these habitats, bird communities have evolved in response to the limitations and opportunities of life in ever-changing landscapes shaped by the most fundamental of all human industries – the production of food.

Agricultural land covers over 70% of the land surface of Britain, and its effect is perhaps best appreciated from the air. Early in the gestation of this book, one of us flew from London to Inverness on a cloudless May morning, and the full variety of Britain's agricultural landscape unfolded below. Initially, large arable fields dominated the scene; mostly the 'nitrogen green' of winter cereal and silage grass, but with frequent yellow patches of flowering oilseed rape, earth-brown fields of germinating spring-sown cereals and sugar beet, scattered patches of woodland and clusters of farm buildings. Further north, and the mosaic of fields became more complex; smaller fields with more hedgerow trees and more irregular outlines; pastures were more frequent – difficult to distinguish from cereal, but in some cases obvious by their more complete stock-proof hedgerows. Over the Pennines and the southern uplands of Scotland, both arable land and hedgerows disappeared altogether in many places to be replaced by a pocket-handkerchief landscape of grassland fields in the valleys, giving way to unenclosed sheep-walk, and heather moorland on the high ground. On some moors, geometrical patterns in the upland vegetation marked 'muir-burns' where patches of heather have been burned to encourage regeneration for red grouse, and some large areas were covered in stands of dense, dark green conifer plantation. Over the central lowlands of Scotland, the lowland patchwork of grass and arable fields returned briefly – the brown spring-sown fields more in evidence here – before sheep-walk, muir-burned heather moorland and conifer plantations again marked the Scottish highlands. During the final descent, lowland agriculture reappeared, clinging to the coastal fringe around the Cromarty Firth; here there was a diverse mix of grass and arable fields, with turnips and freshly germinating spring barley much in evidence. Stone dykes and straggling lines of gorse bushes replaced hedgerows as field boundaries, and the few fields of oilseed rape had yet to flower. In short, from the air (or indeed from the computer screen via Google Earth), agricultural activity of one kind or another is seen to dominate the British countryside. Such is the diversity of the habitats – from arable fields and lowland meadows, to lowland heaths and moorland – that it is very difficult to present a list of 'birds of agricultural land'. Nonetheless, we have attempted to do so in Appendix 1, to provide the reader with a quick reference guide and to demonstrate the range of bird species that make use of agricultural habitats, and may therefore either benefit or suffer from changes in the way in which land is managed.

By virtue of its dominance, the habitats that agriculture provides (and removes) are of pivotal importance to birds. Some of the breeding and wintering populations are of significance in European and even global contexts. The 'dull' farmland that many birdwatchers pass quickly through on their way to watch birds at a local 'hotspot' is thus important for this reason alone. However, it is now all too evident that many of the most serious declines in wildlife populations in Britain, and more widely across Europe, have happened, and in many cases continue, in agricultural landscapes. In the lowlands of the south and east of Britain, intensively managed farmland is the sea in which are scattered the islands of scarce, semi-natural habitats that we strive to preserve. Amongst them, hay meadows, lowland heaths and chalk downland are themselves survivors of grazing systems now no longer economically viable. Yet it is in this 'sea' of enclosed farmland where current declines of birds and other wildlife are most evident. Elsewhere, this analogy is inappropriate. The long-established, low-intensity pastoral farming systems of the uplands and maritime north and west of Britain have moulded the landscapes and habitats of some of our remotest regions. As a result wildlife communities have arisen that are highly dependent upon the continuation of those systems. Yet wildlife losses have been evident here too. Declines of corncrakes, breeding waders and corn buntings in the crofting lands of the highlands and islands of Scotland are cases in point.

Our perspective is largely contemporary, but we recognise that today's agricultural land and its bird populations are only the latest stage in the long history of an ever-changing countryside. The production of food has always been essential for the survival and prosperity of individuals, communities and states, and so social structures at all these levels coupled with technological advances have all played their roles in determining agricultural activity and, through this, the appearance of our countryside. Our first aim, therefore, is to portray the bird communities which we find on farmed land today, and to place these in historical context. Chapter 1 is a brief overview of this history, concentrating on the dramatic changes to lowland agriculture wrought throughout the twentieth century, to help set the remainder of the book in context. With the historical context in place, we move on to introduce briefly the main agricultural habitats and their birds. Enclosed farmland is dealt with in Chapters 2 and 3, focusing on the open field and the field boundary respectively, whilst Chapter 4 considers the lower-intensity grazed systems of the uplands, downland and heaths.

Second, we seek to understand the patterns of distribution and abundance of the individual species that make up these communities, as a product of the effects of agricultural land-use and practices upon each species' essential resources – food and nest sites. In particular, we concentrate on the dramatic technologically and politically driven increases in productivity of agriculture since the Second World War, and explore how these changes have brought about some of the losses of birds, and other wildlife, that have made biodiversity conservation in agricultural systems a pressing priority. The impacts of agricultural change on populations of birds and other wildlife – for example see New's (2005) *Invertebrate Conservation in Agricultural Ecosystems* in this series – have been the subject of a wealth of detailed research in the last two decades. This has usually been motivated by conservation concern for declining species, but in a few cases by economic concerns over the impacts of a minority of species that have succeeded spectacularly on modern farmland. These relationships are explored in detail in Chapters 5–9. Chapter 5 reviews recent trends in the populations of birds associated with agricultural habitats. This review is only possible because, in Britain, we are fortunate in being able to draw upon the evidence of trends in numbers, range, breeding success and survival rates of many of our bird species. These derive from the surveys and monitoring schemes of the British Trust for Ornithology (BTO) and its skilled volunteers, and the Royal Society for the Protection of Birds (RSPB), both supported by the statutory conservation agencies. Chapter 6 focuses on studies of the strength of association between agricultural and bird population change, including consideration of two cases where recent agricultural intensification has, locally, been reversed. These are the growth of organic farming, and the introduction of set-aside of arable fields to curb over-production of cereal and protein crops. In this chapter we also review evidence of recent change in the fortunes of other wildlife on farmland in order to put bird population change in a broader context. Chapters 7–9 describe the more detailed research on distribution and demography (Chapter 7) and ecology (Chapter 8) of individual bird species, or agricultural processes (Chapter 9) that together have shed light on the mechanisms linking agricultural change to bird population change; the evidence of cause and effect. Chapter 8 focuses entirely on a series of case studies of individual species to illustrate the wealth of understanding that has emerged in recent years.

Thirdly, we review what some have called the 'agri-environment era'; that period since the late 1980s when increasing recognition of the environmental damage wrought by intensive agriculture, coupled with overproduction of foodstuffs in European markets, led to gradual redirection of agricultural subsidy from production to the delivery of environmental goods, including wildlife. Partly because of the huge research effort to understand the relationship between agricultural change and decline of bird populations, halting and reversing those declines has been a key aim of many agri-environment schemes and the land management options that they fund. Chapter 10 shows how this improved understanding of relationships between agricultural practice and bird populations, coupled with progressive 'greening' of the Common Agricultural Policy (CAP), and the development of agri-environment schemes has already helped to reverse some of the losses. This chapter concludes by looking to the future. What are the prospects for a more general reversal in the fortunes of populations of birds and other wildlife in habitats increasingly moulded by the effects of European and even global agricultural and wider environmental policies? Will the conservation of rural landscapes and wildlife continue to play an ever more central role in the formulation of agricultural policy, or will these policies revert to the economics of supply and demand as the pressures from global markets and human population growth increase? How can our increasing understanding of the ecology of birds in agricultural habitats help us to find practicable measures to create a countryside which reconciles the need both to produce adequate, high-quality food, profitably, and to maintain landscapes rich in wildlife? We also identify some of the remaining gaps in our knowledge of the ecology of individual bird species, effects of agricultural practices and ecological processes affecting bird populations on agricultural land. For example, how might inevitable global climate change over the next century determine the farming systems that can be practised and the types of crop grown in Britain, and hence the opportunities for wildlife? What might be the impacts of novel technologies such as genetic modification of crops? What might be the challenges and opportunities for bird conservation in managing agricultural land to minimise pollution, conserve soil carbon and water, or provide recreational access?

1 · *The history of agriculture and birds in Britain*

In this chapter we briefly reprise the long-term history of farmland in Britain up to the Second World War and then look in more detail at the sweeping changes since then. There are several excellent accounts of the history of farming in Britain (Symon 1959; Mellanby 1981; Briggs & Courtenay 1985; O'Connor & Shrubb 1986; Rackham 1986; Grigg 1989; Stoate 1995, 1996; Shrubb 2003) and much of the information presented here is summarised from these. In particular, Mike Shrubb's (2003) *Birds, Scythes and Combines* is a detailed and absorbing review of the changing relationships between birds and farming since the mid eighteenth century, written by a farmer with a lifelong interest in birds and bird conservation. Here we attempt no more than an overview of the most important changes from the perspective of bird populations.

Pre-eighteenth century

When Neolithic man colonised Britain around 3000 BC the lowland landscape was certainly more dominated by deciduous woodland than today. The conventional view is of grassland restricted to temporary open areas where forest cover had been destroyed (for example due to fire or tree-fall), though with considerable areas of bog and marshland, especially in floodplains and low-lying coastal areas. However, Vera (2000) has suggested an alternative vision of a shifting mosaic of woodland, scrub and grassland maintained by wild herbivores such as deer, moose, wild boar and the ancestors of now-domesticated cattle and horses. Whichever of these perspectives is closer to the truth, Neolithic settlers brought with them knowledge of arable agriculture and started to clear areas of woodland and scrub in order to grow cereals. Cattle and sheep were also grazed, particularly on the chalk and limestone areas such as Salisbury Plain, the South Downs and the Cotswolds. Successive invasions throughout the Bronze and Iron Ages brought increasingly sophisticated methods of cultivation and an increase in the population, so that by the time of the Roman occupation of AD 43 it numbered around 500 000. In this sense, agriculture can be seen as beginning to replace natural disturbances such as wild grazing, fire, wind and flood in creating open habitats and early successional conditions (Sutherland 2004).

It is interesting to note that even at this early stage agriculture was so successful that production more than met the needs of the local community and grain was exported to what is now France. Development continued slowly over the next millennium, although the introduction by the Saxons of the heavy plough, drawn by oxen was a notable technological advance. From the Norman invasion to the Black Death (1348) the population increased fairly rapidly, and with it the area of land under arable production to almost 4 million hectares. This approaches two-thirds of the total area of land in cultivation today, but between a third and a half of this was left fallow each year to help restore soil fertility and yields were only about a tenth of those achieved in modern times (Mellanby 1981; O'Connor & Shrubb 1986).

During this period open-field agriculture dominated the landscape in many areas. Villages had blocks of land for common grazing, for hay as winter feed and either two or three for growing crops: winter cereal, spring cereal and fallow in the case of the three-field system (O'Connor & Shrubb 1986). As the human population increased, so did pressure on the remaining woodland, and many new, hedged fields (assarts) were created (Muir & Muir 1987). Most of these were in private, rather than communal, ownership. Whilst widespread, the open-field system was by no means universal (Briggs & Courtenay 1985). In parts of the south-east and south-west, land had often been enclosed directly from woodland, and pastoralism dominated. On exposed hills, especially chalk downland, fields were cropped for a few years before being allowed to revert to grassland. In the uplands, on poor soils, and almost universally in Scotland, the infield–outfield system was adopted; the outfield, often higher on the hill, being given over to rough grazing, whilst the lower infield was cropped annually and fertilised. This was a key difference from the three-field system in that it involved no fallowing, and thus no opportunity to clean the land and prepare it fully for a return to cropping. As a consequence of the continuous cropping of the infield (often on a three-year rotation of bere – a variety of barley – followed by two successive annual crops of oats), land often became exhausted and overrun with weeds (Symon 1959).

The Black Death precipitated a relatively sudden change in the farmed landscape as the human population declined, perhaps by as much as a half. Villages became depopulated and large areas of land previously in arable production were abandoned and became overgrown with scrub. The removal of the population pressure meant that assarting ceased, but it also meant a shortage of labour and this, coupled with increasing wool prices, meant that many landowners found it more profitable to turn the entire estate over to sheep farming. This created a requirement for more hedgerow and resulted in evictions of villagers and often the annexing of land previously in common ownership (Muir & Muir 1987). By the early eighteenth century, it is therefore likely that the agricultural countryside was very diverse, with an intimate mixing of areas characterised by small, enclosed fields, hedgerows, patches of woodland and scattered farmsteads and hamlets – the Ancient Countryside of Rackham (1986) – with the open-field landscape in areas best suited to arable agriculture (Grigg 1989; Shrubb 2003). In Scotland, religious and political conflict had held agricultural progress back, and in Symon's (1959) view little had changed between mediaeval times and the end of the seventeenth century. As he put it, 'so long as Scottish farmers were content to work undrained, unenclosed, weed-ridden, over-acid and exhausted lands with inefficient tools . . . progress was out of the question'.

Throughout all these landscapes there remained large areas of what we would now term semi-natural habitat; marshes, fens, downs, heaths and rough grassland, collectively referred to as 'waste' (without the modern, pejorative connotations of that term), and all subject to extensive management under systems of common rights. Such land tended to remain wherever soils were too infertile or difficult to drain to permit cultivation and provided grazing land as well as a source of fuel (e.g. gorse, turf, wood, peat), bedding (e.g. bracken) and thatch (e.g. reed and sedges).

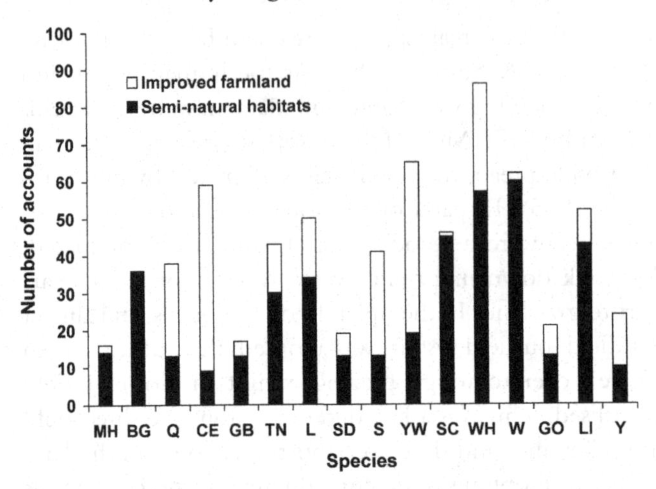

Fig. 1.1 Nesting habitat use of birds of agricultural landscapes in the eighteenth and nineteenth centuries, as summarised from accounts in avifaunas. Data from Shrubb (2003). MH, Montagu's harrier; BG, black grouse; Q, quail; CE, corncrake; GB, great bustard; TN, stone curlew; L, lapwing; SD, stock dove; S, skylark; YW, yellow wagtail; SC, stonechat; WH, whinchat; W, wheatear; GO, goldfinch; LI, linnet; Y, yellowhammer.

The bird communities of these landscapes are poorly known except from the often anecdotal accounts of early avifaunas. However, Mike Shrubb's (2003) analysis makes three important points. First, seed resources, and hence seed-eating birds, must have been very abundant indeed in the weedy arable and grassland systems where cultivation and fallowing were common but effective weed control absent. For example, in a letter to Pennant in 1768, Gilbert White in his *Natural History of Selborne* (1789) comments that 'towards Christmas vast flocks of chaffinches have appeared in the fields' and also that 'we have, in the winter, vast flocks of the common linnets – more, I think, than can be bred in any one district'. Second, the large areas of fallow present in the three-field system would have allowed excellent nesting opportunities for ground-nesting species such as lapwings, stone curlews, skylarks, corncrakes and quail. Third, the large areas of semi-natural habitat meant that species that are now either rare or simply no longer found in the lowland agricultural landscape were then widespread. These include hen and Montagu's harrier, stone curlew, quail, black grouse, great bustard, wheatear and whinchat, as well as still-familiar farmland species such as stock dove and linnet which were seemingly then more widespread in such habitats than on land brought under cultivation (Fig. 1.1).

The mid eighteenth to late nineteenth centuries: 'high farming' and the effects of enclosure

Today it is recognised that 'technology transfer' is a vital component of scientific innovation; that is finding practical application for the technology and disseminating that information to its users. Throughout the eighteenth century the rate of agricultural development was undoubtedly limited by the speed by which technology transfer could occur; farmers learned mainly by word of mouth and by the examples of their neighbours. This changed in 1813 with the establishment of the Royal Agricultural Society of England, which served to promote both research and development and the implementation of technological advance and, the publication of the first county-by-county surveys of agriculture across Britain between 1793 and 1816. In Scotland, the first *Statistical Account of Scotland* published between 1791 and 1799 described the agricultural conditions of every Scottish parish, including accounts of livestock, cropping, cultivations and modes of improvement. These improvements in communication speeded the diffusion of knowledge of developments such as the mechanisation of seed drilling, harvesting and threshing, and advances in animal breeding. In Scotland, the Union of Parliaments in 1707 improved communication between Scotland and England and, gradually, Scottish agriculture began to make up lost ground. By the middle of the eighteenth century, consolidation of common, 'run-rig', arable ground, was proceeding apace. New crops such as turnips and potatoes, and fallow-based crop rotations were being introduced, liming and drainage were improving the agricultural value of land, and pastures were being enclosed to provide hay for winter keep (Symon 1959). Arguably the greatest impact of technology was the development of mechanical pumping systems which permitted the drainage of wetland and reclamation of fen and marsh. In the mid nineteenth century mass-produced clay tile drains for field drainage appeared, and by the end of the century approximately 5 million hectares of land had been drained. At the same time the infant science of agricultural chemistry began to develop (Briggs & Courtenay 1985).

These technological advances permitted significant agronomic developments, with the introduction of new crops such as turnips, potatoes, clover and ryegrass and new farming systems devised around them (Briggs & Courtenay 1985). Clover was particularly important because its nitrogen-fixing ability helped improve soil fertility which had often been depleted under the three-field system. Although not universally adopted, the model farming system that incorporated the new crops was the Norfolk four-course rotation. This comprised a year of a root crop, then spring cereal, a legume, and finally a second cereal course. In the first year, the root crop (typically mangolds, turnips or swedes) acted as a 'cleaning crop', allowing weeds to be hoed from between the widely spaced rows. Sheep, grown for the sale of wool and mutton and use of their manure on arable crops, were then 'folded' on to the full-grown crop. Spring-sown oats (for horse feed) or barley (for the brewing industry) were sown in the second year, and undersown by broadcasting a mix of legume (e.g. usually clover or sainfoin) and grass. This provided summer grazing or a hay crop in the third year. The rotation was completed with a further cereal crop, usually wheat. The key features of this rotation were the introduction of a nitrogen-fixing grass–clover ley to improve fertility, the ability to hoe arable weeds in the regularly spaced rows of the root crop, and the value of this root crop in allowing much larger number of sheep or cattle to be maintained on-farm and fattened over winter.

This rotation formed the basis of the 'high farming' system of the late eighteenth and nineteenth centuries that came to dominate much of lowland Britain. Its drivers were the Industrial Revolution coupled with a doubling of population over the course of a century, and the emergence of growing urban populations that established a need for a large-scale agricultural industry. For the first time, as Mike Shrubb (2003) puts it, 'the land was increasingly adapted to suit the farming system rather than the land use adapted to the nature of the land'. Initially, high farming developed only on the lighter soils, with clay soils generally avoided due to the continuing difficulties of adequate tillage and drainage of heavy soils for the growing of root crops. However, the advent of widescale field drainage using clay tiles from the middle of the nineteenth century onwards allowed a much wider spread of high farming rotations. Whilst the common rights of the old open-field systems limited such change, the huge economic impetus for adoption of high farming, driven on by growing population and technological advance, eventually led to the final enclosure by Parliamentary Act (the 'Enclosure Acts') of remaining areas of open-field agriculture in eastern England, as well as large areas of common 'waste'. Hedgerow creation accelerated with the enclosures, and it has been estimated that some 200 000 miles of hedgerow were planted between 1750 and 1850, increasing the total resource by about 50% (Muir & Muir 1987). In Scotland, the growing profitability of the wool industry, powered by the Industrial Revolution, was reinforced by the realisation that sheep could be held year-round on moorland to provide much improved profit from both wool and meat, compared to the cattle held by the crofting tenants. Sheep farming arrived in the Highlands in the 1760s and the large-scale Clearances of crofting communities followed, adding to an emigration already under way, driven by growing food scarcity and risk of famine (Symon 1959).

Shrubb's (2003) analysis suggests that the impacts of the arrival of high farming and the enclosure of former open-field in the existing farmed landscape may not have been huge and, for some birds, were probably beneficial. Certainly, the widespread planting of hedgerows would have allowed species more typical of woodland and scrub to colonise open farmland, just as we are familiar with today. Field drainage took place primarily on arable land rather than on grassland – in part because a high water table was then considered beneficial for the quality of cattle pasture and hay meadows – and tended in any case to be only partially successful and less than permanent. Populations of species such as sedge warbler, reed bunting and reed warbler may have increased on farmland in association with the habitat provided by arterial ditches. Corncrakes spread into southern and eastern England in the nineteenth century, probably associated with the grass–clover ley course of the rotation, which was often cut late in the summer for hay and so provided the ideal nesting habitat. The increase in fertilisation of grassland from farmyard manure as stocking rates increased, from the growing of leguminous crops in rotations, and from increased use of purchased fertilisers such as lime and bone dust to reduce acidity and correct phosphate deficiency all helped to reverse a long-term decline in fertility of grasslands experienced through the eighteenth and early nineteenth centuries. As Shrubb points out, this recovery in fertility may have increased populations of generalist soil invertebrates such as earthworms and craneflies. This might account for a widespread recovery in starling populations around that time, after they had declined and disappeared from much of the northern part of their range during the late eighteenth century (Alexander & Lack 1944). On the other side of the coin, enclosure, the loss of fallows from crop rotations and the more continuous control of weeds in many crops by hoeing during the spring and summer would all have been detrimental to ground-nesting species such as stone curlews, lapwings, quail, corncrakes, skylarks and yellow wagtails.

Much more damaging, however, was the enclosure and conversion to cultivated farmland of the common 'wastes'. Many of these downland, heathland and wetland habitats lay over light but infertile soils (e.g. peats, sands and chalks) which the rapidly mechanising techniques of high farming laid open to agricultural improvement. The loss of these habitats was on an enormous scale. Hoskins & Stamp (1963) estimated that over 850 000 hectares of 'wastes' were enclosed in England during the eighteenth and nineteenth centuries, with over 80% of this taking place between the 1760s and 1840s. Mike Shrubb (2003) provides some detailed and evocative accounts of the drastic changes to the character of the landscapes wrought by enclosure of these habitats, and catalogues the equally drastic effects that enclosure had on the populations of six ground-nesting species characteristic of open, semi-natural habitats: hen harrier, Montagu's harrier, black grouse, great bustard, stone curlew and woodlark. The scale of the losses of these species and their causes are summarised in Table 1.1.

Drainage and enclosure for agriculture of the extensive lowland wetlands of Britain took place over roughly the same timescale and had equally catastrophic effects on breeding bird populations. In an analysis of those areas of England and Wales with extensive arterial drainage channel systems, Marshall *et al.* (1978) estimated that around 800 000 hectares were drained between the mid eighteenth and mid nineteenth centuries. Bitterns, marsh harriers and spotted crakes were all widespread species which became extinct or virtually so as British breeding birds and species such as avocet, ruff, black-tailed godwit, black tern, Savi's warbler and bearded tit disappeared from the once extensive fenland habitats of eastern England (Shrubb 2003; Brown & Grice 2005).

In summary, the nineteenth century spread of high farming had large and permanent effects on the British landscape, but its main impact on our bird populations was the huge loss of semi-natural, grazed habitats (heaths, downs, moors and mosses) and wetland in the lowlands through conversion to agriculture. The future of very many of the species dependent on these habitats in Britain now depends on the management of the remaining fragments of these habitats or, in cases such as the hen harrier and black grouse, on the management of semi-natural, grazed landscapes in the uplands, to which the ranges of these birds have retreated.

Table 1.1 *Probable effects of enclosure and conversion to agriculture of 'wasteland' in the eighteenth and nineteenth centuries on six ground-nesting species*

Species	Original habitat	Current habitat	Original range	Current range	Cause of decline
Hen harrier	Lowland heaths, downland, fens, bogs, marshes and moorland	Moorland and young conifer plantations	Scot; most of Eng and Wales except extreme S	Upland Scot with small number in Wales and N Eng	Habitat loss and persecution
Montagu's harrier	As above plus crops and rough grassland	Cereal crops, rough grassland and young conifer plantations	S and E Eng	<10 pairs scattered over S Eng	Habitat loss and nest destruction by harvesting/weeding operations
Black grouse	Heathland, moorland and woodland edge	As previously but no longer in lowlands	Throughout Britain except E Midlands and E Anglia	Widespread but declining in Scot; very small populations in N Eng and Wales	Habitat loss, drainage and fragmentation of habitat remnants
Great bustard	Extensive downland and sheepwalk	Extinct	E and S Eng	Extinct	Habitat loss and fragmentation, nest destruction and hunting
Stone curlew	Dry stony, heathland, downland and fallows	As before, but also spring-sown arable crops and specially created nesting plots	E and S Eng	Relict populations in E Anglia and central, southern Eng	Habitat loss and fragmentation, loss of mixed farming, nest destruction in tilled fields and egg collection for food
Woodlark	Wooded parkland and meadows; heathland and downland edge	Heathland and forestry clearfells; locally arable fields	Most of Eng and S Wales	S and E England	Habitat loss

Source: Summarised from Shrubb (2003).

Recession: the late nineteenth century to the Second World War

By the middle of the nineteenth century, the scene was set for extremely rapid development and intensification of agricultural practice. Circumstances, however, contrived to send the industry in entirely the opposite direction, culminating in the Great Depression. Perhaps the most important single change was the repeal of the Corn Laws in the 1840s. Since 1660 the grain market had been protected against imports by a sliding scale of duty which increased to prohibitive levels as home prices fell (Tracy 1989). In 1815 the law was altered to prohibit imports until the price reached a predetermined (high) level. This meant that bread prices were held at artificially high levels until the point was reached at which imports became viable and then the price crashed. Such violent fluctuations in the price of a basic commodity made life very hard for the working classes. As a result the Anti-Corn Law League was formed in 1838. After a decade of effort all restrictions on imports of grain were removed; cereal farmers were then faced with competition from the world market. For some time this fundamental change to the markets had surprisingly little effect. Harvests were good in the late 1850s, imports were limited during the early 1860s by the American Civil War and prices generally remained buoyant. During this period, the last main enclosures of land for arable agriculture took place, and the rate of wetland drainage for agriculture reached its zenith. Whittlesea Mere, the last great fenland mere, was drained in 1851 (Stoate 1995). From the mid 1870s the situation changed rapidly as a series of bad growing seasons in 1875–79 and 1891–94 resulted in very poor harvests. Whereas in the past a poor harvest in a protected market resulted in high prices, now poor home supply was associated with increased imports that pushed prices ever lower. By 1890 the price of wheat was perhaps half of that in 1850. A full account of the prevalent conditions leading to the Great Depression, as it came to be known, can be found in Tracy (1989).

These changing economic circumstances had a profound effect on the agricultural landscape (and hence bird habitat) that was to last for 50 years. As the bottom fell out of the grain market, land was simply taken out of production. Despite a brief respite during the First World War, when wheat production was stimulated through temporary price support, the area of land in tillage fell by a third. The arable land that was taken out of production was often simply abandoned and allowed to revert to poor-quality rough grassland rather than being put down to properly established pasture. In England alone, over 2 million hectares of land was taken out of arable production by 1932, the nadir of the Great Depression. As a consequence farmers increasingly laid off labour. Livestock prices also fell, but not nearly as severely as those in the arable sector. Although sheep numbers declined, the national herd was maintained as dairying rose to prominence as an agricultural sector, supplying a perishable commodity which could not be imported and for whose supply there was no external competition. In many parts of the Scottish highlands farmers turned to vegetables and potatoes as cash crops to maintain farm incomes and sugar beet became established as a key commercial crop, especially in East Anglia. In the uplands, the decline in the profitability of sheep farming was exacerbated in some areas by overexploitation of grasslands at the expense of fertility, and in others by a rising demand for moors for sporting purposes, notably the stalking of red deer and the driving of grouse. In many areas of the Highlands a second round of clearances – this time of sheep – followed as landowners faced by declining rents for their sheep farms, and uncertainty in finding new tenants, let out their moors for the exclusive use of sport shooting (Symon 1959). Overall, the effect of the Depression on the farmed landscape, as well as on local economies and communities, was profound.

Despite the decline in agriculture during this period, effects on the bird communities dependent on its habitats and food sources were probably not marked. Most areas with arable farming retained some cropping as farms still tended to grow their own fodder crops, and weed levels in much of the remaining arable land probably increased as husbandry standards declined in association with the decline in farm labour. Nonetheless, declines and range retractions of both grey partridges and corn buntings in the early decades of the twentieth century may reflect the loss of arable land (Shrubb 1997, 2003). In contrast, the expansion of grassland habitats during the Depression was also associated with gradual deterioration of nineteenth-century tile drainage systems. The spread and increase of redshank, snipe and curlew into inland areas of lowland England during this period probably attests to this (Shrubb 2003). In parts of the uplands, red grouse populations would have benefitted from the targeting of moorland management for this species as quarry on the newly established sporting estates that had ousted sheep farming. Agricultural abandonment also had effects. The 'scrubbing up' of agriculturally marginal downland and heathland areas led to declines of species dependent on open, short-grazed habitats, such as stone curlews and wheatears, but may have benefitted scrub-associated birds such as whinchats, willow warblers and linnets.

The Depression reached crisis point in 1929 when, against the background of general economic recession, world grain prices slumped and Britain was flooded with imports; in 1931 the volume of food imports was 35% greater than normal. Emergency action was required and Government intervention materialised in the form of the 1932 Wheat Act. This once again protected home wheat producers from the vagaries of world markets by guaranteeing prices. This legislative measure had the desired effect and wheat production started to rise sharply. At the same time technological innovation in the form of petrol-driven tractors started to appear. Agriculture had begun to climb rapidly out of the Depression by 1939 when the political and economic upheaval of the Second World War arrived, and large areas of land were returned to cultivation to counter the effects of the German naval blockade on food imports. This conflict was set to have a profound effect on the speed of development of the agricultural industry. British agriculture was about to enter a period of rapid evolution that would change irrevocably the farmed environment, with consequent and often severe ramifications for bird populations.

The Second World War to the present day

It is clear from the account above that politics and policy have long played a major role in shaping British agriculture. Since 1939 agriculture as a science and an industry has advanced extremely rapidly, fuelled by the twin accelerants of policy solutions for political demands and rapid technological advance. These two are inextricably interwoven and, because the fortunes of farmland wildlife have been and remain largely dependent upon policy, it is appropriate to give a very brief outline of the issues involved. A detailed account of how agriculture has been shaped and driven by governments in Western Europe during the twentieth century can be found in Tracy (1989).

Policy changes

Historically, being an island has brought Britain military advantages and disadvantages; both were evident during the Second World War. In 1940, the English Channel saved the country from invasion by German forces; instead it found itself under siege. Not being self-sufficient in terms of food production, the nation had to rely on imports from North America. These imports were carried by merchant ships in convoys that became increasingly the target of attack by dedicated surface ship and submarine units. Merchant shipping losses were so great that national food shortage became a serious Government concern. This clear exposure of an Achilles heel in Britain's defences spurred the post-war Government to ensure that in future the country was self-sufficient in terms of food production. There is no doubt that agricultural intensification would have happened anyway, but there is also no doubt that the graphic illustration of the strategic importance of self-sufficiency accelerated the process.

The Government's commitment to 'a stable and efficient agricultural industry capable of producing such part of the nation's food and other agricultural produce as in the national interest it is desirable' is evident from the Agriculture Act of 1947 (Tracy 1989). Essentially this meant that the Government undertook to buy, at guaranteed prices, all the grain, potatoes and sugar beet produced in the UK. This policy stimulated increased production; by the mid 1950s the UK was producing more milk than required, occasionally more barley, and was self-sufficient in eggs. All this was against a background of relaxed import restrictions, lower prices and therefore increased payments (from the taxpayers' pocket) to bridge the gap between market and guaranteed prices. The Government's post-war promotion of farming also helped promote the image of farmers to the public; food production was clearly a vital function and farmers had an important role to play in the future of the country. Farmers' attitudes were to change too. The Weeds Act of 1949 was to engender a virtual obsession for weed control and a pride in the neatness and tidiness of holdings. Farmers began to realise that their profession really was an *industry* and many began to view their farms as a 'factory floor' devoted to maximising production.

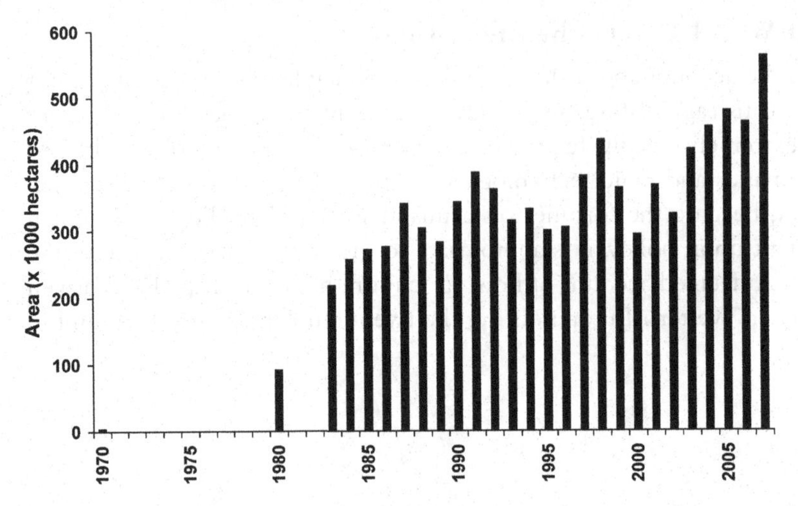

Fig. 1.2 Growth in oilseed rape area in England, 1970–2007. Data from www.defra.gov.uk.

The policy of price support survived the UK's accession to the European Economic Community in 1973 and the formation of the Common Agricultural Policy (CAP). Indeed price support was so successful that, when coupled with the enormous increases in yields brought about through new technologies, 'grain mountains' and 'milk lakes' became commonplace expressions to describe gross overproduction relative to the needs of European markets. The absurdity of this situation was brought into focus in the public eye in the 1980s when the corpulent excesses of production in the European Union (EU) were contrasted starkly with deeply disturbing images of people suffering mass starvation brought about by African famine. To try to stem overproduction in the arable sector, price support payments were switched from a tonnage to an area basis in the 1992 reform of the CAP. Moreover these Arable Area Payments were made conditional on farmers taking between 10% and 15% of their arable land out of production; so-called 'set-aside'. The media were quick to depict this policy as 'paying farmers to do nothing'. Differentials in levels of price support promoted the planting of different crop types. For instance oilseed rape attracted one of the highest area payments and thus the crop proliferated throughout much of the UK (Fig. 1.2) as a break crop in arable rotations. Linseed attracted such a high payment that it was commonly grown for the subsidy alone; the profit margin on the product was so low that it was often not worth harvesting. Public opinion was now very much less favourably disposed to the agriculture industry, and this attitude probably became more acute when the full extent of the damage caused to farmland wildlife by the process of intensification began to emerge in the early 1990s.

Technological advances

Policy can clearly have a profound effect on the agricultural landscape but when a particular policy has achieved its objective, it can always be changed or reversed.

Technological development has similarly had a huge influence on farming practice and hence bird habitats. Unlike policy, however, it is much more difficult to reverse or modify the effects of technological development. Often a single advance will confer huge competitive advantage and the genie cannot easily be put back in the bottle.

Most of the technological developments of the twentieth century can be divided into three categories: mechanisation, agrochemicals and plant and animal breeding. In the last decade or so, a further revolution in agriculture has been spawned with the advent of technology that enables scientists to modify the genotypes of organisms and thereby engineer agriculturally desirable changes with immediate effect. The possible environmental repercussions of the introduction of the first generation of genetically modified crops – modified to confer herbicide tolerance – are discussed in Chapter 9. Here we look back at the more conventional technologies that have been developed over the past 60 years.

Mechanisation

We have already mentioned the arrival, in the 1930s, of the petrol-driven tractor. At this time it was reckoned that to produce an acre of maize by hand took 300 man-hours; with machinery this was reduced to 3.6 hours (Briggs & Courtenay 1985). Thus tractors had two immediate effects. First, they greatly reduced the need for labour and so a sizeable workforce left the land. Second, they reduced the need for horses to work the land and consequently released that land reserved for growing oats as horse feed. Moreover, tractors aren't simply vehicles, they are mobile power sources so that many other labour-intensive tasks could be mechanised. From the point of view of birds, mechanisation had four major consequences:

(1) The replacement of the horse by the internal combustion engine removed the necessity to devote some land to cereal production on all farms (as feed for horses) and this began the marked phase of polarisation of lowland agricultural land use, with grassland-dominated landscapes to the north and west and arable-dominated landscapes to the south and east (Lack 1992). In areas already dominated by grassland farming, the loss of small-scale cereal production for stock feed represented the loss of the single major agricultural source of seed production (both grain and associated annual weeds) for birds.

(2) The speed and efficiency of individual operations markedly increased. More numerous field operations (e.g. multiple cultivations, rolling of spring crops and repeated cutting of silage fields rather than single cuts of hay meadows) reduce the chance that ground-nesting birds can breed successfully. It is not uncommon on intensive arable farms today to see a combine harvester completing the harvest of a wheat field in late summer whilst a plough begins cultivation at the other end of the same field. The period during which seed-rich stubble is present is reduced to zero. The advent of combine harvesting also removed the period during which harvested crops were stored in

stackyards and threshing yards, and so disappeared another rich seed source for birds.

(3) High-technology, powerful machinery is expensive to buy and demands large capital investment. Alternatively, machinery with skilled operatives can be contracted in. Whichever option is chosen, the machinery is also expensive to operate and the logistics of moving it around the farm or turning it within a field carry an associated cost. O'Connor & Shrubb (1986) estimated that doubling of field size from 6 to 12 ha would save 17% in working time. The net result of these changes in the farm economy has been a trend towards specialisation of individual holdings in either arable or livestock production; a switch made easier by the advent of agrochemical production. The economies of scale have also driven a rationalisation of the land use on individual farms. Fields have been enlarged with the consequent (and until relatively recently, grant-aided) destruction of field boundaries. Further, there has been a marked increase in the size of individual holdings as large farms have proportionately lower capital requirements and running costs. Overall there has been a simplification of the farmed environment and a loss of habitat diversity, both at the farm and landscape scale.

(4) Many non-farmed features have suffered, either through destruction under the economy of scale arguments above, or through neglect, partly because of the loss of labour from the land and partly because the function of those features has been lost as a result of specialisation. We have already mentioned the rate of hedgerow removal but many more kilometres of hedgerow have been lost as they have fallen into disrepair. Traditional methods of managing hedgerows such as coppicing and laying were extremely labour intensive and have largely given way to annual trimming with tractor-mounted flails. With the loss of livestock from the east of the country, hedges were no longer needed for their stock-proofing function and were allowed to grow 'leggy' and 'gappy'. Ponds originally established on many lowland farms as a reliable water source for livestock were typically neglected when livestock enterprises disappeared, and many were drained or became silted or choked with scrub (Rackham 1986).

Agrochemicals

Two of the keys to maximising output from arable agriculture are increasing nutrient supply directly to the crop, and efficient control of other organisms (pests and diseases) that either compete with the crop for nutrients or impede harvesting (weeds), cause physical damage to the plant through herbivory, or spread disease (fungi, invertebrates, birds and mammals). The effects of use of these products on the capacity of birds to survive, feed and reproduce are considered in detail in later chapters. Here we briefly review the rapid development in their use.

(i) Fertilisers We have seen how pre-twentieth-century farming systems were constrained by the need to maintain soil fertility. Manuring by livestock, crop rotations

involving a clover ley for its nitrogen-fixing ability, or application of natural products such as crushed bone or lime were employed. The beginning of the twentieth century saw the advent of synthetic inorganic fertilisers based on nitrogen, potassium and phosphate compounds. These freed arable production from the constraints of using a legume crop to enhance fertility. In many arable areas in the east, livestock had been kept purely as a source of organic nitrogen in the form of farmyard manure. When the need for this disappeared so did the stock. In livestock areas, inorganic fertilisers were so efficient at promoting grass yields that they quickly became universal. Improved yields permitted higher stocking rates, and promoted grass growth to such an extent that hay-making has become relatively rare, the crop instead being cut earlier and more frequently for the production of silage.

(ii) Herbicides The first herbicides to be used on a wide scale were the dinitro compounds (such as DNOC) introduced in the 1930s. These were contact herbicides used to kill broadleaved weeds in cereal crops but were highly toxic to animals (including humans). These were largely replaced from the mid 1940s by the phenoxyacetic acids, such as the widely used MCPA. These were much safer for non-target animals but still highly effective in controlling broadleaved weeds in cereal crops. The widespread use of these products had two effects; a substantial reduction in the abundance and distribution of some broadleaved plants such as charlock, poppy, corn buttercup and corncockle, but an increase in the incidence of some grass weeds, especially wild oats and black grass. Towards the end of the 1950s the bipyridylium herbicides, including paraquat, appeared. Paraquat is a highly effective contact herbicide killing most exposed green vegetation. It is also potentially toxic to vertebrates. From the 1960s the number of products available proliferated, many of which were highly specific in terms of their target species. With the swing towards autumn-sown cereals in the 1970s came the development of pre-emergent herbicides capable of controlling grass weeds. In the 1980s, the relatively benign glyphosate (marketed under the brand name 'Roundup') became available. This is a broad-spectrum systemic product that is highly effective on most higher plants and is widely used for post-harvest weed control and to kill off vegetation cover on set-aside before returning the land to production.

(iii) Insecticides Insecticides are applied to crops, either as seed dressings or sprays. The first products to be widely used were the organochlorines and organophosphates developed during the Second World War. The organochlorines such as DDT, dieldrin and aldrin were highly effective, cheap and relatively safe for operators and thus became highly popular. However, they were destined to be remembered for their highly detrimental effect on wildlife, especially seed-eating birds and raptors (Carson 1963). The first organophosphates to be introduced were highly toxic to humans. These were replaced fairly rapidly with more benign products including several systemic products (these are absorbed by the plants then ingested by insects feeding upon them). These carbamates are similar in action to organophosphates.

Perhaps the biggest recent change in insecticide use has been the advent of summer spraying of cereal crops with aphicides in the mid 1970s.

(iv) Fungicides and Molluscicides Cereals and both root and fruit crops all suffer from serious diseases caused by parasitic fungi; indeed fungal infection is far more important than animal pests. Infection by take-all can result in total loss of yield. It is not surprising, therefore, that a wide range of fungicides have evolved to help farmers protect their crops from fungal attack. A number of fungal diseases attack cereal seed and mercury-based seed dressings have helped control these. Mildews and rusts are controlled with systemic fungicides. Broadleaved crops can be vulnerable to damage by slugs, and hence molluscicides have become widely used.

The manner in which weeds or pests affect farm profitability is not necessarily linear. There may be a threshold below which infestation has little effect on yield and profitability is not badly affected, but above which the entire crop is worthless. This makes any cost–benefit analysis of pest control difficult because risk has to be taken into account. Certainly the costs of certain infestations are very high and against a policy context of output maximisation, this encouraged farmers in the 1970s and 1980s towards prophylactic (insurance) spraying regimes rather than treating crops as particular thresholds are approached. More recently, the 1992 switch in price support mechanisms from tonnage to area payments and a fall in grain prices encouraged farmers to look closely at their input costs and adopt more prescriptive and targetted regimes of pesticide application.

Plant and animal breeding

The past 60 years has seen dramatic advances in plant breeding with the aim of improving yields and developing specific phenotypic traits. Modern varieties of barley have higher grain yields, shorter straw (to reduce the risk of lodging) and a higher survival rate of tillers (Briggs & Courtenay 1985). It has been estimated that between 1943 and 1963 improved varieties increased wheat yield by 35% and between 1953 and 1980 spring barley yields increased by 0.84% per year. Similarly, new, hardy varieties of wheat were developed that allowed the crop to be sown and germinate in autumn or early winter. Coupled with the use of seed dressings and pre-emergent herbicides, this precipitated a huge swing in the timing of agricultural operations (Fig. 1.3).

One of the most serious consequences for birds has been the loss of over-winter stubble fields, a crucial foraging habitat for many seed-eating birds. Not every spring-sown cereal was preceded by an over-winter stubble field; often the ground was tilled and prepared the previous autumn, particularly on heavier land where it was uncertain that machinery would be able to operate in the damp conditions prevalent in spring. It is clear, however, that when a winter crop is sown in autumn the option for leaving the preceding stubble for all or part of the winter is removed (Evans *et al.* 2004). Even where stubbles were ploughed early in the winter, the switch to autumn sowing deprived a number of ground-nesting species, including

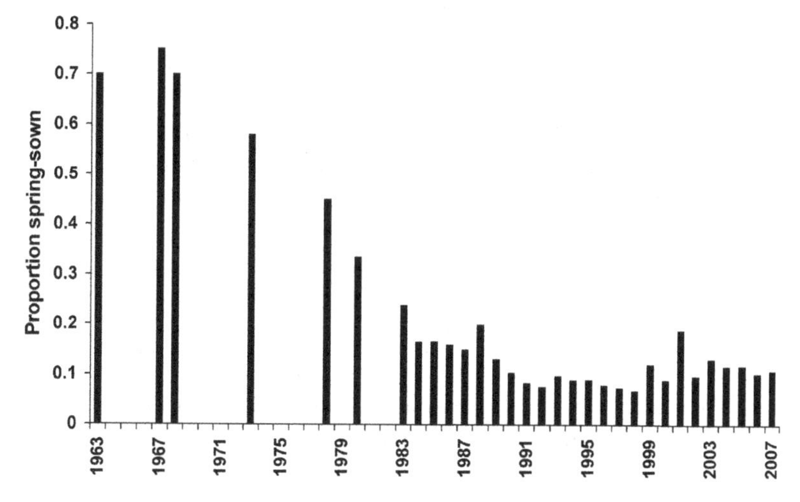

Fig. 1.3 Proportion of total cereal area sown in spring. Data for 1983–2007 from www.defra.gov.uk for England. Data before 1983 estimated from Fuller (2000) for England and Wales.

lapwings and skylarks, of an important breeding habitat. This seemingly insignificant change in the timing of cereal cultivation therefore dealt farmland birds a double blow (see Chapter 9).

Similar advances have been made in varieties of grass since 1945, with both yields and digestibility being improved through breeding. New varieties of ryegrass have increased winter hardiness and are more resistant to trampling (e.g. Moore 1966; Hopkins 1999). These now form the basis of most improved pasture in Britain.

Animal breeding techniques have also improved (Briggs & Courtenay 1985). The Friesian became the most popular dairy breed in the 1950s and 1960s because it produces a high milk yield with low butterfat content. Friesians are often crossed with other breeds such as Aberdeen Angus and Herefords and since 1962, Charolais and Simmental, in order to produce beef from dairy herds. This process was been made possible by the introduction of artificial insemination in the late 1940s. Not all technological advances have proved to be without cost. The development of high protein feeds for cattle based on processed sheep carcasses led directly to the explosion in the incidence of bovine spongiform encephalitis (BSE) in cattle and ultimately to mortality from Creutzfeldt–Jakob disease of humans who consumed infected beef (Anderson *et al.* 1996; Ghani *et al.* 1998). The repercussions of this unfortunate saga were enormous; the bottom fell out of the beef industry in Britain, and given the incubation time of the disease in humans, it may be decades before the full cost to human health is known. Moreover, following outbreaks of the viral disease foot-and-mouth in 2001 and 2007 which resulted in culls of sheep, cattle and pigs, especially in northern England, Devon and south-west Scotland in 2001 (Ferguson *et al.* 2001), and in Surrey in 2007, further economic damage was caused to the livestock industry. The arrival of the highly pathogenic avian influenza strain

H5N1 in the UK poultry flock in February 2007 risks doing the same. Both disease outbreaks were fuelled by modern transportation practices in the livestock industry (e.g. Gauthier-Clerc *et al.* 2007). The knock-on effect that such damage to the industry might have on the environment through changing animal husbandry and patterns of grazing is not yet clear. Undoubtedly, these episodes have created a new atmosphere of concern and suspicion amongst consumers about the health implications of modern intensive livestock production systems.

Summary

Agriculture in Britain has always been in flux. In this chapter we have seen how the nature of farmed landscapes has changed continually, and occasionally rather dramatically and abruptly, under the conjoint influences of changing governments and their policies and technological advance.

Britain's agricultural landscapes are essentially artificial in nature, with their mix of habitats that are actively managed, primarily for food production. Despite this, most agricultural habitats have been in existence long enough to develop their own, specialised wildlife communities. For example, the arable weed flora has a history of 5000 years or more and is probably as old or older than the plant communities of habitats often regarded as semi-natural (Wilson 1992). However, the tremendous acceleration in both technological and public policy development experienced during the second half of the twentieth century brought fundamental changes to the food production industry based on the principle of maximising food production from an agricultural environment seen increasingly as a factory floor. From an industry perspective, this quest to increase yields and achieve self-sufficiency was incredibly successful; truly a green revolution in the agricultural sense. As we shall see these changes have been far too rapid and comprehensive for wildlife to adapt and have consequently caused widespread and severe damage to agricultural ecosystems.

Repairing this damage is seen as one of the biggest challenges facing both conservation organisations and the UK government today. Devil's advocates question the point of restoring 'artificial' habitats and accuse conservationists of adopting a 'time machine' mentality in attempting to recreate past landscapes. It is sometimes suggested that the logical, if spectacularly impractical, conclusion to this approach is the restoration of the wild deciduous woodland that covered much of the country following the last ice age. This is hardly a pragmatic view, however, and most conservationists who can remember corncrakes, red-backed shrikes or cirl buntings in lowland England, corn buntings or grey partridges in Wales or those currently watching the demise of lapwings or black grouse have a more modest aim. They simply wish to see the declines of these species and others halted and reversed. The challenge is to achieve this aim in ways that are compatible with the needs of twenty-first-century agriculture, and do not require a time machine approach to turn the clock back to a former age.

The remainder of this book is devoted to an account of the effects of this post-war agricultural revolution on the birds that inhabit farmed landscapes, to assessing whether an understanding of these effects is now allowing policy-makers, farmers and conservationists to reverse past losses, and to considering how the future of farming in Britain can develop to help farmers to deliver food production, wildlife conservation and other benefits to society on the same land at the same time.

Plate 1.1 Lowland heathland formed as a result of long-term livestock grazing on acidic, dry, nutrient-poor mineral soils. Its area was much reduced and fragmented by enclosure for cultivation in the eighteenth and nineteenth centuries, and more recently further losses have been caused by urbanisation and conifer afforestation. Characteristic breeding birds include nightjar, woodlark, tree pipit and Dartford warbler. © RSPB.

Plate 1.2 Male red-backed shrike feeding nestlings. Formerly a well-known breeding bird of lowland heathland and unimproved grasslands in England, this species declined to extinction as a breeding species in Britain during the second half of the twentieth century. © RSPB.

Plate 1.3 Drainage and enclosure of lowland wetlands were extensive during the eighteenth and nineteenth centuries. Here at Otmoor in Oxfordshire, the RSPB is now restoring these habitats. © RSPB.

Plate 1.4 A juvenile sparrowhawk. Early generations of insecticides proved highly toxic to this species and other raptors (e.g. peregrine falcon). Range and numbers declined dramatically during the 1950s and 1960s, and these species all but disappeared from many arable areas of Britain. However, a full recovery has since taken place following the banning of the insecticides concerned. © Jeremy Wilson.

Part I

The habitats and their birds

2 · *Fields*

In this chapter we briefly introduce enclosed fields, their management and their bird communities as they stand today. We have divided the discussion to consider first tilled fields growing annual crops and second, fields under either permanent or rotational grass cover and used as pasture for grazing stock or to harvest forage crops for over-winter feeding. Fields set aside from agricultural production under CAP initiatives to reduce overproduction are considered separately, and in detail, in Chapter 6.

Tillage

Introduction

Here we refer to tillage as farmland cultivated on an annual basis to grow annually harvested crops. Overall, tilled land can be regarded as the most intensively managed and regularly disturbed of all agricultural land. Most of it is ploughed at least once annually; to this may be added several minor cultivations and other treatments of the soil surface (e.g. rolling), and a regime of inorganic fertiliser and pesticide applications to improve crop establishment and growth, and control weeds and invertebrate pests.

Traditionally, tillage crops were grown in rotation with grass–clover leys in order to maintain soil fertility and help to control arable weeds and crop pests. However, the advent of agrochemical crop protection from the 1950s greatly reduced reliance on diverse rotations, and tillage crops can now be grown in limited, grass-free rotations or as continuous monocultures, as now happens in many intensive cereal growing areas in eastern England. The result has been the gradual loss of the characteristic 'mixed' farming landscapes formed from a patchwork of tilled fields, pastures and leys. Lowland farming practice has become polarised, with a predominance of tillage cropping, and loss of grassland in the south and east, and a dominance of pasture-based stock-rearing and loss of tillage in the west. As Fig. 2.1 shows, although all counties tended to show surges in arable production immediately after the two World Wars, western counties tended to return to predominantly pastoral agriculture with small proportions of arable cropping, whilst arable production continued to grow elsewhere, this causing a more marked polarisation between arable and grassland counties.

In Britain in 2006, tillage crops covered 23% of the total agricultural area and are the main land-use in the southern and eastern lowlands where climatic conditions and soil fertility provide the best conditions for their cultivation. Tilled farmland occupies more than two-thirds of the total area of crops and enclosed grassland throughout eastern England from Humberside south to Essex, with between one- and two-thirds tilled throughout most of the rest of southern and central England, and further north in the Lothian and Fife regions of Scotland. On all tilled farmland, cereals are the dominant crop sown, by area. Wheat (at 1.82 million hectares in

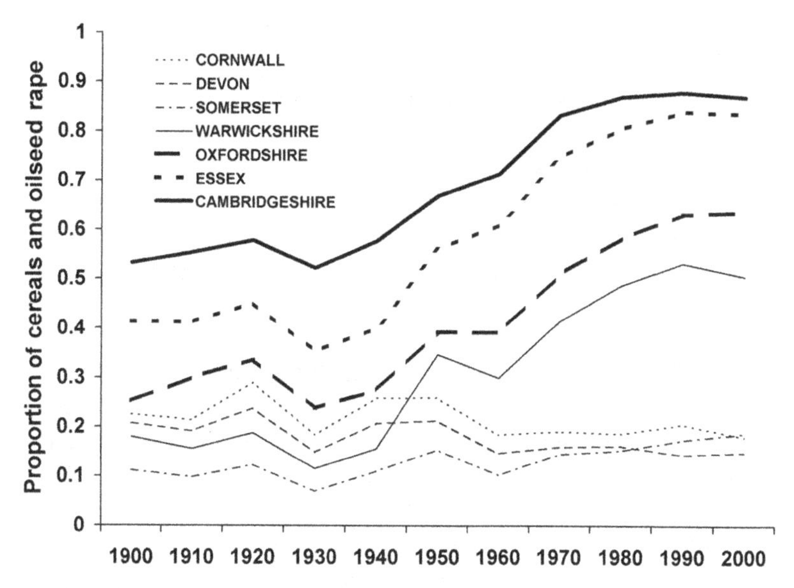

Fig. 2.1 Change in area of the main arable crops (cereals and oilseed rape) as a proportion of the total area of cereals, oilseed rape and enclosed grassland in three western, two midland, and two East Anglian counties, 1900–2000. Data from www.defra.gov.uk.

2007[1]) and barley (0.88 million hectares) account for almost 95% of the total cereal area. Both wheat and barley are grown for livestock feed, and barley is also grown to be malted by the brewing and distilling industries. Wheat, however, is the main cereal grown for human consumption. The recent change in the relative dominance of the two crops reflects technological advances. New 'short-strawed' wheat varieties coupled with the arrival of more efficient fungicides and grass herbicides allowed wheat to be sown in the autumn and its yields began to outstrip those of barley. Furthermore, the development of growth regulators reduced the risk of lodging (when the crop grows too tall and falls over in wet and windy weather) and thus allowed wheat to be sown on fertile soils under high-input regimes. Widespread as they are, wheat and barley do have their particular preferences for climatic conditions and soil types. Wheat has a strong rooting system and tends to grow well on heavier soils, but grows less well in areas of high rainfall. Barley, by contrast, can be grown under a wider range of rainfall conditions and prefers lighter, calcareous soils. Other cereals grown in much smaller quantities are oats (approximately 129 000 hectares in the UK in 2007), and rye and triticale (26 000 hectares in total). The area of oats sown has declined dramatically. In 1938, the total area in Britain was 853 000 hectares. By far the majority was grown to provide feed for cattle and horses and much of the decline can simply be attributed to the mechanisation of cultivation.

[1] All crop area figures from data available from the UK Department for Environment, Food and Rural Affairs (Defra) (www.defra.gov.uk).

Oats differ from wheat and barley in being more tolerant of acid soils, and favouring high rainfall and low sunshine levels. As a consequence, the main areas of oat cultivation have been in Scotland, Wales and north-western England. These are the same regions that have seen the greatest specialisation in pastoral farming systems, thus further contributing to the rate of decline of the oat crop.

A wide variety of other tillage crops are grown in Britain, although with one exception (the 11% of Norfolk sown to sugar beet in 2004), no single crop exceeds 10% of the total agricultural area of any county or region. In addition to Norfolk, beet grown for sugar extraction is mainly confined to Suffolk, Cambridgeshire, Lincolnshire, Nottinghamshire, the former Humberside district and Shropshire, with a total area of 125 000 hectares in England in 2007. Along with a wide variety of other broadleaved crops (brassicas, field beans, peas and maize), beet is also grown as a fodder crop for livestock. By area, field beans and maize dominate this category, and are grown primarily in lowland England, so that stock-feeding crops as a whole reach a maximum extent of 7% of the total agricultural area in Bedfordshire, Cambridgeshire and Dorset. Field beans are an increasingly popular stock-feed break crop in cereal rotations, grow well on heavy, fertile soils and occupied over 179 000 hectares of England in 2006, an increase of over 80% since 1970. Maize is a relatively new crop to Britain, being first recorded in the June agricultural census of 1970. By 2007, over 145 000 hectares were sown in the UK, primarily in southern and eastern counties of England. The vast majority is grown for stock-feed, as an alternative to silage, with the remainder threshed for human consumption. Small areas of other stock-feeding crops are, however, grown in all counties and regions, and the predominant crop varies geographically. For example, turnips, swedes and fodder rape once characterised Wales, the south-west peninsula, northern England and Scotland but have declined in importance as pasture-based farming has become dominant in these areas. Kale and cabbage are most frequently grown in south and south-west England but had also declined from an area of over 60 000 hectares in England and Wales in 1970, to marginally over 10 000 hectares by 1995.

Rape grown for oilseed occupied over 462 000 hectares of England in 2005. Virtually absent from the agricultural landscape except as a fodder crop in 1970 (only 4000 hectares were sown in that year for oilseed), the years since saw a dramatic growth in the popularity of this crop as a break in cereal-based rotations, largely due to subsidy support from the CAP. It is now grown widely throughout Britain, with over 10% of the cropped area sown to oilseed rape over a wide swathe of southern and eastern England in 2006 (Fig. 1.2).

Potatoes occupied around 139 000 hectares in 2007, a decline of approximately one-third since 1970. Production is now concentrated in a cluster of eastern English counties from Norfolk north to Humberside, in Shropshire and Cheshire and, in Scotland, in Fife. The main centre of production from Humberside south to the fens is explained by the fact that potatoes grow best in deep, well-drained and stone-free soils. The silt and peat soils of these areas provide ideal conditions.

Horticultural crops comprise a mixed bag of vegetables grown both in the open and in glasshouses for human consumption, orchards, fruit, nursery stock, bulbs and flowers, and hops for the brewing industry. The total UK area was 165 000 hectares in 2007, a decline of well over 50% from a post-war peak of over 360 000 hectares in 1948. Despite the decline in horticultural enterprises, traditional areas of production remain. Concentrations of commercial orchards are in Kent, Herefordshire and Worcestershire; vegetable growing is still centred in eastern England from Humberside south to Essex, and flowers and bulbs are still found primarily in Lincolnshire and on the Isles of Scilly.

Bird communities of tilled fields

Tilled fields might be expected to provide a relatively inhospitable habitat for birds, nowadays consisting of a more or less uniform expanse of a single plant species, and being routinely disturbed by a sequence of crop husbandry operations from initial tillage, sowing and rolling through a sequence of pesticide and fertiliser applications to eventual harvest. Accordingly, fewer than 20 bird species nest regularly in tillage crops in Britain, most of which are ground-nesters already adapted to natural mosaics of bare ground and short vegetation. The tiny British population of Montagu's harriers nests almost exclusively in fields of cereals or oilseed rape, primarily on low-lying coastal farmland around The Wash, but with a scattering of pairs elsewhere in southern England. Marsh harriers are also nesting increasingly in cereal crops, with over 20% of the English population spreading from reed-beds to nest in arable crops by the mid 1990s (Underhill-Day 1998). Quail, grey partridges and red-legged partridges commonly nest in cereal crops and agricultural grasslands, although grey partridges are more typically found nesting in densely vegetated field margins and hedgebanks than well out into the open field (Potts 1986). Stone curlews are traditionally associated with sparse, rabbit-grazed grass swards on stony, free-draining soils from the Breckland heaths to the southern English downlands. With the loss and fragmentation of these semi-natural grasslands as a result of reduced grazing pressure, afforestation and conversion to arable agriculture, an increasing proportion of the population has now taken to nesting at much lower densities in sparsely vegetated fields of spring-sown crops such as sugar beet, carrots and barley. Here, however, the breeding success of pairs is often put at risk by agricultural operations such as rolling and hoeing. Ringed plovers and oystercatchers are also typically associated with stony substrates such as shingle beaches and gravel pit margins. However, small numbers also breed on stony, sparsely vegetated heathlands and have also taken to nesting on spring tillage and fallowed arable fields in some areas, notably in East Anglia and the Trent Valley, with oystercatchers nesting commonly in cereals in lowland Scotland. Both this species and ringed plover nest at high density in spring cereal strips on the machair shell-sands of the Western Isles (Fuller *et al.* 1986; Jackson *et al.* 2004). Here, small colonies of Arctic and little terns can also be found nesting in cereals. Lapwings are much the commonest wader breeding in tillage crops, with large populations also nesting in meadows and pastures both in the lowlands and uplands. Lapwings nesting on arable land prefer sparsely vegetated fields as generally provided by spring-sown crops and fallow land. Curlews also nest in these habitats in scattered areas of Britain (e.g. the East Anglian Breckland).

Turning to songbirds, skylarks nest on the ground in a wide variety of arable crops and agricultural grasslands but, although they will tolerate taller, denser crop cover than will lapwings, nesting attempts are rarely initiated in crops taller than 50–60 cm, especially where ground cover by vegetation is complete (e.g. Wilson *et al.* 1997). Woodlarks nest similarly, though very locally, in cereal crops and set-aside in parts of their range in southern England (Wright *et al.* 2007). Yellow wagtails also nest commonly in arable crops (Mason & Lyczynski 1980). Early nests tend to be in cereals, and later nests in potatoes, beans, peas and other crops that provide sparse cover at ground level (Gilroy 2006). Given intensification of management of both grazed pastures and forage grass meadows, nests in arable crops may now be more secure than those in the damp grassland habitats with which yellow wagtails are more usually associated. Corn buntings are also ground-nesters, or almost so. Nests are usually either on the ground, or within 20 cm of it, in the tangled vegetation of cereal fields, hay and silage meadows, set-aside or rank grassland along field margins. In the isolated Scottish population on the machair of the Uists, Hartley & Shepherd (1994) found that nests in uncultivated land were more common than those in crops, with birds showing a particular partiality to sites at the base of hogweed plants and in clumps of marram grass. The remaining four species, reed warbler, sedge warbler, whitethroat and reed bunting, have been included simply because in recent years, they have all taken to nesting in fields of oilseed rape in the exceptionally dense crop canopy (e.g. Burton *et al.* 1999). The habit seems to have been taken up only patchily and many rape fields will not contain nests of any of these species. Nonetheless, the crop clearly does provide a dense enough canopy to attract species more usually associated with scrub or reed-beds. Other species such as greenfinch, linnet and yellowhammer have also occasionally been recorded nesting in rape fields.

Although, in total, we have listed 23 species above, the vast majority will be found nesting in arable fields only locally or occasionally or at the field edge. Over much of Britain, the only species that could be considered widespread nesters in open, arable crops are lapwing, skylark, yellow wagtail (patchily in England), reed bunting (in oilseed rape) and corn bunting (patchily).

A much wider variety of species will exploit tilled fields as a foraging habitat. These can be divided roughly into two groups; first, grazers of the green material of the crop and any associated weed species and, second, species that forage either on the ground or in the crop canopy for invertebrates and seeds. The first group consists mainly of swans, geese, dabbling ducks, gamebirds (especially grey and red-legged partridges and pheasant), moorhen, coot, pigeons (especially woodpigeon, stock dove and rock dove) and larks (especially skylark). The second group is much larger and overlaps partially with the above. It includes gamebirds, stone curlew, plovers (especially golden plover and lapwing), curlew, gulls, pigeons, larks, pipits (especially meadow pipit), wagtails (especially yellow wagtail), dunnock, thrushes, warblers (especially sedge warbler, reed warbler, whitethroat and lesser whitethroat feeding in oilseed rape), corvids, starling, sparrows, finches and buntings.

It is amongst these lists of species that most of the significant conflicts between the interests of birds and agriculture persist. For example, the UK has international responsibility for the conservation of large proportions of the global wintering populations of several species of geese. However, populations of some of these (especially pink-footed, greylag, brent and barnacle geese) winter in coastal areas where their grazing of crops and grassland can have impacts on the livelihoods of those farmers whose land they graze (e.g. Vickery & Gill 1999). Similarly, flocks of mute swans, woodpigeons, rooks and house sparrows can have economically significant local impacts on cereals and oilseed rape due either to grazing of the growing crop (Woodpigeons) or consumption of sown or ripening grain (e.g. Inglis *et al.* 1989). Skylarks have historically been the subject of studies designed to measure and ameliorate the economic effects of their grazing of sugar beet seedlings (Green 1980). On the other side of the coin, many of the species which nest or feed in tillage crops are declining in Britain due to a reduction in the availability of invertebrate or seed resources on tilled land during the post-Second World War period of agricultural intensification. These include grey partridge, house sparrow, tree sparrow, linnet, yellowhammer, cirl bunting, reed bunting and corn bunting.

Grassland

Introduction

Trees are the naturally dominant vegetation cover in most of the temperate regions of the world. They are replaced by grassland only in areas too dry, too windswept and cold, or too heavily grazed to allow tree cover to develop. In Britain, most grassland cover, other than in coastal habitats such as saltmarsh and sand dunes, results from human removal of tree cover, followed by grazing of domestic livestock. Where grassland cover has developed uninterrupted by cultivation it is termed 'semi-natural grassland' and is characterised by a stable community of herbs and grasses growing in a competitive equilibrium whose diversity depends mainly on soil fertility. The vast majority of such grasslands are found in the uncultivated uplands, above the enclosure line, but lowland calcareous grasslands (downland) and saltmarsh are also important examples subject to agricultural management. These semi-natural grasslands are considered in Chapter 4, whilst here we focus on the meadows and pastures of enclosed farmland.

Scattered patches of floristically diverse, semi-natural grassland survive on enclosed farmland where soils are poor or where agricultural improvement is otherwise difficult, as on steep slopes, or in flood-prone river valleys. These areas may be roughly classified as either meadows or pastures. Meadows have livestock excluded during the spring and early summer and are cut for hay, with stock returned later to graze the regrowth or 'aftermath'. In the lowlands, meadows are typically found along river valleys and marshlands, historically subject to annual flooding and fertilisation by flowing water. Pastures are used solely for grazing by domestic stock, and grazing may take place at any time during the growing season. Drainage and conversion to arable land, geographical polarisation of tillage and livestock farming, increased reliance on artificial fertilisers rather than organic manures and the replacement of horsepower by tractors have all contributed to the dramatic loss of unimproved hay meadows. Even where old meadows and pastures do survive, almost all have now been agriculturally 'improved' with herbicides and inorganic fertilisers or ploughed and reseeded in order to favour the few highly competitive, nitrogen-loving grasses and herbs such as ryegrasses and clovers that characterise agriculturally sown grasslands. Fewer than 8000 hectares of unimproved, uncultivated hay meadow and pasture now remain (Fuller 1987), yet in contrast to the long-term loss of downland (see Chapter 1), most of the ploughing and improvement of our meadows and pastures dates only from the Second World War.

The characteristic vegetation of meadows differs with location. In the Pennine Dales, meadowsweet, wood cranesbill, false oat-grass and sweet vernal-grass are typical, whereas meadow foxtail and great burnet are common in lowland flood meadows, and crested dog's-tail and greater knapweed in drier meadows. Pastures tend to be dominated by more grazing-tolerant or unpalatable species such as hoary plantain, wild thyme and mat-grass.

Ploughing and reseeding, herbicide, lime and fertiliser applications and drainage all contribute to what most agriculturalists would term the 'improvement' of grassland – the production of a species-poor sward dominated by a few nitrogen-responsive, highly competitive species. Regular inputs of organic manures or inorganic fertilisers promote production of a structurally dense, luxuriant grass crop either as winter fodder (silage) or to allow stock to be grazed directly at higher densities. In 2007, agricultural grasslands (excluding rough grazings) occupied 7.1 million hectares in the UK, a decline of 2.2 million hectares since the agricultural depression of the 1930s (Chapter 1). Most of that decline is accounted for by the resurgence in tillage during the Second World War. In Britain, grasslands (including both enclosed grasslands and upland sheepwalks) are concentrated in Wales, western and northern Scotland, northern England (especially west of the Pennine chain), and the south-west peninsula. Here, the topography, generally high rainfall and wetter, more acidic soils are less suited to growing tillage crops. In all these areas, permanent pastures, long-term leys and sheepwalk constitute the great majority of this grassland. In the uplands sheep are the dominant agricultural grazing animals, with beef and dairy cattle and smaller numbers of sheep in the lowlands. In south-west England a relatively high proportion of the total grassland, peaking at 20% in Cornwall, has been sown within the past five years. Historically, most young grasslands would have been grass–clover mixes sown in rotation with tillage crops to restore soil fertility and provide hay. Nowadays, the majority are simply recently reseeded long-term leys sown to maintain high levels of grass productivity for grazing or silage on intensive dairy and beef farms.

Adjoining these pasture-dominated areas, in eastern Scotland, the Pennine fringe and western England from Cheshire south to Dorset and the Isle of Wight, grassland occupies a lesser proportion of the agricultural land area, generally between one- and two-thirds of the total, but may still dominate locally on heavier, poorer soils and at higher altitudes. Other areas with a similar proportion of grassland occur in Buckinghamshire, Surrey and Sussex. Again, permanent pastures and old leys are predominant, but young, reseeded leys reach 20–30% of the total grassland area in the Grampian region of Scotland, the Welsh Marches, and in Wiltshire and Dorset.

In eastern England tillage crops are the main agricultural land-use and grasslands occupy less than one-third of the total agricultural land area, reaching a minimum of 9% in Cambridgeshire. Further north, from Cleveland north to Fife and the Buchan Plain, outposts of tillage-dominated agricultural land occur along the eastern coastal fringe of England and Scotland wherever light, free-draining soils and relatively low rainfall provide suitable conditions.

Agriculturally improved grassland swards are dominated by sown, nitrogen-responsive, competitive, productive species such as perennial and Italian ryegrass, meadow fescue and red and white clovers; all at the expense of low-productivity, naturally occurring species such as bents, red fescue and many broadleaved herbaceous species (Fuller 1987). Historically, cocksfoot and timothy were also preferred, but are less often sown today. Overall, botanical diversity is usually drastically reduced through the loss of broadleaved species. However, a minority of competitive species that respond well to high nutrient levels, and frequent cutting, grazing or trampling thrive. Examples include annual meadow grass, creeping and spear thistles, dandelion, broadleaved and curled docks and common chickweed.

Bird communities of enclosed agricultural grassland

Whether grassland is harvested as a forage crop of hay or is grazed, the structurally and floristically diverse vegetation cover of unimproved meadows and pastures may be used by a variety of ground-nesting species. Drier fields may be used by red-legged and grey partridge, quail, pheasant, oystercatcher, lapwing, skylark, meadow pipit, yellow wagtail, whinchat and corn bunting; wetter fields may be occupied by mallard, oyster-catcher, lapwing, snipe, curlew, redshank, moorhen, yellow wagtail, sedge warbler, grasshopper warbler and reed bunting. Shelduck may also nest on wet grasslands where these are coastal grazing marshes. By contrast, in the vast majority of intensive silage meadows, skylark may be the only species attempting to nest and, likewise, improved and heavily grazed pastures may support little except the occasional pair of lapwings, oystercatchers or skylarks.

Winter-flooded meadows and pastures and unimproved hay meadows also support three of our rarest breeding birds. On the winter-flooded grasslands adjacent to the Rivers Nene and Ouse in East Anglian fens, tiny populations of black-tailed godwit and ruff nest, and in late-cut hay meadows in western Scotland (mainly the Hebridean islands), corncrakes continue to breed. All three species were once much more widespread in Britain and suffered catastrophic population declines as a consequence of grassland improvement during since the twentieth century. Many of the species that remain more common show evidence of similar, if less severe, responses to the same changes in grassland management as we shall see in Chapters 8 and 9.

A much wider variety of additional bird species make use of meadows and pastures as foraging habitats. Swans (notably mute and whooper), geese (notably brent goose, barnacle goose, pink-footed goose and white-fronted goose), and ducks (notably mallard, wigeon and teal) graze meadows and pastures on their wintering grounds and can have locally significant impacts on grass yields (e.g. Vickery & Gill 1999). Other grazers include gamebirds (partridges, black grouse and pheasant), moorhen and coot (usually near watercourses), pigeons (mainly woodpigeon and stock dove) and skylark. Many more species visit meadows and pastures to feed on aerial, surface- or soil-dwelling invertebrates, and seeds, or to hunt small mammals in the sward. In addition to many of the species listed above, these include hobby, kestrel, golden plover, jack snipe, woodcock, gulls, collared dove, barn owl, little owl, green woodpecker, hirundines, pipits, wagtails, chats, thrushes, red-backed shrike, corvids, starling, sparrows, finches and buntings. Many of these species may be influenced by the same changes in grassland structure and management associated with agricultural improvement that have had such dramatic effects on the populations of birds nesting in grassland. For example, in Britain the chough nests mainly on sea cliffs on the coasts of Wales and the Inner Hebrides of Scotland. Historically, choughs were much more widespread around the coastline of Britain, and populations bred inland in some areas. However, a long-term decline throughout the twentieth century has left the current restricted breeding range and a total British breeding population of only around 430 pairs. At the same time, populations of other corvids (e.g. carrion crow, jackdaw and magpie) have increased dramatically (Gregory & Marchant 1996). Changing grassland management may have played a substantial part in these contrasting fortunes, and the relationships between agricultural grassland management and the birds which forage on those grasslands will be considered in detail in Chapter 9.

Plate 2.1 Winter cereals are harvested as early as July (barley) or August (wheat), and modern combine harvesting leaves little spilled grain for birds. In many areas the field may be ploughed and a new seedbed prepared within days, leaving little or no period of stubble as a food source for seed-eating species. © RSPB.

Plate 2.2 Use of herbicides and insecticides on arable crops reduces the diversity of the weed flora and associated invertebrate populations. The grey partridge was the first species whose decline was shown to be caused, in part, by this reduction in prey availability. Effects on the breeding success of other species, including turtle dove, yellowhammer and corn bunting, have now also been shown. © RSPB.

Plate 2.3 Agricultural improvement of grasslands has made species-rich hay meadows such as this one at West Sedgemoor in Somerset a rare sight. A single, late-summer cut of hay allowed these meadows to support a range of breeding birds, including corncrake, curlew, skylark, yellow wagtail, whinchat and corn bunting. Their modern replacement, the ryegrass silage meadow, cut two or three times each summer, hosts a much reduced species diversity and the only bird attempting to nest – often unsuccessfully – is likely to be the skylark. © RSPB.

Plate 2.4 Corn stooks on Barra, Western Isles, harvested using a reaper–binder (background). Such small-scale arable cultivation was once common in many grassland areas where crops were grown as stock feed, and allowed grain-eating bird species such as corn buntings to persist in grassland landscapes. Modern reliance on baled silage (foreground) as stock feed has seen cereal growing, and granivorous birds, disappear from many northern and western regions of Britain. © Jeremy Wilson.

3 · *The field boundary*

The patchwork quilt of fields that is so characteristic of British farmland is rarely uninterrupted by other features. Even in the open landscapes of downland or the East Anglian fens, fields are often enclosed by farm tracks and uncultivated grassy margins, ditches, or shelterbelts of deciduous or coniferous trees. Elsewhere, hedgerows, stone walls and dykes, fences and earth banks are all common field boundary features. Often, regional landscapes are characterised by their field boundaries. The drystone walls of the Cotswolds and Pennines, the beech hedges of Lothian, and the hedgebanks and sunken lanes of Devon are examples. The hedgerow is perhaps the best-known field boundary feature, usually consisting of a linear strip of woody shrubs, often punctuated with standard trees and associated with other boundary features such as an earthen bank, ditch or uncultivated grassy field margin. The range of woody species associated with hedgerows is considerable. Hawthorn, blackthorn, elder, hazel, roses, wych elm and field maple are all common in lowland England. Elsewhere, 'hedges' of gorse develop along earthen or stone banks in parts of Scotland, Wales and south-west England, whilst planted hedges consisting solely of beech are common in some areas of Scotland, the retained leaves serving as a windbreak through the winter months. Hedgerows vary enormously in their origin, age, botanical composition and current management. They may have arisen by planting, or by natural regeneration along banks or fence lines protected from grazing, or have been left as natural woodland vegetation at the margin of woods cleared for agricultural purposes. Though far from an infallible guide, the botanical composition of a hedge can reveal much about its age and origin (Pollard *et al.* 1974); older hedges generally contain a higher diversity of woody species, and those containing characteristically woodland species (e.g. small-leaved lime) may betray a woodland edge origin. In England, hedges form one of the most visible components of a fundamental difference in the history of the lowland, rural landscape – the distinction between what Rackham (1986) refers to as the 'Ancient' and 'Planned' Countrysides. The main historical and current differences between these are summarised in Table 3.1.

The Ancient Countryside is the product of a continuous influence of human inhabitants on the landscape for at least a thousand years. The hedgerows are botanically rich, sinuous and reflect a long history of use to enclose stock, provide fuel wood, and establish and mark boundaries. Lowland landscapes characteristic of the Ancient Countryside still remain in western England, from Somerset north to Lancashire, and south and east of a line from Suffolk to the New Forest. The Planned Countryside predominates between these two areas from Yorkshire, Lincolnshire and Norfolk, south-west through the Midlands to Wiltshire and Dorset. It reflects the influence of the eighteenth- and nineteenth-century Enclosure Acts (Chapter 1); the rapid imposition of enclosure upon a landscape of open-field, strip cultivation and common grazing systems, interspersed with woodland blocks, that had been established since mediaeval times. Enclosure was achieved mainly by hedge-planting, and most hedges in Planned Countryside are thus relatively young, straight and botanically simple, being based on the planting of one or few species (notably hawthorn). As Rackham (1986) noted, the distinction between these two landscapes

Table 3.1 *Key differences between the 'Ancient' and 'Planned' Countryside as summarised by Rackham (1986)*

Ancient Countryside	Planned Countryside
Open-field agriculture absent or limited in extent	Open-field agriculture well established until effects of Enclosure Acts
Many small woodlands; much heathland	Fewer, larger woods; heathland rarer
Many ponds	Few ponds
Hamlets and small towns	Villages
Hedges mainly species-rich and sinuous	Hedges straight, and mainly of hawthorn
Roads many; often sunken; many footpaths	Roads few, straight and on surface; few footpaths

often remains very sharp; for example, 'it bisects each of a dozen parishes on the Cambridgeshire–Suffolk border'.

As well as field boundaries, the agricultural landscape of Britain is punctuated by other patches of habitat that influence bird communities. Streams and rivers, ponds, woodlands, villages and their gardens and farm buildings all appear as islands in a sea of agriculturally managed land. Indeed the complexity and variety of these non-agricultural habitat patches probably account for a far greater proportion of the overall density and number of species of birds present in an agricultural landscape than the matrix of fields which lies between them. Yet, historically and ecologically these habitat patches are far from being islands; woodlands, for example, have long been managed as a source of wood fuel and building materials by agricultural communities, just as many today are managed to support populations of game. Equally, wildlife populations on farmland, including those of birds, are influenced by the presence and distribution of these habitat patches just as the wildlife of those habitats may in turn be influenced by the management of the agricultural 'sea' in which they sit. For example, in analyses of Common Birds Census (CBC) data, O'Connor & Shrubb (1986) and Whittingham *et al.* (2008) found that the habitat features most consistently correlated with the abundance of common breeding bird species on farmland were hedgerows, woodlands and woodland edge habitats, standard trees in field boundaries, and ponds. In other words, of the four most general correlates of bird abundance, none involved the nature of agricultural land-use in the landscape. On the other side of the coin, a recent analysis of bird species diversity in patches of deciduous woodland in East Anglia (Hinsley *et al.* 1995) found that both the size and isolation by distance of woodland patches from other woods were the strongest correlates of the probability of presence of woodland birds such as nuthatch and great spotted woodpecker. At first sight we might think that this result has rather little to do with agriculture, until we realise that it is the expansion of agricultural land-uses that is largely responsible for the size and distribution of remaining woodland fragments.

Hedgerows

Hedgerows are a well-studied wildlife habitat in Britain, not least because of the rapidity of their loss or neglect during the second half of the twentieth century as increasing mechanisation has favoured larger field units, and the need for hedgerows to provide stock-proofing in arable areas has declined. Field boundaries comprising a hedgerow with standard trees, ditch (wet or dry) and adjacent rough grass margin can support a diverse breeding bird assemblage. The hedgerow itself is effectively linear scrub and its breeding birds reflect this. Where the hedge is regularly but not severely trimmed, and not permitted to grow beyond 2–3 m tall, it retains a dense, thorny structure with a rich herbaceous ground layer. Periodic laying or coppicing of the hedge is particularly effective in helping to retain a stock-proof hedge structure. This management supports breeding species such as red-legged and grey partridges, pheasant, wren, dunnock, robin, blackbird, song thrush, lesser whitethroat, whitethroat, chaffinch, greenfinch, goldfinch, linnet, yellowhammer and, locally, cirl bunting. Cuckoos are also found in well-hedged farmland as brood parasites of dunnocks. Historically, red-backed shrikes also nested in farmland hedgerows in parts of southern England but have long since been lost. Where a hedgerow ditch is permanently wet, mallard, moorhen, grasshopper warbler, sedge warbler, reed warbler and reed bunting may also nest. A lack of trimming results in gradual succession of the hedgerow towards a linear wood, and the breeding bird assemblage changes. Species requiring a dense thorny structure or thick, herbaceous ground flora such as dunnock, lesser whitethroat, whitethroat, greenfinch, goldfinch, linnet and yellowhammer tend to be lost and to be replaced by woodland species such as woodpigeon, turtle dove, mistle thrush, blackcap, titmice, bullfinch and, occasionally, nightingale. Where neglect of hedgerow management is associated with a failure to prevent grazing stock gaining access to the hedge, then little may remain except a line of small blackthorn or hawthorn trees with virtually no ground flora. Relict hedges such as these support very few breeding birds; perhaps only chaffinches and the occasional pair of blue or great tits. Where standard trees occur in an otherwise well-managed hedgerow, other canopy or hole-nesting species more typical of mature woodland may occur, including kestrel, stock dove, little owl, barn owl, green, great spotted and lesser spotted woodpeckers, redstart, spotted flycatcher, treecreeper, jackdaw, starling and tree sparrow.

Additional species may visit hedgerows and grassy field margins as foraging habitats. These birds may be breeding birds from adjacent habitats, wintering species or passage migrants. Sparrowhawks nesting in woodland frequently hunt along hedges, flying low and flicking from one side of the hedge to the other in pursuit of songbirds. Magpies and carrion crows will search hedgerows for songbird eggs and young during the breeding season, whilst rank grass cover along field edges is an important habitat for barn owls hunting voles and mice. Barn swallows and other hirundines frequently hawk insects along hedge lines, finding the leeward side of hedges an especially valuable foraging habitat in poor weather conditions

(Grüebler *et al.* 2008), and in insect-poor arable farmland (Evans *et al.* 2003a). Fieldfares, redwings and, in some winters, waxwings flock in hedgerows to take fruit. Species from nearby habitats such as woodland and farmyards may also visit hedgerows on open farmland to feed. These include warblers, goldcrest, treecreeper, nuthatch, jay, tits and house sparrow, plus winter visitors such as redpoll, siskin and brambling.

With such a rich bird fauna, management of hedgerows has considerable potential to influence the abundance and species composition of the assemblage of birds found on lowland farmland. The effects of hedge removal, management, structure and botanical composition on the bird populations that exploit them for nesting or feeding will be discussed in detail in Chapter 9.

Woodland and gardens

The history of woodlands and forests and their associated bird communities are described in detail by Fuller (1995), so here we give only a brief outline of the bird community of farm woodlands and its influence on bird populations in agricultural landscapes.

Scattered patches of deciduous woodland are as much a feature of lowland agricultural landscapes as hedgerows, and have been so since the widespread clearance of the natural, post-glacial forest in Neolithic times. Until the early twentieth century woodland played a pivotal role in rural and industrial society. Woodlands were managed as a source of timber trees for large-scale building purposes, rods and poles for fencing, tool handles and wattle-work, and logs for fuel. Wood fuel was used for domestic purposes, but also in pottery kilns and for iron smelting. Before the Industrial Revolution, woodlands were therefore central to much industrial activity. Most British tree species (pines are an exception) regenerate well after cutting, and this allowed woodlands to be managed as self-renewing resources. Woods were bounded by ditches, banks and hedges to exclude grazing animals, and pollarding and coppicing were employed to harvest 'underwood' renewably on a regular basis whilst allowing a small proportion of trees to grow on to maturity as a source of timber. Although many woodlands are now managed for game or for conservation, aesthetic or amenity values, their former industrial importance has declined, traditional management practices have been abandoned, and timber production has become dominated by plantations, often of exotic conifers, and often planted at the expense of ancient woodland. Approximately half of England's ancient woodlands have been lost since the Second World War to agriculture, forestry, roads and housing. On the other side of the coin, large areas of secondary woodland have grown up on railway and motorway embankments, neglected heath and downland, old orchards and abandoned agricultural fields, and the advent of set-aside during the 1990s provided the long-term financial incentive for some farmers to plant new woodlands on less productive arable land. The result is that the woodlands breaking up agricultural landscapes today are a hugely diverse mix of ages, management practices (or lack of them) and tree species compositions.

Typically, patches of mature deciduous or mixed woodland on a lowland English farm might be expected to support a cross-section of the following breeding species: sparrowhawk, kestrel, pheasant, woodcock, stock dove, woodpigeon, cuckoo, little and tawny owls, green and great spotted woodpecker, wren, dunnock, robin, blackbird, song and mistle thrush, blackcap, garden warbler, chiffchaff, goldcrest, spotted flycatcher, long-tailed, marsh, willow, blue, great and coal tits, nuthatch, treecreeper, jay, magpie, jackdaw, carrion crow, starling, chaffinch and bullfinch. Redwings, fieldfares, siskins, redpolls and bramblings and, in some winters, waxwings visit woodlands in autumn and winter. The common practice of feeding of pheasants with grain on game-rearing estates may also attract into woodland species more usually associated with open farmland such as yellowhammer and reed bunting. Smaller numbers of woodland sites may hold a rookery or heronry, and species such as red kite, buzzard, hobby, lesser spotted woodpecker, redstart, wood warbler, firecrest, pied flycatcher, siskin, redpoll, hawfinch and crossbill occur more scarcely or in localised areas. Where felling and replanting or coppicing occur, or large numbers of mature trees fall naturally (as in the severe gales of October 1987), species more associated with scrub and thicket stages of woodlands and plantations may be found, including tree pipit, whitethroat, willow warbler, goldfinch, linnet, yellowhammer, reed bunting and, locally, nightjar, woodlark or nightingale. The precise mix of species found in any given wood depends of course on its size, geographical location, tree age and species composition, understorey structure and management, and degree of isolation from other woodlands. All these issues are dealt with in detail by Fuller (1995) and are beyond the scope of this volume. For our purposes it is largely sufficient simply to draw attention to the fact that the sheer size of the above species list means that presence of woodland patches can clearly have an enormous impact on the abundance and total number of bird species found within the agricultural landscape.

However, it is not simply a matter of pointing out that much of the bird species diversity and biomass in a farmland landscape is actually found in the woodland patches within it; birds in woodland and open farmland and the management of these interspersed habitats interact in ecologically significant ways. For example, some unenclosed woodlands, notably the New Forest and the sessile oakwoods of western Britain, have a long history of grazing by agricultural livestock and, as 'wood-pastures' are recognised as a distinctive form of ancient woodland both in the lowlands and uplands (Kirby *et al.* 1995; Stiven & Holl 2004; Hopkins & Kirby 2007). High numbers of grazing animals – usually sheep – tend to result in a sparse field and shrub layer, with notable reductions in palatable species such as bramble and little regeneration (Palmer *et al.* 1994). Such woodlands tend to have few species dependent on shrub layer habitats (e.g. thrushes, dunnock, robin, wren, blackcap and garden warbler) with hole-nesters such as woodpeckers, titmice, redstart and pied flycatcher as well as tree pipit and wood warbler well represented (e.g. Simms 1971; Williamson 1972; Irvine 1977; Fuller 1982).

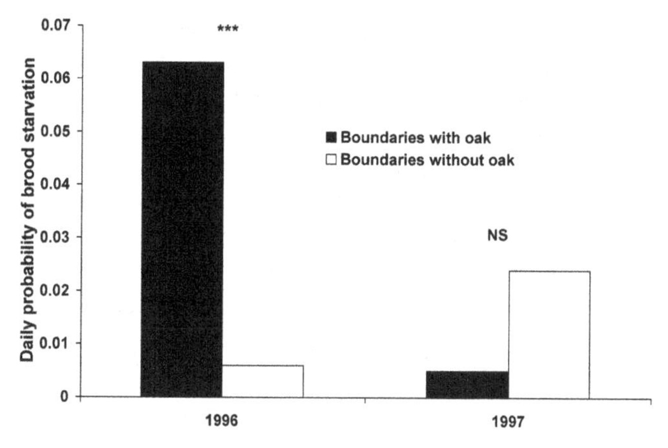

Fig. 3.1 Daily probability of starvation of chaffinch nests with oak trees present or absent from the field boundary in which the nest is located. Years of low (1996) and high (1997) abundance of defoliating caterpillars are compared. Data from Whittingham *et al.* (2001a). *** denotes a difference statistically significant at $p < 0.001$. NS denotes a difference not statistically significant at $p < 0.05$.

Many individual bird species exploit both woodland and farmland habitats. The rook is a good example. This omnivorous species relies heavily on pastures and cultivated land for its main food sources, soil-dwelling invertebrates and grain, yet it nests colonially in woodland trees. Although generally common on arable fields and pastures, Rooks are significantly scarcer in areas such as the Cambridgeshire and Lincolnshire Fens where woodland cover is exceptionally low. This is a pattern which applies to many other species which are otherwise commonly seen on farmland, such as robin, mistle thrush, chaffinch, bullfinch and titmice (Gibbons *et al.* 1993). For example, a recent detailed study of the breeding and foraging ecology of chaffinches on mixed farmland in Oxfordshire found that although the species nested at high densities in farmland hedgerows, it relied heavily on foraging arboreally, especially in standard oaks and willows in hedgerows and woodland to secure caterpillars as food for nestlings (Whittingham *et al.* 2001a). In a year when abundance of defoliating caterpillars was low (1996), chaffinch nests in field boundaries containing oak trees suffered lower brood starvation rates than those in field boundaries without oak trees. However, this difference was not apparent in 1997, a year of 73% higher caterpillar abundance (Fig. 3.1).

There is also some evidence that populations of many species found nesting in hedgerows on farmland can be regarded as overspill from preferred woodland habitats. For example, Williamson (1969) found that after the severe winter of 1962/63 when wren numbers fell catastrophically, subsequent population recovery resulted initially in recolonisation of woodland, riparian and garden habitats, with birds only returning to farmland hedgerows in 1966 and 1967. Similarly, Osborne (1982) showed that hedgerow chiffchaff territories were almost all within 250 m of copses when overall population levels were low, but that birds could be found up to a kilometre from woodland patches in years of high population densities. In an experimental study, Krebs (1971) found that great tits holding territories on farmland hedgerows moved to occupy vacancies created in woodland by the removal of the resident territory holders, although this result was not found to be consistent when the study was repeated (Webber 1975; Krebs 1977). In one case the vacancies were not filled at all, and in another they were filled by other non-territorial woodland birds. In other cases, however, it seems that even farmland hedgerows may be 'full', with non-territorial 'floating' individuals awaiting territorial vacancies. Thus, Edwards (1977) experimentally removed territorial birds of seven species from Dorset farmland, and observed subsequent patterns of recolonisation. Although blackbirds, song thrushes and dunnocks showed incomplete recolonisation of the vacancies, wrens, robins, great tits and yellowhammers had all more or less reoccupied the vacant territories within a month.

Gardens vary greatly in size, but larger gardens often represent a fine-scale mosaic of 'scrub' (garden shrubs or hedges), open grassland (lawns), herbaceous vegetation with patches of bare earth (flower beds and vegetable patches), water (the garden pond) and the occasional mature tree. Add to this our habit of growing crops which may also be favoured by birds, our propensity to provide seed food and kitchen scraps through the winter months and our frequent addition of predators (domestic cats), and the clusters of gardens found in rural villages and around the suburban edge of towns and cities represent a 'habitat' that is probably as uniquely distinctive to birds as it is to us, and very likely to influence wider bird community composition. Perhaps the most striking example of the potential impact of garden management on a bird population is the recent increase in the population of wintering blackcaps in Britain. These birds originate from breeding populations in central Europe, and it is believed that the new migration route to wintering grounds in Britain (as opposed to south-western Europe and north Africa) may have evolved as a consequence of the provision of winter food sources in gardens (e.g. Berthold *et al.* 1992).

Common breeding bird species in smaller gardens include wren, dunnock, robin, blackbird, song thrush, blue tit, great tit, coal tit and chaffinch. Larger gardens with mature trees and shrubbery may also hold such species as stock dove, woodpigeon, collared dove, tawny owl, little owl, great spotted woodpecker, mistle thrush, blackcap, garden warbler, chiffchaff, willow warbler, goldcrest, spotted flycatcher, long-tailed tit, treecreeper, magpie, jackdaw, starling, goldfinch, bullfinch and tree sparrow. Effectively, the larger the garden the more its breeding bird community comes to resemble that of woodland or parkland. A wider variety of species will also visit gardens to feed even though they do not nest there. Hunting sparrowhawks and kestrels, green woodpeckers visiting lawns to feed on ants, and winter visits of redwings or fieldfares to feed on berry-bearing shrubs or fallen fruit are all examples. However, the range of species that visit winter-provided food is perhaps the best-studied facet of garden bird communities, and where the effects of habitat changes on nearby farmland may be felt. Analysis of long-term data from the British

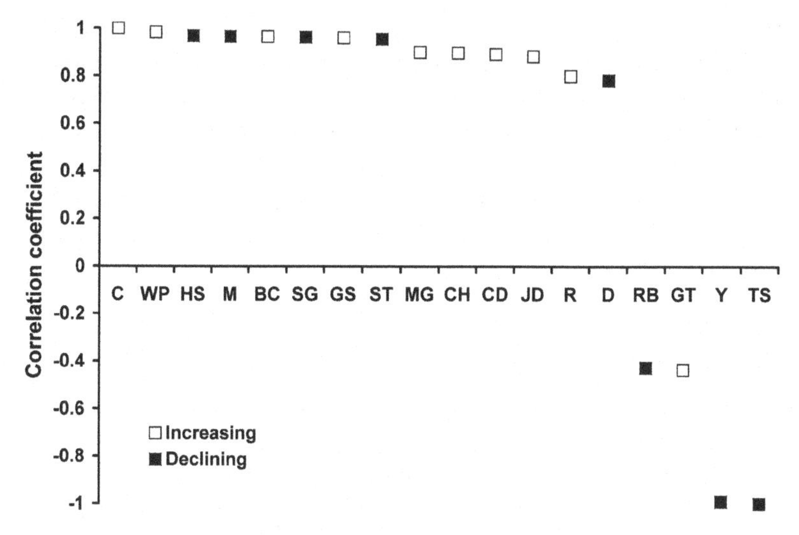

Fig. 3.2 Correlation between national breeding population trends and trends in use of garden feeding stations for 18 species. Data from Chamberlain *et al.* (2005). C, carrion crow; WP, woodpigeon; HS, house sparrow; M, mistle thrush; BC, blackcap; SG, starling; GS, great spotted woodpecker; ST, song thrush; MG, magpie; CH, chaffinch; CD, collared dove; JD, jackdaw; R, robin; D, dunnock; RB, reed bunting; GT, great tit; Y, yellowhammer; TS, tree sparrow.

Trust for Ornithology's (BTO) Garden Bird Feeding Survey showed that the most frequently recorded garden bird species are robin, blackbird, blue tit, greenfinch, dunnock, song thrush, great tit, starling, house sparrow and chaffinch (Chamberlain *et al.* 2005). Of 41 species for which data were analysed, 21 increased in gardens between 1970 and 2000 (notably collared dove, magpie, greenfinch, woodpigeon, jackdaw and chaffinch – all of which have shown concurrent increases in national breeding populations; see Chapter 5), and only five declined (house sparrow, pied wagtail, song thrush, mistle thrush and starling – which, other than pied wagtail, have all shown national population declines). Typically, correlations between trends in use of gardens and national breeding population trends were positive; in other words, species increasing nationally such as great spotted woodpecker tended to use gardens more, and species decreasing nationally such as house sparrow tended to use gardens less (Fig. 3.2). Notably, however, three seed-eating species whose populations have declined markedly on lowland farmland in recent decades (tree sparrow, yellowhammer and reed bunting) showed increasing use of gardens as their populations declined (Fig. 3.2). Might these increases be a response to shortages of seed foods on farmland?

Peak reporting rates for these species and others (e.g. long-tailed tit, brambling, goldfinch and siskin) tend to be in late winter and early spring (February–April; see www.bto.org). This strengthens the argument that the increase in garden-feeding by declining seed-eating species may be a response to the long-term decline in seed availability in the wider countryside. This probably has its greatest impacts in late winter – a time of year when depletion and germination of seed and grain, coupled with the need for birds to attain breeding condition are always likely to have imposed a period of food limitation on populations. The possibility that supplies of food in gardens may be contributing to the maintenance of local populations of seed-eating birds on farmland has not been studied specifically but is certainly worthy of further investigation. The fact that chaffinches and greenfinches are two of the commonest visitors to garden seed-feeders and are maintaining stable or increasing populations on farmland may be no coincidence.

Farmsteads, farmyards and walls

Farm buildings offer nesting opportunities to a variety of traditionally hole- or cliff-nesting species, especially where there is easy access to the interior of the building and where thatch is used for roofing. Nesting species include kestrel, rock dove/feral pigeon, stock dove, collared dove, tawny owl, barn owl, little owl, swift, barn swallow, house martin, wren, grey wagtail, pied wagtail, robin, blackbird, spotted flycatcher, blue tit, great tit, jackdaw, starling, house sparrow and tree sparrow. Shrubb (2003) even lists shelduck. Modern farm buildings tend to reduce such opportunities, not least because legislation now requires those used for storage of food products to be bird-proof. This, coupled with the frequent conversion of redundant farm buildings for human occupation can markedly reduce local populations of species such as barn owl and barn swallow which depend on such sites for nesting (Ramsden 1998; Toms *et al.* 2001; Evans *et al.* 2003b). At some farms, nest boxes and platforms may be provided to assist some of these species. In upland areas, kestrel, barn owl, barn swallow, house martin, starling and, locally, chough may nest in abandoned buildings, with pied wagtails, wheatears and wrens commonly finding nest sites in drystone walls. Traditionally, farm yards, stackyards and threshing yards have provided an abundant source of weed seed and spilt or stored grain as a food for seed-eating birds over the winter months (Shrubb 2003). Woodpigeon, collared dove, jackdaw, starling, house sparrow, tree sparrow, chaffinch, brambling, greenfinch, goldfinch, linnet, twite, yellowhammer, reed bunting and corn bunting are all species to have benefitted in the past. They may do so to a lesser extent nowadays as modernisation of buildings, indoor wintering of cattle and increasingly stringent hygiene regulations have encouraged less spillage and cleaner farmyards and have prevented bird access to grain stores. The relative cleanliness of the modern farmyard may be another reason for the recent increase in late-winter recording of a number seed-eating species at garden feeding stations. Traditional farmyards also attracted larger numbers of wintering brown rats that provided an abundant food source for barn owls and, locally, other wintering raptors such as hen harrier and short-eared owl. The shift in diet of barn owls to smaller mammals (e.g. mice, field voles and shrews) following the advent of the combine harvester and the disappearance of stackyards after the Second World War (Love *et al.* 2000) may have rendered this species more susceptible to the effects of severe winter weather (Shrubb 2003).

Ponds and watercourses

Smaller farm ponds with shallow water may support breeding species such as little grebe, mallard, water rail, moorhen and coot, with snipe, wren, sedge warbler, grasshopper warbler and reed bunting in rank fringing vegetation, scrub and wet grassland. Where areas of reed have colonised or been planted, breeding reed warblers may also occur. Larger ponds, especially those with islands relatively secure from predators, may also attract great crested grebe, teal, gadwall, mute swan, greylag and Canada goose and, where deeper water occurs, tufted duck. Earthen banks, where present, may be used by nesting kingfishers or sand martins. A wide variety of other species will visit ponds to feed, and this may be where more unusual bird visitors to a farm are most likely to turn up. Cormorants, grey herons, gulls, swifts, hirundines, meadow pipits, and pied and grey wagtails are among the most likely, but hobbies hawking for dragonflies, or a wintering jack snipe, water pipit or green sandpiper might be seen, along with the possibility of other species of wildfowl or terns. Overall, the presence of a farm pond can add substantially to habitat and bird diversity but, because of the declining use of ponds as a water source, especially in arable areas, numbers of farm ponds have decreased rapidly during the twentieth century either through deliberate in-filling, drying-up after local field drainage, or simple neglect (Rackham 1986; Shrubb 2003). For example, in a study of moorhens near Huntingdon, Relton (1972) found that the number of farm ponds in the study area had declined from over 150 in the 1890s to under 70 by the late 1960s.

Plate 3.1 The presence and management of hedgerows on lowland farmland throughout Britain makes a critical contribution to the variety and abundance of birds supported. Here the protection from agricultural operations afforded by a green lane will offer added benefits; the periodically trimmed, shrubby hedgerows offer nesting opportunities for a wide range of scrub and woodland species, and the wide, uncultivated verges provide a rich foraging habitat and opportunities for ground-nesting species. © RSPB.

Plate 3.2 Standard trees allowed to grow as part of the hedgerow network increase the range of species found on lowland farmland. They provide opportunities for hole-nesting birds such as kestrels, stock doves, woodpeckers, little owls and tree sparrows, whilst breeding birds dependent on defoliating caterpillars when feeding nestlings (e.g. tits and chaffinches) will forage in the canopy. © RSPB.

Plate 3.3 Farm buildings can provide a wide range of nesting and feeding opportunities for birds. Species such as collared doves, barn owls, barn swallows, pied wagtails and house sparrows may be more likely to be found around farm buildings than anywhere else on the farm. © RSPB.

Plate 3.4 Many farm ponds have become disused and neglected, especially in arable areas where they are no longer needed as a source of drinking water for livestock. Their restoration can provide opportunities for a range of wetland bird species on farmland. © RSPB.

4 · *Semi-natural heathlands and grasslands*

As we saw in Chapter 1, Victorian high farming, powered by the Industrial Revolution and a rapidly growing and urbanising human population, led to the enclosure and agricultural improvement of huge areas of semi-natural lowland heathland and downland in Britain. The fragmented remnants of these habitats remain and continue to support the highly distinctive bird communities that we introduce below. However, very large areas of grazed heathland and moorland remain in the uplands of Britain. Here the impacts of agricultural grazing are an important determinant of vegetation cover and bird communities, alongside competing upland land uses such as forestry and sport shooting of red grouse and red deer (Sydes & Miller 1988). *Countryside Survey 2000* estimated a total stock of 6.05 million hectares of terrestrial, semi-natural habitat (acid grassland, bog, dwarf shrub heath, fen, marsh and swamp, bracken and calcareous grassland) of which 5.22 million hectares (86%) was in the uplands (Haines-Young *et al.* 2000). This included 87% of acid grasslands, 94% of bog, and 89% of dwarf shrub heath – the three habitats that make up most of the grazed upland heaths and moorlands of Britain.

Heathland

The word 'heath' refers to woody undershrub species of the family Ericaceae – the heaths and heathers – or to land dominated by this form of vegetation. Of these heather or 'ling' is by far the most common, with bell heather and cross-leaved heath also common, and Dorset heath and Cornish heath locally dominant in parts of southern England. Tracts of 'heathland' are not necessarily dominated solely by this ericaceous vegetation. Gorse (alias furze or whin), bracken, lichens and grasses may also dominate large areas. Heathlands are of two fundamental types. The first forms over wet, acid peats, usually above the enclosure line (Pearsall 1971). This type is discussed below as just one of a complex of vegetation communities characterising the British uplands. In the lowlands, heath vegetation most often develops on heavily grazed land over dry, drought-prone, mineral soils and is generally believed to be the consequence of human clearance of woodlands followed by subsequent use for grazing and supply of fuelwood, litter and thatch, rather than for cultivation (Webb 1986). At first sight it may seem odd that heather-dominated vegetation develops to characterise two such radically different landscapes and soil types. However, heather requires acidic, low-nutrient environments with limited humidity variations and high light levels, and hence both anaerobic upland peats and dry, mineral soils can provide these basic requirements. Nowhere in the world is heather-dominated vegetation found more extensively than in Britain, and beyond the Atlantic coastal fringe from Norway south to Spain, the limited frost tolerance of heather constrains its distribution (Webb 1986).

Upland heath and moorland

Vegetation communities characteristic of 'uplands', the uncultivated hill-land above the permanent enclosure line, occupy almost 30% of the land area of Britain. Almost

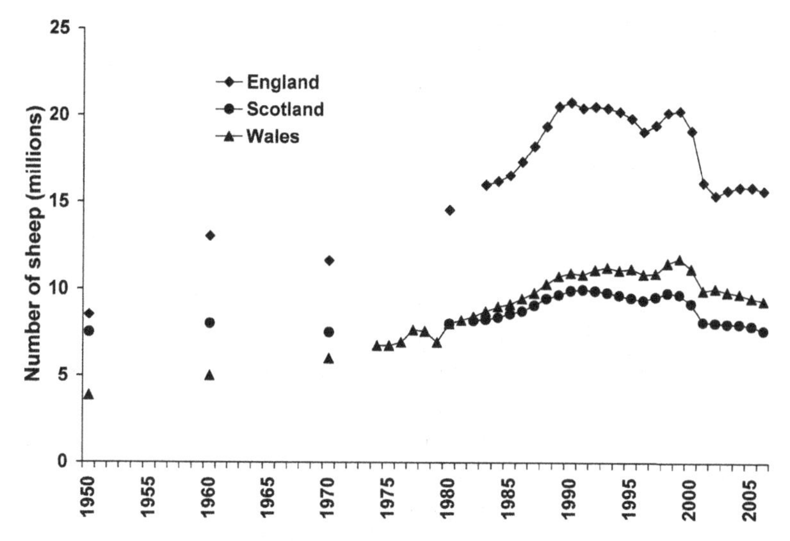

Fig. 4.1 Trends in sheep numbers in England, Scotland and Wales, 1950–2006. Data from www.defra.gov.uk, www.scotland.gov.uk, www.statistics.gov.uk and Fuller & Gough (1999).

all of this area has been influenced by human activity. Deforestation, burning, grazing and sport-shooting of red grouse and, in the Scottish highlands, red deer have produced 'sheepwalks' (upland grasslands), grouse moors (upland dwarf shrub heaths) and deer forest (mixtures of upland heath and grassland) as the predominant upland habitat types (Ratcliffe 1990). The combination of oceanic climate, mountainous terrain and widespread human impacts of native forest clearance, grazing, burning and, recently, acidic deposition have combined to produce a unique, anthropogenic landscape characterised by plant communities that are either absent from or rare elsewhere in the world (Ratcliffe & Thompson 1988). Just a few of the least fertile areas above the tree line on the montane, 'high tops', some peat bogs (notably in the Flow Country of Caithness and Sutherland) and inaccessible sites such as lake islands and cliffs may retain the only genuinely natural vegetation cover to be found in Britain. Our oceanic climate and prevailing Atlantic winds generate dramatic climatic gradients from east to west which mean that characteristically 'upland' vegetation communities may be found at sea level in parts of northern and western Britain where high rainfall and wind-speed and low sunlight levels produce climatic conditions that only occur at higher levels further south and east. Significant upland deforestation by grazing and burning has occurred in the uplands for well in excess of 2000 years, although complete woodland clearance of many areas has only happened in the past 100–300 years (Birks 1988; Ratcliffe 1990). Grazing and burning favour the gradual replacement of all woody vegetation by bracken and grasses (e.g. Anderson & Yalden 1981; Hester & Sydes 1992; Thompson *et al.* 1995), and upland heath has been in retreat for at least the past 50 years as the national sheep flock has increased from 19.5 million in 1945 to over 40 million in 1990 (Fig. 4.1). Sheep numbers more than doubled in England and Wales between 1950 and 1990 (Fuller &

Gough 1999). They had once been almost as high, at the zenith of Victorian high farming in the late nineteenth century, but then a much higher proportion of the flock was held in the lowlands – especially East Anglia and south-east England – and hill ground was spared grazing pressure over-winter because of the difficulty of providing supplementary food. Nowadays, the relative ease of conveying supplementary food to the hill allows more sheep to be over-wintered on high ground and grazing pressure to be maintained year-round (Hudson 1984). Although sheep numbers increased less spectacularly in Scotland (by 32% between 1950 and 1990), red deer numbers almost doubled between 1959 and 1989 (Staines *et al.* 1995).

In heavily grazed areas, blocks of heather often do remain, but are in poor condition, and healthy stands are increasingly restricted to cliff ledges or other areas inaccessible to sheep. The strongholds of true upland heath are those areas managed primarily to maintain stocks of red grouse for shooting. Grouse moors are now most characteristic of the drier eastern uplands of Britain including the Grampians, Lammermuirs, eastern Pennines and North York Moors. Here sheep densities are kept low, predators of grouse such as foxes and crows are killed, and heather cover is managed by rotational burning to maintain a mosaic of heather patches of varying age in order to provide the ideal combination of nesting cover and fresh heather shoots as food for the grouse (Watson & Miller 1976; Hudson 1992). Further west, grouse moors do remain, but have been in continuous decline over the years (e.g. Redpath & Thirgood 1997) for several reasons. These include difficulty in maintaining heather cover in the face of sheep-grazing, more vigorous competition from grasses on wetter soils, falling economic values of grouse moors relative to sheep and forestry and the loss of gamekeeping and associated heather management practices. Where grouse management has been abandoned, the highest remaining red grouse densities may occur in areas where pure stands of heather are mixed with patches of blanket bog dominated by a mix of heather and cotton-grass (Ratcliffe 1990).

Since the late 1990s, sheep numbers have declined again in England, Scotland and Wales (Fig. 4.1) as EU CAP reforms have replaced headage-based payments (i.e. a subsidy paid per animal) by area-based payments, thus reducing incentives to maximise flock sizes. This trend was exacerbated by the outbreak of foot-and-mouth disease in 2001 which reduced the national flock directly by culling in some areas, and has had a longer-lasting impact on the confidence of farmers in restocking. An increase of 25% in *Countryside Survey 2000*'s 'fen, marsh and swamp' vegetation category of semi-natural habitats in the Scottish uplands (Haines-Young *et al.* 2000) may reflect expansion of soft rush across upland pastures experiencing management neglect as stocking densities have begun to fall in recent years.

As well as the upland heaths dominated by heathers, moorlands may also be dominated by other woody undershrubs such as bog myrtle, crowberry and bilberry, or by bracken and grasses such as bents, fescues, wavy hair-grass, vernal-grass, mat-grass and purple moor-grass, with rushes, mosses and cotton-grasses in wetter areas (Pearsall 1971; Ratcliffe 1990; Thompson *et al.* 1995). At the lower moorland edge, heathland may intergrade with complex mosaics of woodland, scrub and bracken cover as in the birch–heath mosaics of north-east Scotland (Gillings *et al.* 1998) or the ffridd habitats of Wales (Woodhouse *et al.* 2005; Fuller *et al.* 2006). On open moorland, woody cover of heather and bilberry predominates in the wetter, more acid areas, where sheep densities also tend to be lower. Over neutral or calcareous rocks, where soil fertility is higher, grassland dominates, largely because of prolonged, preferential and heavier grazing. Upland grassland sheepwalks are now by far the dominant vegetation cover in much of the uplands of Wales, Lakeland, the Pennines, and the Southern Uplands and Breadalbane Hills of Scotland. Across this great climatic and altitudinal range, and diversity of vegetation mosaics in the British uplands, breeding bird communities are highly variable. Indeed, Britain probably hosts a diversity of upland bird communities that is not matched anywhere else in the world in a similar geographical area, and here we try to summarise them briefly.

The bird communities of upland heathland and moorland

Well-drained, inland upland heath or grassland, above approximately 300 m above sea level, has a breeding bird assemblage which includes teal, red grouse, black grouse, hen harrier, merlin, golden plover, lapwing, curlew, cuckoo (here parasitising meadow pipits), short-eared owl, skylark, meadow pipit, wren, stonechat, whinchat, wheatear, twite and crow. On steeper slopes, where crags, rocky bluffs and scrub-filled gullies occur, there are nesting opportunities for additional species, notably golden eagle, buzzard, peregrine, kestrel, rock dove, stock dove, pied wagtail, ring ouzel, jackdaw and raven. In wetter areas and blanket bogs, many of these species will be less abundant or absent, and others may be found, including red-throated diver, greylag goose, mallard, wigeon, red-breasted merganser, dunlin, snipe, redshank, greenshank, black-headed and common gull and, in a very few places in Scotland, northern rarities such as common scoter, wood sandpiper and Temminck's stint. Alongside fast-flowing streams and rivers in the uplands, common sandpiper, dipper and grey wagtail are the characteristic breeding species, with goosander, red-breasted merganser and, very locally, goldeneye found, especially in Scotland, on the larger rivers and lakes. Where moorland intergrades with scrub woodland at its lower edge as in the highland birch woods of Scotland, open-ground species such as meadow pipit may be found alongside tree pipits, willow warblers and chaffinches, with more characteristically woodland species such as wrens and tits occurring in older-established birch woodland. Similarly, in the complexes of unimproved grassland, bracken, dwarf shrub heath and gorse scrub that make up the Welsh ffridd, an unusually diverse mix of species characteristic of open and scrub habitats may be found in close association, including chough, wheatear, whinchat, stonechat, mistle thrush, meadow pipit, tree pipit, yellowhammer and linnet. Where characteristically 'upland' vegetation types extend almost to sea level as in parts of northern and western Scotland, the Outer Hebrides and the Northern Isles, maritime species such as Manx shearwater, eider, white-tailed eagle, oystercatcher, ringed plover, whimbrel, Arctic and great skuas, herring, lesser black-backed and great black-backed gulls, Arctic tern and rock pipit may be found nesting. In this context, the coastal bogs and moors of Orkney, Shetland and the Outer Hebrides and the Flows of Caithness and Sutherland are unique in Britain with colonies of skuas, gulls and terns alongside the more usual range of upland breeding species. Finally, above approximately 900 m altitude, in the few areas of Britain (mostly in the Cairngorm massif) where a true montane vegetation community is found, three further breeding birds occur: ptarmigan, dotterel and snow bunting. Here, ptarmigan occur where montane heath is interspersed with screes and boulders and may be the commonest vertebrate species, with densities higher than reported from elsewhere in the species' largely Arctic breeding range. Snow buntings nest under rocks in boulder-strewn areas, and dotterels on the dry ground of rounded spurs or flat summits.

Across this diverse range of topographies, altitudes and vegetation communities, the three species that are perhaps the most consistently abundant on open moorland are red grouse, skylark and meadow pipit. A fourth species, wheatear, is also common and widespread, but more locally distributed than the other three because it prefers short-grazed grassland or other sparsely vegetated areas and may be scarce on much heather moorland. Colonial species such as black-headed gull may also attain very high population densities locally, but not across wide areas of the uplands.

During the winter months, upland heath and grassland is an inhospitable habitat occupied by few bird species. On a good day, winter birdwatching in the uplands may reveal red grouse, wren, kestrel, meadow pipit and crow, but little else. Only snow bunting can be regarded as a true winter visitor to upland areas of Britain. Small flocks occur in many upland areas, often feeding on seeds of purple moor-grass or seeds and insects blown on to the surface of snow patches, but are unpredictable and infrequent in their occurrence in any one place. At other times of the year, a variety of additional species may use upland grasslands and heathlands as foraging habitats. These include sparrowhawk, jack snipe, woodpigeon, swift, barn swallow, blackbird, song thrush, fieldfare, redwing, mistle thrush, magpie, rook, chough, starling, chaffinch, greenfinch, goldfinch, linnet and reed bunting.

As discussed above, the main impact of agricultural activity in its broadest sense, on both the vegetation and bird communities of upland heathland and grassland, has been the effect of the shifting balance of heather and other woody vegetation covers, and grass, sedge and rush cover, as a function of the interactions of grazing and grouse management against the backcloth of the prevailing geology and climatic conditions. These issues will be discussed in more detail in Chapters 8 and 9. Changing agricultural management at the upland edge – the margin between enclosed farmland and the open moor – will also be considered in more detail insofar as it has affected a number of species whose populations are concentrated at this interface – for example, golden plover, curlew, black grouse and twite. In addition, a huge area of former sheepwalk, especially in the southern and eastern uplands of Scotland, has been afforested during the twentieth century, mostly in large monocultural blocks of non-native species such as Norway and Sitka spruce, and lodgepole pine. The contrast between the breeding bird community of a dense, mature conifer plantation (perhaps goldcrests, coal tits, siskins, chaffinches, carrion crows and the occasional pair of tawny owls, sparrowhawks, goshawks or crossbills) and that of the upland heath or grassland it replaced is stark. Mature conifer forestry excludes the internationally significant upland bird assemblage in favour of a limited range of common and widespread woodland species (e.g. Thompson *et al.* 1988). In areas such as Galloway and parts of the Scottish Borders where both grouse moor management and hill sheep farming became ever more uneconomic and conifer afforestation has been particularly extensive, the direct loss of moorland breeding bird populations may be substantial. For example, Ratcliffe (2007) estimated that the afforestation of some 2500 km^2 of moorland across the Southern Uplands of Scotland may have caused the loss of 5000 breeding pairs of curlew, 300 pairs of golden plover and virtually all of the small population of dunlin.

During the early stages of succession of a new plantation, a variety of upland species may benefit, if temporarily, from the increase in nesting and feeding opportunities provided by the exclusion of grazing, and the associated flush of vegetation growth and increase in invertebrate and small mammal populations (Evans *et al.* 2006a). These include hen harrier, black grouse, short-eared owl, nightjar, stonechat and whinchat, along with a number of species more often associated with lowland or wooded habitats such as barn owl, tree pipit, grasshopper warbler, whitethroat, willow warbler and redpoll. In Scotland, young conifer plantations in the uplands are now the main habitat of the small breeding population of redwings that has established itself in the twentieth century. However, once the plantation reaches the age at which the canopy closes and shades out ground vegetation, then all of these benefits are lost and birds of open moorland, including raptors and black grouse, are excluded (e.g. Avery & Leslie 1990; Madders 2000, 2003; Rebecca 2006; Pearce-Higgins *et al.* 2007). Some of this bird community may be retained in areas where more enlightened restocking policies have created a sustainable successional mix of growth stages from clear-fell to mature, pole-stage plantation. The impacts of conifer afforestation and forest management on breeding bird assemblages in the uplands are discussed in detail by Avery & Leslie (1990) and are beyond the scope of this volume, although Chapter 9 will consider how the ecological impacts of afforestation may interact with those of grazing and grouse moor management in influencing the breeding birds of unafforested ground.

Lowland heathland

Areas of lowland heathland probably peaked in Britain during the sixteenth and seventeenth centuries. Since then a range of factors has combined to reduce the area of lowland heathland in Britain to a few percent of its former extent (Webb 1986). These include improvements in agricultural technology enabling the cultivation and improvement of the poor soils, the eighteenth-century Enclosure Acts, urbanisation, natural reversion to woodland, afforestation, and agricultural reclamation during the Second World War. The vast majority of lowland heathland is found in the southern, coastal counties of England from Cornwall to Sussex, and in East Anglia. Elsewhere in northern and western Europe, the destruction of lowland heathland has been even more complete than in Britain. Very few heaths in Britain retain their former complement of human uses, the best remaining examples being in the New Forest. Many of the rest are threatened by the difficulty in maintaining the traditional forms of management that are essential to maintain heathland vegetation communities (Tubbs 1991). For example, lack of grazing results in invasion of trees and the gradual replacement of ericaceous vegetation by grass and bracken. This has happened to considerable areas of British heathland over the last 200 years. Birches, pines and oaks are common invaders of open heathland, so that remaining areas are often a mix of open heath, interspersed with patches of scrub and more mature woodland. Accidental summer fires (and those started by arsonists) on heaths heavily used for recreational purposes have also stripped soils, killed shrubby vegetation and led to replacement by stands of bracken and grasses such as wavy hair-grass and purple moor-grass.

The bird communities of lowland heathland

Where open heathland is invaded by scrub and woodland, its bird communities come to reflect a mix of woodland species associated with the basic heathland vegetation types; heather, gorse, bracken, grassland and wet flushes of grasses, sedges and rushes. Here we focus on the latter habitats whose characteristic breeding species comprise lapwing, snipe, redshank, curlew, grasshopper warbler and reed bunting (in the wetter areas), stone curlew, ringed plover, nightjar, woodlark, skylark, tree pipit, meadow pipit and wheatear (on dry, bare or sparsely vegetated ground), wren and Dartford warbler (in dense heather), and whinchat, stonechat, dunnock, robin, blackbird, song thrush, linnet, goldfinch, yellowhammer and, formerly, red-backed shrike (in areas dominated by gorse or bracken). In addition, the cuckoo is frequent on heathland as a parasite of meadow pipit nests. This relatively limited species diversity is matched by low breeding densities of many species. For example, in a detailed study of Hartland Moor in Dorset, Bibby (1978) found that only four breeding species bred commonly (meadow pipit, stonechat, wren and Dartford warbler) with smaller numbers of linnets, yellowhammers, willow warblers, whitethroats, dunnocks, robins and blackbirds.

Additional species that will visit heathland vegetation types as a foraging habitat include a wide range of species that feed on invertebrates or seeds on the ground or in over herbaceous or scrubby vegetation. Few of these, however, could be regarded as specialising in heathland foraging habitats. Green woodpeckers searching heathland turf for ants and hobbies hawking dragonflies in summer over heathland are two examples, and heathland with scattered shrubs and small trees as hunting perches is frequently chosen as a feeding habitat by small numbers of great grey shrikes which visit Britain from the continent in winter, subsisting on small mammals and large beetles. Two birds of prey that are characteristic of heathland in the winter months, though very localised in their occurrence, are the short-eared owl and hen harrier, both of which may be seen quartering open heath searching for birds and small mammals and may roost, sometimes communally, in deep heather (e.g. Clarke *et al.* 1997). Four rather more widespread resident birds of prey – kestrel, sparrowhawk, barn owl and little owl – may also be seen over heathland hunting birds, mammals and large invertebrates.

Historically, there have been catastrophic long-term declines in heathland specialists such as stone curlew, nightjar, woodlark, red-backed shrike and Dartford warbler due largely to destruction of habitat through cultivation, afforestation or urbanisation (Webb 1986). More recently, further losses have accrued for several reasons. These include under-grazing of heaths either by removal of domestic stock or the effects of myxoma- tosis on rabbit populations, and replacement of heather or gorse cover by coarse grasses and bracken after human-caused fires, or as a consequence of increasing atmospheric nitrogen deposition (Tubbs 1991; Marrs 1993a; Blackstock *et al.* 1995; Rose *et al.* 2000). The response of stone curlews to loss of open, stony ground on the Breckland grass heaths due to afforestation and under-grazing has been to colonise adjacent areas of spring-sown crops such as carrots and sugar beet (Green & Griffiths 1994). The effects of agricultural change on stone curlews nesting on both heathland and arable land are discussed as a detailed case study in Chapter 8. In the case of nightjar and woodlark, harvesting and replanting of conifers has provided new areas of suitable nesting habitat on former heathland in areas such as Breckland, the heaths of the Thames basin, Weald and east Devon, the Suffolk Sandlings and the North York Moors. Nightjars increased by 50% between 1981 and 1992 to a total of over 3000 'churring' males (Morris *et al.* 1994), mainly as a consequence of the colonisation of large (>10 hectares) clear-felled areas and restocked, young conifer plantations up to eight years old (Ravenscroft 1989; Scott *et al.* 1998). Numbers have continued to increase since then, with a further survey in 2004 finding a further 36% increase in the number of churring males (Conway *et al.* 2007). The breeding population of woodlarks is also increasing; from 241 pairs in 1986 to about 350 in 1993, 1552 by 1997 (Wotton & Gillings 2000), and 3084 in 2006 (S. Wotton pers. comm.). Populations in Breckland and along the Suffolk coast have benefitted from the clear-felling of conifer plantations, occupying similar habitats to those favoured by nightjars (e.g. Bowden 1990). Those on the heaths of Hampshire and Surrey have probably benefitted from improved heathland management, including removal of encroaching scrub and woodland, and restoration of grazing and burning regimes, and populations are also expanding beyond heathland habitats to nest on adjacent arable farmland, notably on rotational set-aside fields (Wright *et al.* 2007) (Fig. 4.2). However, densities and reproductive output on these fields are low and, in the long run the future of both woodlark and nightjar populations in Britain probably remains dependent upon a combination of appropriate grazing management of heathland coupled with forestry strategies that ensure maintenance of areas of clear-fell and pre-thicket-stage plantation within forest blocks. Recent studies of both species suggest that high levels of human, recreational

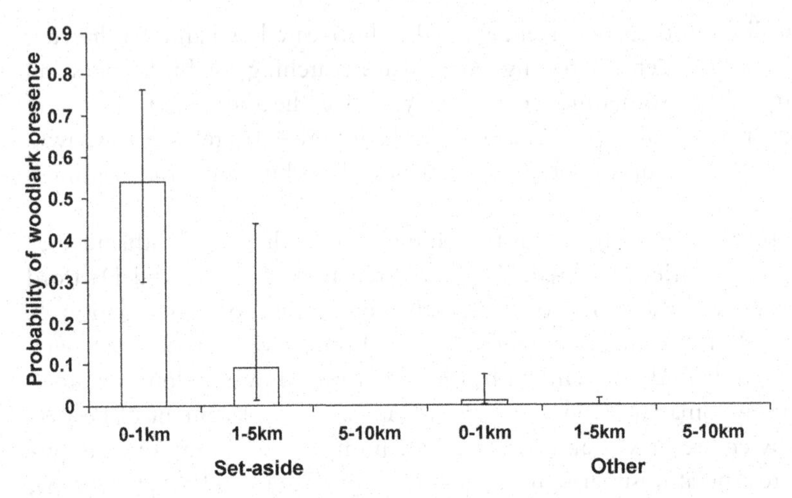

Fig. 4.2 Probability of encountering woodlark territories on farmland (set-aside and other field types) as a function of distance from the edge of heathland and forestry habitat in Thetford Forest. Error bars are 95% confidence intervals. Data from Wright *et al.* (2007).

disturbance may also limit densities and breeding success (e.g. Liley & Clarke 2003; Mallord *et al.* 2007).

In contrast, the Dartford warbler remains dependent on areas of heather and gorse cover, free from significant encroachment by scrub and trees (Westerhoff & Tubbs 1991). Here, its numbers fluctuate mainly because of variation in winter weather conditions. The successive hard winters of 1961/62 and 1962/63 reduced the population to just 11 breeding pairs. Numbers have since increased to 423 pairs by 1984 (Robins & Bibby 1985), an estimated 1890 pairs by 1996 (Gibbons & Wotton 1996) and to 3208 pairs in 2006 (S. Wotton pers. comm.). The population has also spread from its traditional strongholds on the Dorset and Hampshire heaths to Surrey, Sussex, the coastal heaths of Devon, Cornwall, south Wales and Suffolk, Cannock Chase and even on to heathland of more upland character on Dartmoor, Exmoor and the Brecon Beacons. However, Dartford warblers remain vulnerable to the effects of severe winter weather, and will benefit from restoration and recreation of heathland, especially if this can reverse previous fragmentation to create larger, contiguous blocks (van den Berg *et al.* 2001). Locally, the species' preference for foraging in gorse (it holds a richer invertebrate fauna than heather: Bibby 1979) may cause problems as territories richest in gorse are often those closest to main roads where Dartford warblers are vulnerable to being killed by traffic (Catchpole & Phillips 1992).

Amongst our heathland species, the red-backed shrike is perhaps the exception to the general story of long-term historical decline, followed by stabilisation and recovery due to a combination of improved management of remaining heathland, and colonisation of new habitats. Red-backed shrikes were once widespread across southern England in farmland and heathland habitats which combined scrub or hedgerows to provide nest sites and hunting perches, with areas of open, herbaceous cover as a source of grasshoppers, large beetles and other insects that are the species' favoured prey. By the late 1960s, the species had become confined to such habitat patches on lowland heathland in the New Forest, Surrey and East Anglia. 1989 was the first year in recorded ornithological history when no red-backed shrikes bred in England, and there has been no regular breeding since (Brown & Grice 2005). The possible causes of the spectacular and rapid demise of this species in Britain are discussed in Chapter 9.

Lowland calcareous grassland: the 'downs'

The sheep- and rabbit-grazed 'downs', characteristic of low-fertility, calcareous soils throughout southern and eastern England, form the largest areas of unenclosed semi-natural grassland in the lowlands. Measured by floral and invertebrate diversity, calcareous grasslands are some of the richest wildlife habitats in Britain, and are extremely varied (Oates 1994), with the National Vegetation Classification recognising no fewer than 14 main types (Rodwell 1992). The species composition of chalk grassland reveals much about its age. For example, juniper, pignut, milkworts, horseshoe vetch, felwort and pasque-flower occur only in swards that are centuries old, and are destroyed immediately by ploughing. The lack of such species may thus indicate cultivation as long ago as the Middle Ages. Approximately 36 000 hectares of unimproved, lowland, calcareous grassland remain in Britain, a pitiful remnant of what once existed. Conversion to arable land during the Enclosure Acts of the eighteenth century, during the high farming period of the late nineteenth century, and during the ploughing campaigns of the Second World War, have claimed the vast majority. Many of the largest remaining areas have survived simply by virtue of being on land secured for military use – as on Salisbury Plain and Porton Down in Wiltshire, and parts of the Norfolk Breckland. Conversion to arable land has not been the only cause of downland loss; abandonment of grazing in areas where arable farming has become predominant plus reduction in rabbit-grazing after the outbreak of myxomatosis in the 1950s have resulted in the invasion of many areas of chalk downland by deciduous scrub and its gradual succession to secondary woodland. Downland scrub may be dominated by a single shrub species such as yew, juniper, box or hawthorn or may be a much more complex mix of species, including privet, dog rose, wayfaring tree, buckthorn and blackthorn.

The bird communities of downland

Downland provides cover for only a limited range of ground-nesting species, although species more associated with scrub such as whitethroat, willow warbler, linnet and yellowhammer may be found wherever successional processes have begun to occur. Stone curlews and lapwings will use bare or sparsely vegetated ground, and hole-nesting wheatears are typically associated with the areas around rabbit

Table 4.1 *Population estimates of selected breeding bird species on the Salisbury Plain Training Area in 2000*

Species	SPTA breeding population (pairs)	% of British population
Quail	36+	12+
Stone curlew	30+	10+
Corn bunting	391	3.2–4.6
Grasshopper warbler	264	2.5
Whinchat	586	2.1–4.2
Barn owl	70	1.75
Stonechat	223	1.0–2.5
Skylark	14 612	0.8
Meadow pipit	8 869	0.5

Source: Data modified from Stanbury (2002) using British population size data from Chapter 5 (Table 5.2).

warrens. Where vegetation cover is longer, other ground-nesting species such as red-legged and grey partridges, quail, curlew, skylark, meadow pipit, tree pipit, whinchat, stonechat, grasshopper warbler and reed bunting can occur. Cuckoos also occur on downland as brood parasites of meadow pipit nests. A much wider variety of species will forage in or over the short turf for invertebrates and seeds. These include pigeons, green woodpecker, hirundines, wagtails, chats, thrushes, corvids, starling, finches and buntings. Most areas of downland are now remnant islands in a sea of arable, agricultural land, and species that nest in adjacent farmland may make particular use of downland habitats as a relatively rich source of invertebrate prey. For example, grasshoppers – almost absent from tilled land – are common in downland grass and may be exploited as nestling foods by yellowhammers and corn buntings. As with lowland heathland, downland is a favoured hunting habitat in winter for raptors such as hen harriers and short-eared owls, and locally breeding species such as hobby, kestrel, sparrowhawk, barn owl and little owl.

A recent survey of the breeding birds of the Salisbury Plain Training Area (SPTA) – at over 30 000 hectares, the last remaining downland landscape in Britain – gives some insight into a bird community that would once have been much more widespread over the semi-natural grassland landscapes of lowland Britain (Stanbury 2002). In total, 76 breeding species were recorded, but the populations of some of these (Table 4.1) are little short of spectacular in the context of lowland England, and show just how rich the bird community of unimproved grassland systems can be.

The site supports over 10% of the British populations of quail and stone curlew annually, more than 1% of the populations of barn owl, corn bunting, grasshopper warbler, whinchat and stonechat, and a staggering 23 500 pairs of skylarks and meadow pipits. Given that meadow pipits (let alone stone curlews, grasshopper warblers, whinchats and stonechats) are effectively absent as breeding birds from improved, lowland agricultural grasslands, and that skylarks typically persist in such grasslands at densities of approximately 18 pairs per km^2 (Donald 2004), as compared with over 40 pairs per km^2 on the SPTA, the likely nationwide impact of agricultural grassland intensification is clear. As Stanbury notes, in the same year that over 250 pairs of grasshopper warblers were recorded on Salisbury Plain, the whole of the adjacent county of Hampshire yielded only four known pairs.

Other than loss through cultivation, the main conservation problem caused by changing management on downland is very similar to that on lowland heathland. Lack of grazing caused by reduced livestock densities and the effects of myxomatosis on rabbit populations has led to longer vegetation and gradual encroachment by scrub and woodland. The current remnant distribution of breeding stone curlews reflects these processes (Green & Taylor 1995). Even where former downland breeding habitats remain as in the Salisbury Plain area of Wiltshire, many breeding pairs are now nesting on spring-tilled fields or specially created tilled plots on adjacent arable land, and visit downland only to feed on earthworms and arthropods in areas where the vegetation remains sufficiently short (Green *et al.* 2000). Similarly, wheatears, which once nested abundantly in the holes provided by rabbit nurseries, are now rare on downland except as migrants in spring and autumn. Salisbury Plain now represents the only significant downland population centre, with nesting wheatears very rare or absent on the South Downs, the Marlborough and Berkshire Downs and on the Chilterns (Gibbons *et al.* 1993).

As one might expect, as succession towards woodland proceeds as a result of withdrawal or reduction in grazing pressure, the breeding bird community becomes increasingly dominated by woodland species, with those species more characteristic of open downland habitats becoming scarcer. These changes may entail the loss of rare breeding birds specifically associated with remaining downland fragments, such as stone curlew, but the overall breeding bird assemblage becomes more species-rich. Where scrub cover remains light and with much open grassland between the individual bushes, skylarks and meadow pipits will remain, but will be joined by dunnocks, whitethroats, willow warblers, linnets and yellowhammers. Where scrub is denser and has a closed canopy, dunnocks and willow warblers will remain but the other species will probably be absent, being replaced by species more characteristic of woodland such as great tit, robin, blackbird, song thrush, nightingale, blackcap and chaffinch. These changes are well illustrated by the study made in 1980 in the Chiltern Hills by Fuller (1982).

Open downland in winter can be a rather desolate habitat with few birds. Over a large area of downland grass on the Berkshire Downs (now largely converted to arable cultivation), Colquhoun & Morley (1941) found only nine bird species, with grey partridge, rook and starling the most numerous. Where scrub exists, however, its fruit may attract large numbers of woodpigeons, blackbirds, song thrushes, mistle thrushes, fieldfares and redwings for as long as stocks last, whilst mixed flocks of tits can be found throughout the winter. Patches of dense scrub may be used by a wide range of species as communal roosts including long-eared owls, woodpigeons, thrushes, linnets, greenfinches, chaffinches and bramblings. Downland scrub also provides a vital food source for migrant species such as blackcaps, garden warblers and whitethroats feasting on berries to fatten up prior to autumn migration.

Saltmarsh

Saltmarshes are natural grass–herb mixes which form above the neap high tide mark of estuarine habitats and develop as existing plants trap further silt and mud brought in by the tides, and this effect is combined with the accumulation of dead plant material from previous years of growth. Saltmarshes are covered partially by most tides and completely by spring tides, and are thus dominated by a shoreward succession of gradually less salt-tolerant plants. Typically, the lower, more frequently inundated parts of the marsh will be dominated by saltmarsh grass, glasswort or the invasive cord-grass. Further up the shore, sea aster, annual seablite, sea-lavender, thrift, and sea purslane or some combination of these may often be dominant, with red fescue and sea rush marking the upper limit of the marsh.

The main historical impact of agriculture on saltmarshes has been the wholesale reclamation of marsh for pasture or arable land through the construction of sea defences. Large areas of intensively managed arable land around the Wash have been reclaimed from the sea since mediaeval times. Here, the total area reclaimed is considerably greater than the total area of saltmarsh and intertidal flats now remaining. Gradual accretion of new saltmarsh does occur at the seaward edge of existing marshes, but this is a minor compensation for losses due to reclamation for agriculture. Its effects in many areas are being outstripped by a combination of increasing rates of sea-level rise coupled with hard sea defences, resulting in saltmarsh erosion in situations where reinundation of areas previously reclaimed is not possible. In recent years, with net rates of saltmarsh loss in the UK of over 100 hectares per year, schemes to create new intertidal habitats through managed coastal realignment have been implemented both for their flood protection and environmental benefits (Atkinson *et al.* 2004), and show some promise in recreating biologically functional saltmarsh and mudflat habitats.

Other than the dynamics of destruction and recreation of saltmarsh through the balance of agricultural reclamation, sea-level change and managed coastal realignment, the main agricultural influence on saltmarshes is extensive grazing by cattle or sheep. Large areas of grazed saltmarsh are found in several of Britain's large estuaries, including the Solway, Morecambe Bay, Severn, Ribble and Dee. Where such grazing occurs, the taller herbaceous components of the saltmarsh community tend to be lost and the vegetation may become dominated by a 'billiard-table' sward of red fescue and saltmarsh grass. In a few areas, this close-cropped grassland may be cut commercially for turf, and managed with additional fertiliser inputs and cutting.

The bird communities of saltmarsh

Structurally, saltmarsh vegetation communities provide the same mix of low shrubs and grassland that characterise the heathland and grassland bird communities discussed above. Breeding birds include shelduck and red-breasted merganser (in burrows, thick vegetation or amongst tree roots at the landward edge of the marsh), mallard, grey partridge, moorhen, snipe, redshank, meadow pipit, yellow wagtail, sedge warbler and reed bunting (wherever rank vegetation provides nesting cover), oystercatcher, lapwing, ringed plover and skylark (on bare or sparsely vegetated areas), black-headed gull and common tern (usually on saltmarsh islands relatively secure from terrestrial predators), and stonechat and linnet (in patches of gorse, bramble, sea buckthorn or shrubby seablite at the inland edge of the marsh). Narrow strips of saltmarsh along the seaward edge of the Hebridean machair (see below) can also hold high densities of breeding Dunlin (Fuller *et al.* 1986). The mixture of saltmarsh vegetation and the maze of muddy tidal creeks attracts a much wider variety of species to forage. These include grazers such as swans, geese, ducks (notably wigeon and teal) and pigeons, species feeding on saltmarsh invertebrates and seeds (e.g. plovers, pigeons, hirundines, pipits, wagtails, chats, corvids, starling, finches and buntings) and a large number of waterbirds visiting the tidal creeks to hunt for fish and invertebrates. This latter group includes grey heron, cormorant and many migrant and wintering waders, gulls and terns.

Saltmarsh habitats are especially important wintering areas for several species of seed-eating passerines (skylark, shore lark, linnet, goldfinch, twite, reed bunting, Lapland bunting and snow bunting). In particular, saltmarsh habitats in eastern England between the Humber and the Thames probably support almost all of the isolated and declining Pennine breeding population of twite (Brown & Grice 2005) during the winter; the birds feeding on the seeds of sea aster, sea-lavender, annual seablite and glasswort (Brown & Atkinson 1996). Late in the winter when seed sources are at their scarcest, wintering twite are strongly associated with patches of seeding glasswort on the lower marsh, and this is one of the reasons why rates of saltmarsh loss to erosion may be sufficient to account for the observed declines in the East Anglian wintering population (Atkinson 1998).

The impact of grazing by cattle and sheep on the suitability of saltmarsh habitats for ground-nesting and foraging birds is of particular interest. For example, twite and other seed-eating species are notably scarce wherever flowering and seeding of favoured plants is inhibited by heavy grazing. This may be one reason why wintering twite are rare in the large areas of saltmarsh in the Solway Firth and on Morecambe Bay, despite the relative proximity of these estuaries to moorland breeding habitats in Scotland and the Pennines (Brown & Grice 2005). In the German Wadden Sea, for example, relaxation of grazing pressure on saltmarsh vegetation is known to allow recovery in seed production of favoured saltmarsh plants, and in wintering populations of twite, shore lark and snow bunting (Dierschke & Bairlein 2004). Amongst breeding species, saltmarshes support over half of the British breeding population of redshank and some of the highest densities recorded (Thompson &

Hale 1993; Brindley *et al.* 1998). The impacts of grazing on saltmarsh-breeding populations of redshank have been particularly well studied (Norris *et al.* 1997, 1998; Brindley *et al.* 1998), and we will return to these studies in Chapter 9.

Machair

As a complex mix of cultivation, dune grasslands and semi-natural grazed grasslands, machair fits none of the categories listed above, yet it represents an agriculturally managed complex of habitats unique to the British Isles for the community of bird species that its supports (Fuller 1982). Machair can be defined as a flat, sandy coastal plain with associated mobile and stabilised dune systems. It is dominated by a natural grass and herb community on a base-rich soil that develops over the calcareous 'shell-sands' of the dunes. In Britain, it is found in many areas of western and northern Scotland, but is most widespread along the western coast of the Outer Hebrides where the drier parts of the machair plain are cultivated in a strip rotation of spring-sown cereals (usually oats and rye as a winter cattle feed), potatoes, grassland and fallows. Seaweed was traditionally used as a fertiliser, but modern inorganic fertilisers and herbicides are now in common use. Similar rotations are now virtually absent elsewhere in the British Isles, and even on the machair, larger field blocks are now replacing strips, and cereals undersown with grasses are replacing the fallow phase of the rotation. Cultivation creates a complex succession of habitats as one traverses the machair plain from its seaward edge to the acidic, 'upland' peats inland of the coastal strip. At the seaward fringe, the unvegetated foreshore gives way to mobile dunes, and then to dune grasslands often grazed extensively by cattle and sheep. Further inland lies the cultivated machair which in turn gives way to grazed wet pastures, hay meadows, lochans and marshes at its inland edge. The land then rises over peat soils, supporting successively drier grazed grasslands interspersed with marshy areas (the 'black land') and these in turn give way to exposed peat and heath–grass mixes of the open moor.

The bird communities of machair

The breeding bird community of this complex of habitats comprises teal, mallard, buzzard, corncrake, oystercatcher, ringed plover, lapwing, dunlin, snipe, redshank, Arctic tern, little tern, cuckoo, skylark, meadow pipit, wren, whinchat, stonechat, wheatear, sedge warbler, linnet, twite, reed bunting and a rapidly declining population of corn buntings. In addition, rock doves, blackbirds, song thrushes, pied wagtails, starlings, greenfinches, house sparrows and a few pairs of barn swallows nest amongst croft buildings and abandoned farm machinery. Amongst this breeding bird community, the populations of corncrakes and breeding waders are of international importance.

The vast majority of the remnant population of corncrakes in Britain nests in the traditionally managed, late-cut hay meadows and yellow iris beds of the Hebridean machair. Here, subsidised agricultural management, specifically to benefit nesting corncrakes, has halted the long-term decline of the species, on the brink of its extinction as a British breeding bird, and a modest recovery is now apparent (O'Brien *et al.* 2006). The history of the decline of corncrake populations in Britain and recent research and conservation action to secure its future is worthy of detailed discussion and is discussed as a case study in Chapter 8.

Six species of breeding waders are abundant on the machair, although their individual distributions vary markedly across the complex of habitats available; these are oystercatcher, lapwing, ringed plover, dunlin, snipe and redshank (Fuller *et al.* 1986; Jackson *et al.* 2004). Dry dune pastures hold high densities of oystercatchers, ringed plovers and lapwings, whilst dunlin occupy the wet dune slacks. On the dry, cultivated machair plain, oystercatchers, ringed plovers and lapwings are again the dominant species, with dunlin occupying the edges of cultivation or patches of grassland within the ploughed area. Further inland, where dry machair gives way to damp machair pastures and fens, dunlin and redshank reach their highest densities, with lapwing and snipe also present in good numbers. In this area, the hay meadows so important for nesting corncrakes support few waders, but as the land rises further inland, the grazing and marshes of the black land support oystercatchers, lapwings, snipe and redshank with dunlin in marshy areas at the machair edge. Roughly, wader habitats classify into those dominated by the calcareous shell-sands closest to the sea (i.e. the machair or 'white land') and those forming the boundary between the machair and the peaty upland (the 'black land'). The wader communities of these two broad habitat types differ substantially, although there is some variation between different machair areas in different parts of the Outer Hebrides (Fuller *et al.* 1986). Overall, nowhere else in the British Isles supports a similar assemblage of breeding wader species, and no land area of comparable size supports such high densities of birds. Supporting around a quarter of the ringed plovers nesting in Britain and perhaps a third of the dunlin, the machair of the Outer Hebrides is amongst the most important breeding grounds for waders in western Europe.

As with the other habitats discussed, machair is visited by a wide range of other species that forage on or over machair habitats, but whose nests are not directly associated with them. These include several birds of prey (hen harrier, kestrel, merlin, peregrine, golden eagle and white-tailed eagle), golden plover, curlew, gulls, short-eared owl, rock pipit, fieldfare, redwing, hooded crow, raven, twite and snow bunting. Counts of wintering birds on the machair by RSPB in 2004 revealed up to 3000 twite (N. Wilkinson pers. comm.), suggesting that the weed-rich fallows of the cultivated machair may be an important wintering area for a significant proportion of the Scottish breeding population.

Plate 4.1 This area of unimproved grassland on a steep valley side in Devon supports breeding cirl buntings. One reason for this species' selection of these habitats is the rich fauna of large invertebrates, especially grasshoppers taken as food for nestlings. The improved, reseeded grassland on the other side of the valley will have lost much of its biodiversity, and its value as cirl bunting habitat. © RSPB.

Plate 4.2 A mosaic of heather moorland with the muir-burn strips characteristic of management for red grouse, and sheep-grazed grassland. Other characteristic breeding birds include merlin, lapwing, golden plover, curlew, short-eared owl, wheatear, stonechat, whinchat and ring ouzel. Sadly, illegal persecution all too often means that hen harriers are rare or absent. © Jeremy Wilson.

Plate 4.3 Twite in typical nesting habitat on heather moorland. This species relies on agricultural habitats on both sides of the limit of enclosure, nesting in dense heather or bracken on open moorland, but commuting distances of up to a kilometre or more to collect seed food for nestlings from hay meadows, rough pastures or arable ground. © RSPB.

Plate 4.4 Cereal and fallow strips on the Uist machair. The intimate mix of arable strips, dune grassland, hay meadows and wet pasture supports some of the highest densities of breeding waders in Europe as well as corncrakes, skylarks, meadow pipits, wheatears, corn buntings and even colonies of Arctic and little terns. Twite breeding on nearby moorland commute to the machair to collect seed for nestlings. © Jamie Boyle.

Part II

Trends and patterns of change

5 · *Population sizes and trends*

Chapter 1 emphasised that the agricultural environment has never been static. The clearance of native woodland, fluctuations in the balance of arable and pastoral land, the widespread planting of hedgerows after the Enclosure Acts, the mechanisation and specialisation of agriculture and the modern use of agrochemicals are all examples. Recent policy and technological advances have accelerated the pace of change. Semi-natural habitats have been through a long period of decline, destruction and neglect and whilst the remnants of these habitats are now more secure because of intense conservation activity, and the protection afforded by statutory site designation (Marren 2002), new conservation challenges have arisen on farmland. So, the character of the countryside has always been shaped by farming practices, and given the scale and rapidity of some of the changes, especially recently, we should expect to have seen major changes in the fortunes of farmland wildlife. Few long-term historical changes in wildlife populations have been documented systematically, although the *Historical Atlas of Breeding Birds in Britain and Ireland: 1875–1900* (Holloway 1996), and *Birds, Scythes and Combines* (Shrubb 2003) provide a thorough review of what can be gleaned from early avifaunas to describe long-term change. We are however extremely fortunate in Britain to have national monitoring schemes in place, which allow us to assess change in bird populations over the past 50 years (Table 5.1). Knowledge of bird population trends in Britain is amongst the best in the world. We owe a great deal to the foresight of the BTO in establishing systematic population monitoring of a wide range of breeding birds in the 1960s, to the efforts of those who continue to design, manage and fund monitoring schemes and, especially, to those who devote their time, on a voluntary basis, to gathering the data. This monitoring effort is central to conservation. Without it, we do not know where to direct scarce resources, nor can we assess the efficacy of any conservation measures we choose to implement. This chapter focuses on a review of trends over the past 40 years for those species associated with agricultural habitats. We also set these recent changes in longer-term historical context in Britain, and the wider European context, as well as introducing the contemporary conservation priority-setting protocols that this wealth of high-quality monitoring data has made possible.

Using abundance and trend information to determine conservation status

'Red data lists' are now a familiar element of global conservation inventories, helping to identify and document changes in population status and allow limited conservation resources to be focused on the most immediate conservation priorities. In Britain, the relative richness of bird survey and monitoring data allowed the then Nature Conservancy Council (NCC) and Royal Society for the Protection of Birds to publish the first Red List for birds in 1990 (Batten *et al.* 1990). This list used a set of largely quantitative criteria to decide the conservation status of each species,

Table 5.1 *Systematic survey and monitoring schemes contributing to our knowledge of the size, trends and habitat associations of bird populations of agricultural habitats in Britain*

Survey	Time-span	Description
Common Birds Census (BTO)	1962–2000	Annual breeding season surveys of farmland or woodland plots selected by observers. Numbers of territories estimated for each species from the clustering of records across visits. Informal mapping of habitat data on survey plots. Very detailed and accurate bird data for each plot, but low sample sizes, and non-systematic plot selection and habitat description.
Breeding Bird Survey (BTO, JNCC, RSPB)	1994–present	Successor to CBC. Based on two visits to two parallel transects across randomly selected 1-km squares ($n = 3604$ in 2007). Birds recorded in distance bands to permit density estimation using distance sampling methods. Systematic habitat data collected. Gives lower confidence in bird population sizes on individual survey squares than the CBC, but the much greater sample of survey squares is more representative of regions and habitats across Britain, and gives greater statistical confidence in trend data.
Wetland Bird Survey (BTO, WWT, RSPB, JNCC)	1960s (for most species) – present	Monthly non-breeding counts of predefined wetland sites, including still waters and estuaries, but also some riverine, marshland, canal and open coast sites. Divers, grebes, cormorants, herons, swans, geese, ducks, rails, waders and kingfishers are all counted systematically.
National Bird Atlases (BTO, SOC, IWC)	1968–72 and 1988–91 (breeding); 1981–84 (wintering)	Atlases of the distribution at the 10-km^2 scale. 1988–91 breeding atlas and wintering atlas also provided an indication of abundance in each 10-km square based on fixed time visits to tetrads or the square as a whole. Estimation of range change available by comparison of the two breeding atlases. A further atlas, combining breeding and winter seasons, is being carried out during 2007–11.
Single-species surveys (RSPB, Statutory Conservation Agencies, BTO)	Ongoing	Periodic (usually a 6- or 12-year cycle) national surveys of the breeding distribution and abundance of individual species that are either too scarce or too localised in their distribution to be monitored satisfactorily by CBC or BBS. For the rarest and most localised species (e.g. stone curlew), surveys may be full population censuses. For others (e.g. hen harrier), survey design also incorporates stratified random sampling of grid squares, usually at the 1-km^2 scale, so that the total population is estimated, with confidence intervals, by extrapolation from numbers in sample survey squares.

Sources: For technical details of the field and analytical methods used to estimate population sizes and trends, see Marchant *et al.* (1990) (CBC), Raven *et al.* (2007) (BBS), Maclean & Austin (2006) (WeBS), Gibbons *et al.* (1993, 2007b) (*Breeding Atlases*) and Table 5.2 for details of recently published single-species surveys.

including rarity, the international importance of populations and population decline. Since then, the criteria and list have been updated twice (Gibbons *et al.* 1996a; Gregory *et al.* 2002), and the most recent of these updates uses seven criteria to determine population status in the UK. These are global conservation status, recent population decline (measured by either numbers or range), historical population decline, European conservation status, breeding rarity, localised distribution, and international importance of the UK population. Species are allocated either to Red, Amber or Green list according to the following thresholds for the various criteria (see Gregory *et al.* (2002) for technical details), with a species listed according to the highest priority criterion that is exceeded. Any species not qualifying against any Amber or Red listing criterion is included on the Green list, and the three lists exclude species recorded only as vagrants, scarce migrants or those whose populations have been introduced by humans (e.g. red-legged partridge, pheasant and little owl).

Global conservation status: Red if, from the perspective the species is considered 'Critically Endangered', 'Endangered' or 'Vulnerable' globally according to IUCN criteria for measuring extinction risk (BirdLife International 2000)

Recent population decline: Red if the breeding or non-breeding population has declined by = >50% in numbers or range over 25 years, Amber if the breeding or non-breeding population has declined by 25–49.9% over 25 years.

Historical decline in breeding population: Red if a severe population decline was recorded for the period 1800–1995 by Gibbons *et al.* (1996a), but Amber if the population has at least doubled in the last 25 years and the species now has a population size of at least 100 breeding pairs.

European conservation status: Amber if the species is considered of unfavourable conservation status in Europe (Tucker & Heath 1994, recently revised by BirdLife International 2004) and breeds in the UK.

Breeding rarity: Amber if the species has a five-year running mean population of 1–300 pairs in the UK.

Localised breeding and non-breeding populations: Amber if = >50% of the population is found on 10 or fewer Important Bird Area (IBA: Heath & Evans 2000) or Special Protection Area (SPA: Stroud *et al.* 2001) sites.

International importance during the breeding or non-breeding seasons: Amber if = 20% of the European population, as assessed by BirdLife International (2000), occurs in the UK.

This assessment categorised the conservation status of 247 species in the UK, with 40 (16%) allocated to the Red list, 121 (49%) to the Amber list, and 86 (35%) to the Green list. Below, we make use of these categories to help place the current conservation status of birds of agricultural habitats in historical and geographical context and to show how this prioritisation has driven the recent growth in research and conservation action for those species of greatest concern.

Changes in breeding populations in Britain

Table 5.2 summarises the recent population trends and population sizes of those species with substantial proportions of their British breeding populations in agricultural habitats (see Appendix 1).

Population size

The abundance of all but the scarcest breeding species (where complete censuses are possible) is usually estimated by extrapolation from densities in smaller sampling units such as grid squares (Baker *et al.* 2006), and distribution of this population between different habitat types is usually even less well known (see below). Nonetheless, we can use the national population estimates from Table 5.2 to review briefly patterns of overall abundance.

As one might expect, scarcer species are more likely to be of high conservation concern (Red or Amber listed) than more abundant species (10 of 13 species with <1000 pairs, 23 of 41 species with 1000–100 000 pairs and 12 of 35 species with over 100 000 pairs). Nonetheless, the declines of starling and house sparrow in England account for a staggering loss of over 70 million breeding individuals since the 1960s, whilst the four biggest population increases (woodpigeon, wren, robin and chaffinch) account for a gain of over 16 million. So, whilst the appearance of collared doves or disappearance of corn buntings in the local countryside may stick in the memory, the changes in relative abundance of many of those species that remain common and widespread have been numerically spectacular.

All 15 species with over 1 million breeding pairs are widespread residents or partial migrants found throughout Britain in a wide range of habitats, and all are passerines except woodpigeon and the artificially sustained population of pheasants. The pattern is similar between 100 000 and 1 million pairs – 19 widespread resident or partial migrants, 10 of which are passerines, and just two long-distance migrants (barn swallow and whitethroat). The 25 species having between 10 000 and 100 000 pairs are a diverse mix that includes the more common raptors (sparrowhawk, buzzard and kestrel), species confined mainly to semi-natural grassland and heathland habitats (e.g. golden plover, common gull, tree pipit, wheatear, whinchat, stonechat and twite) and a preponderance of long-distance migrants (e.g. turtle dove, cuckoo, tree pipit, yellow wagtail, wheatear, whinchat, grasshopper warbler, lesser whitethroat and spotted flycatcher). Here, we find species whose geographical range is limited in Britain, either by latitude (e.g. turtle dove) or altitude (e.g. golden plover), and species whose population declines have left them absent from large areas of habitat within their former distributional limits (e.g. turtle dove, tree sparrow and corn bunting). Two species in this category (grey partridge and tree sparrow) have experienced such severe declines in the past 25 years that they would then have had populations exceeding 500 000 and 1 million pairs respectively. The 28 species with fewer than 10 000 pairs include some whose population size and trends are poorly

Table 5.2 *Current population size and recent changes in range and abundance of breeding birds associated with enclosed farmland and grazed, semi-natural habitats*

Species	Current British breeding population size[a]	Range change (%) in Britain 1968–72 to 1988–91[b]	Trends in abundance (%) based on UK CBC, WBS and BBS data[c]			Recent trends in abundance for scarcer species[d]
			1967–2005	1980–2005	1994–2007	
Red grouse	154 700p	−12.7			NS	
Black grouse	3920–6156m	−28.4				Declining[5]
Grey partridge	70 000–75 000p	−18.7	−87	−82	−39	
Quail	4–315m	98.5				Fluctuating[3]
Red kite	1200p*	150.0				Increasing[1]
Hen harrier	687t[2]	33.5				+42% (1998–2004)[2]
Montagu's harrier	7tf	−36.0				Stable[3]
Kestrel	35 400p	−4.1	NS	21(E)	−29	
Merlin	1300p	15.9				Unknown
Corncrake	1067m[6]	−75.6				+122% (1993–2004)[6]
(Great bustard)			Extinct – subject to reintroduction attempt			
Stone curlew	307p[7]	−41.9				+143% (1990–2006)[3,7]
Lapwing	154 000p	−9.0	−33	−53	−18	
Snipe	52 500p	−19.1			38	
Black-tailed godwit	44–52p	37.2				Fluctuating[3]
Redshank	36 600p	−11.8	−45	−43	NS	
Black-headed gull	127 907p	−18.8				
Common gull	48 163p	−12.7				
Turtle dove	44 000t	−24.9	−82	−80	−66	
Cuckoo	9600–20 000p	−4.9	−59	−59(E)	−37	

Barn owl	4000p	−37.5				Stable[8]
Nightjar	3693–5519m[9]	−51.2				+36% (1992–2004)[9]
Woodlark	3085p*	−62.8				+89% (1997–2006)*
Skylark	1 700 000t	−1.6	−59	−51(E)	−13	
Tree pipit	74 400t	−15.0	−83	−82(E)	NS	
Meadow pipit	1 680 000t	−3.2	−44	−44(E)	−16	
Yellow wagtail	11 500–26 500t	−9.4	−70	−67	−47	
Dunnock	2 060 000t	−3.2	−34	−12	25	
Ring ouzel	6157–7549p	−27.0				Declining[10]
Song thrush	1 030 000t	−2.1	−50	−18	18	
Mistle thrush	205 000t	−2.0	−41	−37	−12	
Grasshopper Warbler	10 500p	−36.7			68	
Dartford warbler	3208p*	60.7				+70% (1994–2006)*
Spotted flycatcher	58 800t	−2.3	−84	−79	−59	
Red-backed shrike	0–5p	−86.5				Extinct[3]
Chough	300–346p	12.8				+45% (1992–2002)[13]
Starling	8 100 000–10 800 000i[11]	−3.6	−83	−79(E)	−26	
House sparrow	6 000 000p[12]	−5.3	−69	−62(E)	−10	
Tree sparrow	68 000t	−19.6	−97	−94(E)	NS	
Linnet	535 000t	−4.6	−71	−44(E)	−27	
Twite	10 000p	−1.1				Declining (E)[14]
Bullfinch	157 700t	−6.5	−49	−32	−18	
Yellowhammer	792 000t	−8.6	−55	−53	−19	
Cirl bunting	697p	−83.2				+511% (1989–2003)[15]
Reed bunting	176 000–193 000t	−11.7	NS	NS	31	
Corn bunting	8500–12 200t	−32.1	−86	−85	−36	

(cont.)

Table 5.2 (*cont.*)

Species	Current British breeding population size[a]	Range change (%) in Britain 1968–72 to 1988–91[b]	Trends in abundance (%) based on UK CBC, WBS and BBS data[c]			Recent trends in abundance for scarcer species[d]
			1967–2005	1980–2005	1994–2007	
Mute swan	5299p	−2.6	216	159	NS	
Greylag goose	18 150p	258.0			220	
Mallard	47 700–114 400p	−1.2	167	37	27	
Golden eagle	442tp[4]	5.4				+3 (1992–2003)[4]
Sparrowhawk	38 600p	19.6	179	87(E)	NS	
Buzzard	31 100–44 000t	8.4	422	219	56	
Hobby	2200p	141.3				Increasing
Red-legged partridge	72 000–200 000	32.1	NS	−30	43	
Pheasant	1 688 000–1 788 000f	0.9	96	92(E)	40	
Moorhen	240 000t	−9.2	29	NS	16	
Oystercatcher	113 000p	11.4	122	55	−17	
Ringed plover	8540p	12.2				Unknown
Golden plover	38 400–59 400p (UK)[16]	−7.7			NS	
Dunlin	9150–9900p	20.6				Unknown
Ruff	37i	200				Fluctuating[3]
Curlew	105 000p	−2.8	NS	NS(E)	−36	
Rock dove	Unknown	Unknown				Unknown
Stock dove	309 000t	−6.8	181	NS(E)	NS	
Woodpigeon	2 450 000–3 040 000t	−2.3	143	84	22	
Collared dove	284 000t	6.7	428	58	27	

Little owl	5800–11 600p	−11.0	−41	−37	−26	
Long-eared owl	1100–3600p	−24.6				Unknown
Short-eared owl	1000–3500p	−15.2				Unknown
Green woodpecker	24 200p	−4.1	211	96(E)	31	
Barn swallow	678 000t	1.1	NS	NS(E)	25	
Pied wagtail	255 000–330 000t	0.3	87	NS	15	
Wren	8 000 000t	−0.3	95	50	25	
Robin	5 500 000t	1.0	38	43	21	
Whinchat	14 000–28 000p	−16.2			−26	
Stonechat	8500–22 000p	−14.6			278	
Wheatear	55 000p	−6.8			13	
Blackbird	4 620 000t	−1.9	17	NS	24	
Fieldfare	1–4p	205.9				Declining
Redwing	2–17p	22.5				Declining
Lesser whitethroat	64 000t	16.2	NS	−25	NS	
Whitethroat	931 000t	−6.7	−62	58	31	
Great tit	1 952 000t	−0.5	100	52	55	
Blue tit	3 333 000t	−1.4	44	26	14	
Magpie	590 000t	1.1	102	32	NS	
Jackdaw	503 000t	−3.0	87	71	40	
Rook	1 022 000–1 304 000p	−0.4			NS	
Carrion/hooded crow	950 000t	2.8	116	48(E)	Carrion 19/hooded–14	
Raven	12 900p	−9.2			134	

(cont.)

Table 5.2 (*cont.*)

Species	Current British breeding population size[a]	Range change (%) in Britain 1968–72 to 1988–91[b]	Trends in abundance (%) based on UK CBC, WBS and BBS data[c]			Recent trends in abundance for scarcer species[d]
			1967–2005	1980–2005	1994–2007	
Chaffinch	5 562 000t	0.7	30	20	14	
Greenfinch	695 000t	−2.7	30	48	27	
Goldfinch	299 000t	5.4	NS	NS(E)	39	

Notes: Species in **bold** (except great bustard) are those currently Red or Amber-listed in the UK (Gregory *et al.* 2002) due to breeding population decline or range contraction, and where the breeding population is strongly associated with agricultural habitats.

[a] p, pairs; t, territories; i, individuals; m, males; f, females. Population size data are for Britain (unless otherwise stated) and taken from Baker *et al.* (2006), apart from where denoted by superscripts to the following references: [1]Wotton *et al.* (2002a); [2]Sim *et al.* (2007b); [3]Brown & Grice (2005); [4]Eaton *et al.* (2007); [5]Sim *et al.* (2008); [6]O'Brien *et al.* (2006); [7]www.ukbap-reporting.org.uk; [8]Toms *et al.* (2001); [9]Conway *et al.* (2007); [10]Wotton *et al.* (2002b); [11]Robinson *et al.* (2005b); [12]Robinson *et al.* (2005a); [13]Johnstone *et al.* (2007); [14]Langston *et al.* (2006); [15]Wotton *et al.* (2004); [16]Thorup (2006). *unpublished results from latest national surveys (S. Wotton pers. comm.).

[b] Range change is the % change in number of occupied 10-km squares in Britain between the 1968–1972 and 1988–1991 *Breeding Atlases*, taken from Gibbons *et al.* (1993).

[c] NS, no significant change during the period. All are UK abundance trends unless followed by (E) which denotes an England-only trend. Trends are derived from CBC, WBS and BBS data, as appropriate to the species. Annually updated species trend graphs can be viewed at www.bto.org/birdtrends, in addition to BBS trends for the four UK countries (where available).

[d] Recent population trend for species not adequately covered by the BBS. Where possible, the trend has been quantified for a period (as stated in brackets) between two recent population censuses.

known (quail, dunlin, long-eared owl and short-eared owl), some formerly more abundant species in continuing decline (e.g. black grouse and ring ouzel), some now responding well to management intervention which we will discuss in greater detail later (e.g. red kite, corncrake, stone curlew, nightjar, woodlark, chough and cirl bunting), a few right on the edge of their European breeding range in Britain (e.g. Montagu's harrier, ruff, redwing and fieldfare) and one for which the British breeding population is probably extinct for the foreseeable future (red-backed shrike). All of these species have breeding ranges limited by habitat (e.g. ring ouzel and Dartford warbler) or latitude (e.g. little owl) within Britain, or have declined within an historically much larger distribution (e.g. black grouse, corncrake and red-backed shrike).

In a wider context, we can compare these breeding population sizes with those elsewhere in Europe to determine for which species the UK hosts a high proportion of the European population (BirdLife International 2004). The UK holds greater than 25% of the European population of four of the indigenous species listed in Appendix 1 – stock dove (42–59%), curlew (30–49%), oystercatcher (25–38%) and wren (21–37%) – and between 10% and 25% of a further 10 species (greylag goose, mute swan, sparrowhawk, moorhen, woodpigeon, meadow pipit, dunnock, robin, blue tit and carrion crow). The international significance of Britain's stock dove population is perhaps the most striking on this list, given the relative lack of research attention that the species has received.

The international importance of Britain's over-wintering populations of migratory waders and wildfowl is well known. The same BirdLife International data set shows that the UK holds greater than 10% of the European population of at least 12 species that are concentrated on agricultural habitats in winter – pink-footed goose (82%), lapwing (48%), common gull (41%), brent goose (39%), Bewick's swan (34%), golden plover (26%), greylag goose (24%), wigeon (22%), teal (20%), barnacle goose (18%), mute swan (14%) and jack snipe (12%).

Habitat-specific distribution and population sizes

One of the key advantages of the Breeding Bird Survey (BBS) over its predecessor, the Common Birds Census (CBC), is that BBS can be used to estimate habitat-specific detectabilities of birds (Buckland *et al.* 2001). This is because BBS methods involve random survey square selection, recording of birds in distance bands along transect lines within these squares, and systematic recording of habitat type along transect lines (see Crick 1992), allowing distance sampling techniques to be used. This, in turn, allows habitat-specific densities and hence population sizes of birds to be estimated. Previously, the problem of different but unquantifiable detectabilities in different habitat types had confounded attempts to make reliable population size comparisons between habitats (e.g. Bibby & Buckland 1987). BBS data have now been used by several studies to estimate the habitat-specific densities and population sizes of common British breeding birds (Gregory & Baillie 1998; Gregory

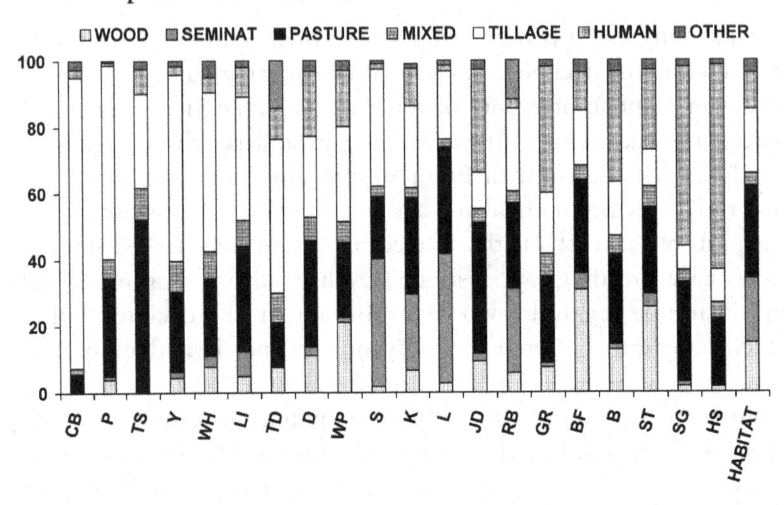

Fig. 5.1 Estimated proportions of the total British populations of 20 bird species occurring within seven habitat classes in 1998. WOOD, broadleaved, coniferous and mixed woodland, and scrub; SEMINAT, semi-natural grassland, heathland and bog; PASTURE, improved or unimproved grass farmland; MIXED, mixed farmland; TILLAGE, tilled farmland; HUMAN, urban, suburban and rural human sites; OTHER, freshwater and coastal habitats, inland rock and other miscellaneous habitats occurring over very small areas. CB, corn bunting; P, grey partridge; TS, tree sparrow; Y, yellowhammer; WH, whitethroat; LI, linnet; TD, turtle dove; D, dunnock; WP, woodpigeon; S, skylark; K, kestrel; L, lapwing; JD, jackdaw; RB, reed bunting; GR, greenfinch; BF, bullfinch; B, blackbird; ST, song thrush; SG, starling; HS, house sparrow. Species are ordered from left to right by the total proportion of their population estimated to be in the enclosed farmland (PASTURE/MIXED/TILLAGE) categories. HABITAT, proportion of total habitat area made up by each habitat type for comparison with species columns. Data from Newson *et al.* (2005). (See Newson *et al.* (2008) for most recent data for a wider range of species, published as we go to press.)

1999), with a recent analysis considering 20 species of predominantly lowland farmland birds (Newson *et al.* 2005). Figure 5.1 summarises these results to show the broad habitat associations of these 20 species during the breeding season. Towards the left-hand end of the *x*-axis is a group of six species whose populations are heavily concentrated on farmland – corn bunting mainly on tilled farmland, and tree sparrow, yellowhammer, whitethroat, linnet and turtle dove across a wide range of farmland types, but with tree sparrow tending to be at highest density on grass-dominated farmland, and the others on mixed or tilled farmland. Lapwing and skylark are found at high densities on both farmland and semi-natural grasslands, whilst dunnock, woodpigeon, kestrel, reed bunting, bullfinch, jackdaw, greenfinch, blackbird and song thrush are all habitat generalists, though with the latter four species found at highest densities in human habitats (villages, gardens, parkland). The remaining two species, starling and house sparrow, are particularly associated with human habitats and pastoral farmland, but tend to be found at much lower densities elsewhere. This illustrates the more general point that many species found

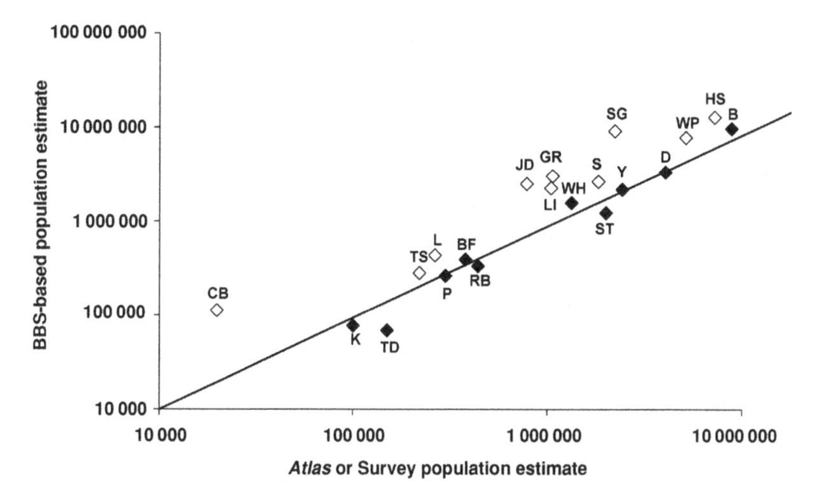

Fig. 5.2 Comparison of *Atlas* or single-species survey-based population estimates and those derived from BBS data by Newson *et al.* (2005). Estimates are given as number of breeding individuals. Pairs of population estimates for a given species would lie on the solid line if equal. Open diamonds show the 10 species for which the BBS-based estimate was the larger of the two by the greatest percentage. K, kestrel; P, grey partridge; L, lapwing; WP, woodpigeon; TD, turtle dove; S, skylark; B, blackbird; ST, song thrush; WH, whitethroat; JD, jackdaw; SG, starling; D, dunnock; HS, house sparrow; TS, tree sparrow; GR, greenfinch; LI, linnet; BF, bullfinch; Y, yellowhammer; RB, reed bunting; CB, corn bunting.

commonly on farmland may also have substantial populations in other broad habitat types, with significant levels of movement of individuals between agricultural and other habitat types (e.g. Fuller *et al.* 2004).

Newson *et al.* (2005) went on to compare UK population estimates derived from these habitat-specific density estimates with previously available population estimates derived from either the 1988–1991 *Breeding Atlas* (Gibbons *et al.* 1993) or more recent single-species surveys of lapwing, skylark and corn bunting (Donald & Evans 1995; Browne *et al.* 2000; Wilson *et al.* 2001). These estimates span four orders of magnitude and at this gross scale, the two sets of estimates accord well, although the BBS-derived estimates tend to be higher than those derived from the *Atlas* or from single-species surveys (Fig. 5.2). This is perhaps because BBS uses distance-sampling to attempt to correct for individuals that are present but remain undetected. Closer inspection reveals that the 10 species for which this difference between BBS-based and *Atlas*/survey-based population estimate is greatest are all species that aggregate during the breeding season. This may be because they forage communally (e.g. woodpigeon), nest colonially (e.g. tree sparrow, house sparrow) or both (e.g. jackdaw, starling, linnet, greenfinch), or because the restricted availability of suitable nesting habitat on most farmland today tends to cause the birds to aggregate (e.g. lapwing, skylark, corn bunting). The abundance of such species would be overestimated if transect routes followed by observers deviated from the ideal straight line (which in reality they do) in such a way that they systematically encountered habitat features around which these species aggregate. As an example, if transect lines tended to follow access routes through or adjacent to farmyards, then observers might be more likely to find aggregations of starlings and house sparrows close to the transect route than would be the case if the route was a straight line, thus overestimating numbers of these species. In this context, it would be helpful to collect more detailed habitat and distance data for at least a sample of BBS survey squares and transect routes in order to check whether densities of aggregated species may be being systematically over- or underestimated.

Trends in breeding populations

Before the 1960s, systematic censusing of bird populations in Britain was the exception rather than the rule, with the BTO's Heronries Census dating back to 1928 being perhaps the most remarkable of the exceptions. Nonetheless, Britain has a long and proud tradition of natural history recording, which allowed Alexander & Lack (1944), Parslow (1973) and Sharrock (1974) to summarise the changing status of our breeding bird populations. These historical reviews were updated by Gibbons *et al.* (1996b) and rendered semi-quantitative by converting previous descriptive accounts of changing status to scores. Thus, over a given time period, a 'huge and widespread decline' such has been the fate of red-backed shrikes would score −5, and a 'spectacular increase' such as that of collared doves would score +5, with species showing no evidence of a clear trend scoring zero. Gibbons *et al.* (1996b) presented these scores separately for the periods 1800–1849, 1850–1899, 1900–1939, 1940–1969 and 1970–1995. We considered that any species scoring less than at least −3 between 1800 and 1940 equated to a Red- or Amber-listed species of its time, and compared this list with those species currently Red- or Amber-listed on the basis of population decline or range contraction (the species listed in bold in Table 5.2). Some striking patterns emerge from this comparison across various broad ecological groupings of species (Table 5.3).

Overall, the number of species Amber- or Red-listed because of population decline has increased, especially amongst the 'seed-eating' species, invertebrate feeders (other than waders), and long-distance migrant species. These categories are of course not mutually exclusive. Two species – the great bustard and red-backed shrike – have become extinct in the UK. On the other side of the coin, the conservation status of raptors has generally improved due to a reduction in human persecution. Indeed two of the five species still listed as of high conservation concern – hen harrier and red kite – have recovering populations in parts of their range, helped greatly by reintroduction programmes in the case of the red kite (Carter *et al.* 2003). The pigeons and corvids, amongst which are the species traditionally of greatest concern to farmers as crop pests, continue to fare well as a group, the notable exceptions being the turtle dove (the only long-distance migrant in the group) and chough (the only species in the group associated with semi-natural habitats).

Table 5.3 *Comparison of conservation status before 1940 and in recent times*

Species group[a]	Number of species	Red-/Amber-listed on basis of 1800–1940 trends	Now Red-/Amber-listed on basis of range contraction or population decline	Change (%)
All species	89	33	45	+36
Long-distance migrants	19	7	13	+86
Resident, short-distance migrants or partial migrants	71	27	34	+26
SEED-EATERS[b] (larks, finches, sparrows, buntings)	14	5	11	+122
INSECTIVORES (cuckoo, nightjar, green woodpecker, barn swallow, pipits, wagtails, dunnock, wren, chats, thrushes, warblers, spotted flycatcher, red-backed shrike, starling)	28	6	15	+150
Raptors (including owls)	12	8	7	−12.5
Waders (including stone curlew)	11	6	5	−17
Geese, swans, ducks, gamebirds and rails	11	4	5	+25
Corvids and pigeons	12	4	2	−50

[a] Excludes Great Bustard and introduced species since these species have no population status coding.
[b] This term describes a group of species that are predominantly granivorous when full-grown through most still feed invertebrates to nestlings.
Source: See text and Table 5.2 for derivation of table content.

The growing proportion of farmland bird species exhibiting contracting breeding ranges and declining populations started to become apparent from inspection of CBC data as early as the late 1980s (Marchant *et al.* 1990). The widespread, generic nature of these declines did not become apparent until subsequent reviews were published (Fuller *et al.* 1995; Siriwardena *et al.* 1998a; Gregory *et al.* 2004). The more detailed trend information available from the CBC and BBS for the last 30 years or so (Table 5.2) shows that 13 species have may have declined by more than half in the 25 years between 1980 and 2005. This list is worth reiterating in full since many of these species will be the focus of more detailed discussion in later chapters: **grey partridge, lapwing, turtle dove, cuckoo, skylark, tree pipit, yellow wagtail, spotted flycatcher, starling, house sparrow, tree sparrow, yellowhammer** and **corn bunting**. To these of course should be added **black grouse, corncrake, stone curlew, red-backed shrike** and **cirl bunting** – five species that have suffered either precipitous historical declines making them simply too rare to be monitored by CBC or BBS, or have become extinct. Of these 18 species, perhaps only black grouse and tree pipit would never have been regarded as lowland farmland species in Britain. BBS trends for the period 1994–2007 are little rosier. Of the first group of 13 species above, only lapwing, skylark, tree pipit, house sparrow and tree sparrow have declined by less than 25% over this period, and **curlew** and **cuckoo** have been added to the ranks of species showing worrying rates of decline. Conversely, populations of a few of the rarest species, notably corncrake, stone curlew, cirl bunting and, in England and Wales, black grouse, have recovered markedly since the early 1990s in response to intensive conservation management (Aebsicher *et al.* 2000; Brown & Grice 2005; Sim *et al.* 2008; Chapter 8).

If we examine Table 5.2 for species whose populations have doubled over the same period (1980–2005) then there are only two – mute swan and sparrowhawk – both species which have shown strong recoveries from previous reductions in numbers caused by environmental pollution and, in the case, of sparrowhawk, illegal persecution (Newton 1986; Sears & Hunt 1991). Looking more broadly across all species that have shown statistically significant increases over this period, they comprise largely common wetland species (mute swan, mallard and moorhen), artificially sustained gamebirds (pheasant), recovering raptor populations (sparrowhawk and buzzard), resident insectivores (e.g. green woodpecker, robin and wren) and a range of species which, either through their exploitation of agricultural crops or garden habitats and food sources (or both), are to some extent commensal with humans (e.g. woodpigeon, collared dove, magpie, jackdaw, carrion crow, great tit, blue tit, chaffinch and greenfinch). When Dolton & Brooke (1999) considered change in populations of terrestrial British breeding birds from the perspective of biomass, they estimated an overall decline of 10% between 1968 and 1988, and a 29% decline if pheasant was excluded from the calculations. They noted too that most species contributing to the decline were strongly farmland-associated whilst those showing increases were a more heterogeneous group containing several commensal species. However, the increasing population of pheasants was by far the greatest contributor to avian biomass increase on farmland and the continuing increase of the very large pheasant population will be the main contributor today. Given that Dolton & Brooke's analysis was based solely on breeding population sizes, yet of the order of 25 million pheasants may now be released for shooting in Britain each year (Tapper 1999), the broader ecological impacts of such high pheasant densities merit detailed investigation (Sutherland *et al.* 2006). As yet studies have only examined the impacts of pheasant release pens on woodland structure. These studies have found

Table 5.4 *Examples of consistent and contrasting changes in breeding populations of selected bird species in UK countries, as reported by the BBS from 1994 to 2007 (Risely* et al. *2008)*

Species[a]	England	Scotland	Wales
(a)			
Dunnock	+19	+40	+38
Blackbird	+21	+25	+47
Great tit	+43	+87	+80
Blue tit	+10	+16	+29
Reed bunting	+19	+56	
(b)			
Curlew	−20	−48	−33
(c)			
Oystercatcher	+47	−27	
Lapwing	+5	−38	
Woodpigeon	+27	−11	+35
Cuckoo	−56	+39	−52
House sparrow	−18	+30	+93

[a] Species are grouped into (a) those showing geographically consistent increases, (b) those showing consistent declines and (c) those showing contrasting trends between countries.

that, provided densities are not excessive, then pheasant management can lead to a richer field and shrub layer in woodland with attendant benefits for other breeding birds such as warblers (Sage *et al.* 2005; Draycott *et al.* 2008).

Overall, rates of decline of lowland farmland birds, so severe from the 1970s to the 1990s, have slowed more recently (Gregory *et al.* 2004). Interpretation is however complicated by the fact that BBS achieves better geographical coverage than its predecessor, the CBC, whose design meant that its distribution of survey plots was always more representative of the more intensively managed landscapes of lowland England than of other parts of the UK (Fuller *et al.* 1985). This means that a trend calculated over the whole of the UK from BBS data is more likely to conceal variation between different regions, variation that only the development of the BBS since 1994 has allowed us to measure. Table 5.4 shows this variation for example species for which there are sufficient BBS data to allow trends to be reported separately for at least two countries in Britain.

Recent population changes are indeed consistent (at least in direction if not in rate) for many species listed above, but the five species listed in section (c) show that this is by no means always the case, with all five showing starkly contrasting trends between at least two countries in Britain. Such contrasts are also likely to exist at yet finer scales. In England, there are sufficient BBS survey squares to allow a breakdown of trend information for many species by Government Office Region and, again, although for many species trends are reasonably consistent across regions this is not always the case. The thrushes (blackbird, song thrush and mistle thrush) provide a good example; all three have shown relatively small overall changes (+21%, +20% and –23% respectively across England as a whole between 1994 and 2007: Risely *et al.* 2008). Yet these overall changes conceal a tendency for increases towards the north and west to be countered by smaller increases or declines in the south and east. Table 5.4 also implies that increases have tended to outweigh declines over the past 10 years, but this in part reflects the fact that many of the declining species are already too scarce or localized for BBS trends to be estimated from the smaller samples of survey squares in Scotland and Wales (77% of the 3604 BBS squares surveyed in 2007 were in England: Risely *et al.* 2008). Trends cannot be calculated outside England for any of the five most rapidly declining species between 1994 and 2007; the corn bunting, to take just one example, is probably now extinct in Wales and reduced to no more than 800–1000 occupied territories in Scotland (Forrester *et al.* 2007).

The species listed in Table 5.4 are noticeably more associated with lowland than upland agricultural habitats, and this simply reflects the relative scarcity of BBS survey plots in the uplands. The problem was even more acute during the period of the CBC from 1962 to 2000, and our knowledge of the population trends of many widespread breeding birds of moorland and other upland habitats is therefore poor. The most comprehensive analysis of CBC data from upland plots that has proved possible was undertaken by Henderson *et al.* (2004a) who examined trends between 1968–1980 and 2000 for 35 breeding species over 13 areas of marginal upland grassland from Dumfries and Northumberland in the north to Powys in the south, with most of the sites in the Pennines. With such a small number of sites and substantial variation in trends between sites, most species showed no overall, significant change in abundance, but 11 species associated with moorland declined – to extinction across all sites in the case of ring ouzel – whilst only five increased (Fig. 5.3).

The CBC is not the only data source on upland breeding bird abundance and, even though not designed for long-term monitoring purposes, many upland bird surveys were carried out during the 1980s and early 1990s by the statutory conservation agencies, RSPB and others. These usually had the aim of identifying the most important upland areas for breeding birds (especially waders) as a precursor to designation of sites as Special Protection Areas (SPAs) under the EU Birds Directive. Between 2000 and 2002, 13 of these areas from Exmoor to Lewis and Harris were resurveyed using the same field methods originally employed, in order to assess population change over the intervening years (Sim *et al.* 2005). The results (Fig. 5.4) show some of the same evidence of marked geographical variation in trend that BBS data has revealed for other species in Table 5.4, but also clear evidence of strong decline amongst some species (notably lapwing, curlew and ring ouzel), and increases in others (stonechat and raven). Snapshot data from two periods separated by many years are of course much weaker evidence of trend than high-quality annual

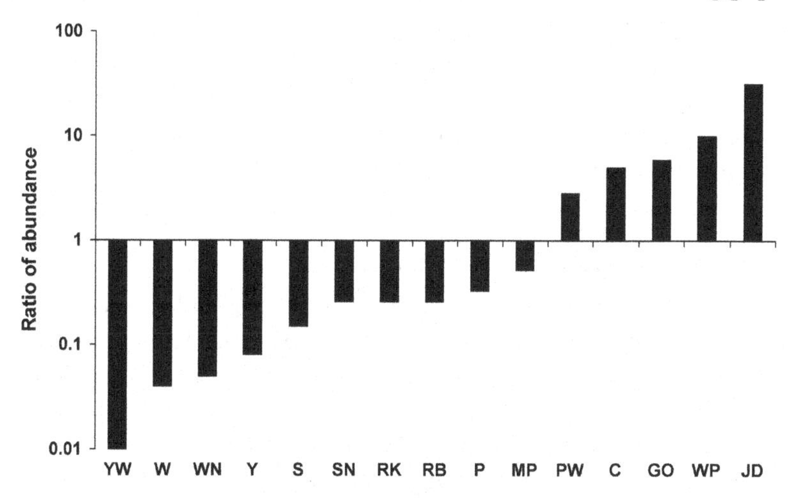

Fig. 5.3 Significant changes in abundance of breeding birds on 13 marginal upland CBC sites between first surveys undertaken between 1968 and 1980, and a repeat survey in 2000. Confidence intervals not shown for clarity, but are large in all cases. Ring ouzel became extinct at all five sites where it was recorded in the period 1968–1980 and is not shown. A further 18 species either showed no significant change in abundance or were not strongly associated with agricultural management of grasslands. YW, yellow wagtail; W, wheatear; WN, whinchat; Y, yellowhammer; S, skylark; SN, snipe; RK, redshank; RB, reed bunting; P, grey partridge; MP, meadow pipit; PW, pied wagtail; C, carrion crow; GO, goldfinch; WP, woodpigeon; JD, jackdaw. Data from Henderson *et al.* (2004a).

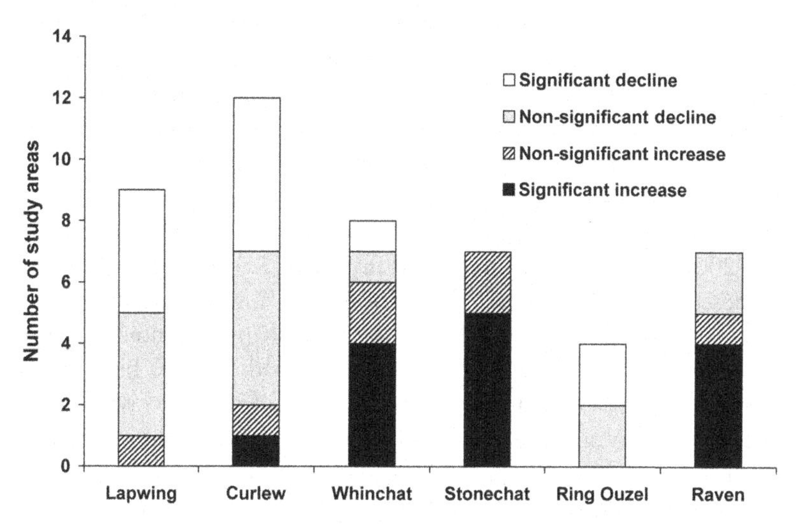

Fig. 5.4 Variation in trends of breeding birds in different upland areas of Britain. Data from Sim *et al.* (2005).

monitoring data. However, 45 of the 48 significant population changes recorded by these resurveys were higher than between-year changes in abundance on BBS plots for the same species, suggesting that at least the statistically significant abundance changes recorded by the resurveys are likely to indicate real long-term trends rather than annual fluctuations in essentially stable populations.

Changes in wintering populations in Britain

Although breeding range and numbers provide the conventional index of conservation status for birds, some species only occur in Britain in the winter, or else relatively small breeding populations are greatly augmented by winter immigration. Britain's only atlas of the distribution and abundance of birds in winter (Lack 1986) does give a valuable snapshot of the situation in the early 1980s, but examples of long-term monitoring of trends in winter bird populations in agricultural habitats are rare. The Wetland Birds Survey (WeBS), run in partnership by BTO, the Wildfowl and Wetlands Trust (WWT), RSPB and the Joint Nature Conservation Committee (JNCC) involves annual counts by volunteers at around 2000 wetland sites (mostly estuaries and still-water bodies), and is able to report trends for many wildfowl and waders since the 1960s at all scales from the individual site to national level. This trend information encompasses a range of wildfowl species listed in Appendix 1 which spend a significant proportion of time during winter exploiting agricultural habitats, usually grazing on grassland, stubbles or other crop waste, or newly sown crops. Figure 5.5 summarises these changes for some of these.

These data show very marked increases for most species, although note that WeBS excludes the large populations of wintering species such as lapwings, golden plovers and gulls which winter on farmland, away from WeBS survey sites. In general, very little is known about the overall distribution of these species across agricultural and wetland habitat types. Wintering numbers of mallards and the European race of white-fronted goose are the exceptions, declining by 63% and 32%, respectively, since the late 1970s. The former is most probably 'short-stopping'; in other words, increasingly wintering in areas further north and east as winters become milder (Maclean & Austin 2006). Rarer species are absent from Fig. 5.5, and perhaps most notable amongst these is the bean goose, a species which was historically much more numerous as a wintering species in Britain, but which now winters largely in Germany, the Netherlands, Hungary and Sweden. Fewer than 1000 birds now winter annually in Britain, and these are heavily concentrated into two wintering flocks; one on the Yare Valley grazing marshes in East Anglia, and the other on mixed farmland on the Slammanan Plateau in Scotland's central lowlands.

Trends in the wintering populations of terrestrial species are much less well known and although for many resident species we might reasonably take the view that these trends will generally mirror breeding trends, there are several species that are either absent as breeding species, or whose breeding populations are greatly supplemented by winter influxes to farmland. Independent data on their winter population trends

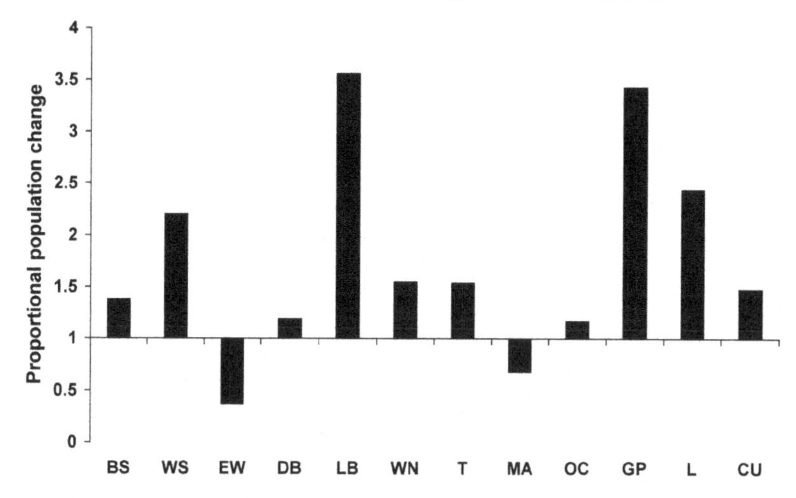

Fig. 5.5 Proportional changes in British population at wetland sites between the winters 1978/79 and 2003/04 for 12 wildfowl and wader species whose wintering populations make extensive use of agricultural habitats for feeding. A value of 1 on the *y*-axis indicates no change in population; 2 indicates a doubling, and 0.5 a halving. Data from Maclean & Austin (2006), which also gives a detailed description of the WeBS data collection and analysis methods. BS, Bewick's swan; WS, whooper swan; EW, European white-fronted goose; DB, dark-bellied brent goose; LB, light-bellied brent goose; WN, wigeon; T, teal; MA, mallard; OC, oystercatcher; GP, golden plover; L, lapwing; CU, curlew.

would be informative. Such species include jack snipe, redwing, fieldfare, brambling and snow bunting, as well as lapwing, golden plover, snipe, long and short-eared owls, starling and chaffinch whose breeding populations are supplemented by winter influxes of birds from northern Europe.

The Banbury Ornithological Society's (BOS) long-term winter monitoring of 1200 km^2 of the south Midland mixed farmland is an impressive exception to this general lack of winter trend data. Data from the 1975/76 to 1995/96 winters were published by Easterbrook (1999) and compared with national trends in breeding populations over the same period. He found strong positive correlations between change on the BOS survey area and both national CBC breeding population change and national range change between the two breeding bird *Atlases* across a sample of 48 species. Of 21 species that declined by over 25% on the BOS survey area, only eight were then Red- or Amber-listed for population decline nationally (Gibbons *et al.* 1996a). These eight (grey partridge, skylark, song thrush, starling, bullfinch, reed bunting, tree sparrow and corn bunting) remained so listed when the population and conservation status of UK breeding birds was revised by Gregory *et al.* (2002). But, of the other 13 species, this revision added dunnock, mistle thrush, marsh tit, willow tit, house sparrow and yellowhammer to the Amber or Red lists on the basis of decline, and changed the listing of starling from Amber to Red. For this group of largely non-migratory farmland birds, these results show how winter trends in the Banbury area did indeed reflect the wider, national reductions in breeding populations that led to their revised conservation status. Similarly, if we look at species recorded as stable or increasing by Easterbrook, we find that of 27 species, all but three (redwing and fieldfare – Amber-listed due to their small UK breeding populations, and linnet – Red-listed due to population decline) were Green-listed by Gibbons *et al.* (1996a). Since then, just two of these locally increasing species have moved to the Amber list (kestrel and goldcrest), again suggesting that the winter trends recorded around Banbury were broadly reflective of national population changes. The linnet counter-example is interesting – a substantial (>25%) increase in the local wintering population over a 20-year period contrasting with a severe national decline in the breeding population over the same time. Linnets are partial migrants with some birds moving south as far as north Africa to winter whilst others remain in the UK (Wilson 2002), so here may be one example where changes in the proportion of birds emigrating mean that winter trend provides information that would not be detected by breeding season monitoring alone. In passing, it is worth noting that Easterbrook's observation of stability in the wintering populations of redwings and fieldfares in the BOS study area over the 20 winters between 1975/76 and 1995/96 is a rare example of robust evidence of the status of these two numerous members of the winter farmland bird community.

The European context

Few nations are fortunate enough to have quantitative data on bird populations and their trends that are as comprehensive as those of Britain, but Tucker & Heath (1994), now updated by BirdLife International (2004), collated available data on population size and trends across the whole of Europe for 524 species, and devised a two-tier system of classification to categorise each species' conservation status. Each species was first assigned a European Threat Status as indicated in Table 5.5.

Species meeting any of the IUCN criteria CR, EN, VU, NT (near-threatened) or DD at a global level are then assigned to Species of European Conservation Concern (SPEC) category 1 (SPEC1), whilst species meeting criteria CR, EN or VU solely at a European level, or meeting any of the additional European Threat criteria (D, R, H, L) are assigned to SPEC2 (if more than 50% of their breeding or wintering population or range is in Europe) or SPEC3 (if less than 50% of their breeding or wintering population or range is in Europe).

We have used this classification to compare (Table 5.6) the conservation status of those farmland, grassland and heathland birds listed in Appendix 1 at UK and European levels.

Overall, there is the expected tendency for species that are secure at European level to be in the non-declining category in the UK (45 of 63 species) and, likewise, for species in the Vulnerable, Declining or Depleted categories across Europe as a whole to be declining in the UK (25 of 37). This confirms that, in general, patterns of population change in the UK are similar to those across Europe for the same set of species. However, the exceptions are interesting. Eight species (ring ouzel, song

Table 5.5 *Classification of European Threat Status*

Criterion	Description
Critically Endangered (CR) Endangered (EN) Vulnerable (VU)	Species meeting IUCN Red List criteria (IUCN 2004) CR, EN or VU at a European level and therefore considered to face an extremely high, very high, or high risk, respectively, of extinction in Europe
Declining (D)	Species not meeting the above criteria, but which has declined by >10% over 10 years (1990–2000) or three generations, whichever is the longer
Rare (R)	Species not meeting the above criteria, but numbering fewer than 10 000 breeding pairs (or 20 000 breeding individuals or 40 000 wintering individuals)
Depleted (H)	Species not meeting the above criteria, but which has not yet recovered from a moderate or large decline during 1970–90 which led to classification by Tucker & Heath (1994) as EN, VU or D
Localised (L)	Species not meeting the above criteria, but is heavily concentrated with >90% of the European population occurring at 10 or fewer sites
Secure (S)	Species not meeting the above criteria, and therefore having a favourable conservation status in Europe
Data Deficient (DD)	Species where there is inadequate information to assess extinction risk based on distribution and/or population status

Source: BirdLife International (2004).

Table 5.6 *Comparison of conservation status of native farmland, grassland and heathland birds (taken from Appendix 1A), at UK and European levels; species are categorised row-wise by UK conservation status (see Gregory* et al. *2002) and column-wise by European Threat Status (Table 5.5)*

Birds of Conservation Concern in the UK Status	European Threat Status[a]				
	Vulnerable	Declining	Depleted	Rare	Secure
Red	3	6	9	0	8
Amber (decline)	1	6	2	0	11
Amber (non-decline) or Green	2	4	4	1	43

[a] None of the species has European Threat Categories of CR, EN, R or L at European level, and only corncrake is graded SPEC1 due to being considered near-threatened at a global level.

thrush, grasshopper warbler, twite, bullfinch, yellowhammer, cirl bunting and reed bunting) are Red-listed in the UK but their European populations as a whole are considered secure. It is perhaps noteworthy that although all four 'farmland buntings' in the UK are in decline, for only one – corn bunting – is this decline widespread across the continent. Conversely, there is a rather mixed bunch of ten species which are classed as either Vulnerable (Bewick's swan, brent goose), Declining (ruff, jack snipe, curlew, wheatear) or Depleted (dunlin, short-eared owl, green woodpecker, barn swallow) at the European scale, but whose populations trends in the UK vary. Of these, the ruff is an extremely rare breeder, but wintering populations of brent geese and breeding populations of green woodpeckers and barn swallows are currently increasing. Breeding dunlin, curlew, short-eared owl and wheatear – all now found mainly on upland farmland and moorland – are either stable or decreasing.

Summary

Britain is fortunate to have a wealth of high-quality bird population monitoring data extending back to the 1960s and in some cases earlier, the great majority of those data having been collected voluntarily by the membership of the BTO. Taken together, these trend data reveal some dramatic changes in the populations and breeding ranges of some species. In particular, there has been a preponderance of severe declines amongst seed-eating species, amongst insectivorous species characteristic of dry rather than wetland habitats, and amongst long-distance migrants. Populations of raptors, corvids and pigeons have generally been on the increase, as have populations of those boreal or arctic-breeding geese, swans and ducks that graze crops and grassland in winter. The wider ecological impacts of the huge population and biomass of released pheasants in Britain requires investigation. Trends of many widespread upland breeding birds remain poorly known, and efforts are currently under way to try to improve upland BBS coverage. This is an important initiative given that the limited data available from the CBC, together with periodic surveys of individual upland blocks suggest that substantial population changes have occurred, including widespread declines of breeding waders and ring ouzels, and increases in stonechats and corvids. Overall, the number of species categorised as of high or medium conservation concern due to range or population decline has increased markedly over the second half of the twentieth century. In a wider geographical context, these trends reflect similar changes across Europe. In agricultural habitats, a greater proportion of bird species are considered to be of unfavourable conservation status at a European scale than in any other habitat. Declines of bird populations in agriculturally managed habitats are a Europe-wide phenomenon.

In the last decade, the BBS suggests that declines of some farmland species in the UK have slowed or reversed and, in particular, there has been a strong recovery in numbers, if not in breeding range, of three rare species (corncrake, stone curlew and cirl bunting), whose remnant populations have been subject to intensive conservation management. The more geographically representative survey coverage of Britain offered by the BBS, and the recent repeat surveys of upland areas both show us that there may sometimes be marked variation in trends between different regions. This is no surprise, but has never before been so well measured. This finer-grained resolution of population trends can help us to understand what may be driving them. For example, what clues may lie in the fact that house sparrow populations are now increasing in Scotland and Wales, but remain in long-term decline in England, or the fact that curlews have declined by over 80% in North Wales whilst doubling in numbers in the South Pennines?

Plate 5.1 The corncrake was once widespread, nesting in hay meadows and arable crops. Earlier harvesting and conversion of late-cut hay meadows to early-cut silage has caused a dramatic decline, with the species now found only in the north and west of Scotland where hay-making persists. Here, management schemes to provide early and late cover and to compensate farmers and crofters for later, 'corncrake-friendly' cutting of meadows has brought about a marked recovery in numbers. © RSPB.

Plate 5.2 Although still a widespread – though declining – breeding bird in upland edge habitats in northern England and Scotland, the lowland populations of lapwings have declined severely. Replacement of spring- by autumn-sown cereals, drainage and improvement of grasslands, high stocking rates and impacts of predation on low-density populations have all played a part. Large numbers of lapwings from British and European breeding populations still winter on arable and grassland throughout lowland Britain. © RSPB.

Plate 5.3 The woodpigeon is one species which has benefitted greatly from intensive, arable agriculture. In particular, the rapid increase in acreage of oilseed rape provides this species with an abundant food source throughout much of the year. © RSPB.

Plate 5.4 Many seed-eating species – finches, sparrows and buntings – have declined in part due to loss of winter seed sources, especially weed-rich over-winter stubbles. The tree sparrow is one example. Rotational set-aside and agri-environment scheme measures directed at restoring seed supplies in arable environments have helped to stem declines, and tree sparrows have shown signs of a modest recovery in numbers in recent years. © RSPB.

6 · *Patterns of associations between agricultural change and wildlife populations*

Chapters 1 and 5 showed us that the agricultural landscapes and bird communities that we now see, and are outlined in Chapters 2–4, are merely a snapshot from a long history of continuous change. Nonetheless, acceleration in technological development coupled with increasing state intervention in agriculture brought an unprecedented pace of change to the food production industry, and to the environments in which food is grown, during the second half of the twentieth century. Over the same time period, the declines in populations of birds and other wildlife on farmland have been severe. A great deal of recent research has been devoted to understanding whether or not these changes reflect cause and effect; whether recent agricultural change has simply been too rapid and comprehensive for wildlife to adapt and has been the main cause of the widespread losses. The burgeoning interest in this field of research now confronts us with a daunting array of studies published in the ecological, agricultural and ornithological literature over the past 15 years (Fig. 6.1), so to organise our discussion we have attempted to separate studies which have focused on *pattern* from those which have focused on *process*. By pattern, we mean studies that have sought to establish whether changes in bird (and other wildlife) populations across wide geographical extents or long durations are associated with agricultural change. By process, we mean more detailed studies (which are thus usually more restricted in duration and extent), which seek to establish the mechanisms linking agricultural change to the fortunes of individual bird species. In this chapter (focusing on patterns) and Chapters 7–9 (focusing on mechanisms), we review the evidence for a cause-and-effect relationship between agricultural change and wildlife loss on farmland, and seek to understand which have been the most important processes affecting bird populations.

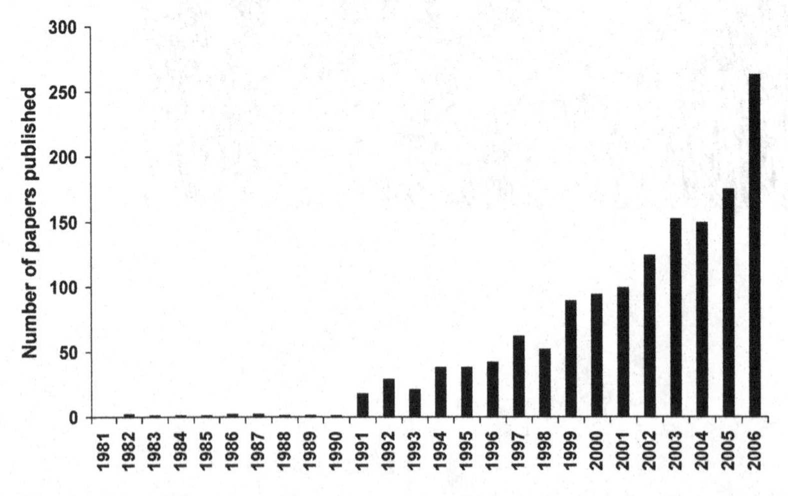

Fig. 6.1 Growth in peer-reviewed papers in which the terms related to agriculture/farming/farmland and birds/biodiversity are in the title, key words or abstract. Data collated from a search of 26 agriculture, conservation, ecology and ornithology journals using Web of Knowledge (last accessed January 2008).

The main elements of post-war agricultural change (Chapter 1) include agrochemical crop protection and nutrition, land drainage, increasing livestock densities, crop breeding, earlier sowing and harvesting, loss of non-crop habitats, simplification of crop rotations, specialisation of farming enterprises and, now, new technological developments such as genetic modification of crops. Together they are often subsumed under the umbrella of 'agricultural intensification'. This term is useful since the above changes are heavily interdependent and all are concerned with increasing agricultural productivity by making as great a proportion of primary production as possible available for human consumption (Krebs *et al.* 1999). To the extent that this is achieved, we might expect that the rest of nature will suffer, and that measures of agricultural intensification will be strongly inversely correlated with measures of the diversity, range and abundance of birds and other wildlife in agricultural habitats. Generally speaking the term 'agricultural intensification' describes a move towards high input/high output systems, and monoculture as a cost-effective goal. Monoculture results in a decline in both habitat quality and diversity and is the antithesis of an ecosystem (Evans 1997).

Spatial and temporal associations between agricultural intensification and bird population trends

The strength of the association between bird population trends and measures of agricultural intensification has been tested in England and Wales by Chamberlain *et al.* (2000). Their first finding was that a wide range of management changes measured between the 1960s and the mid 1990s were not only strongly intercorrelated, but that their rates of change were also highly coincident, with a period of stability to the early 1970s, followed by rapid change between then and the late 1980s, and a further period of relative stability since then. In particular, variables describing autumn cropping, agrochemical usage and intensive grassland management all showed strong increases during the 1970s and 1980s, in contrast to variables describing spring cropping, fodder cropping and low-intensity management which showed marked decreases. Mirroring this result, there was a marked pattern of common population change across an ecologically and taxonomically diverse group of 29 bird species for which CBC trend data were available. Generally speaking these species showed relative stability before the 1970s, followed by a period of very rapid change (mostly characterised by decline) between the mid 1970s and the late 1980s, which then slowed into the 1990s. Taken together these results showed that the timing of broadly synchronous intensification of a wide range of interdependent agricultural management practices matched that of a similarly synchronous set of population changes amongst farmland birds. During this period of rapid agricultural intensification, most of the associated bird population changes were declines (notably tree sparrow, turtle dove, skylark, song thrush, linnet, blackbird, dunnock, bullfinch, corn bunting, starling, grey partridge, reed bunting and lapwing), though a small number (e.g. stock dove, great tit, blue tit, rook, jackdaw and chaffinch)

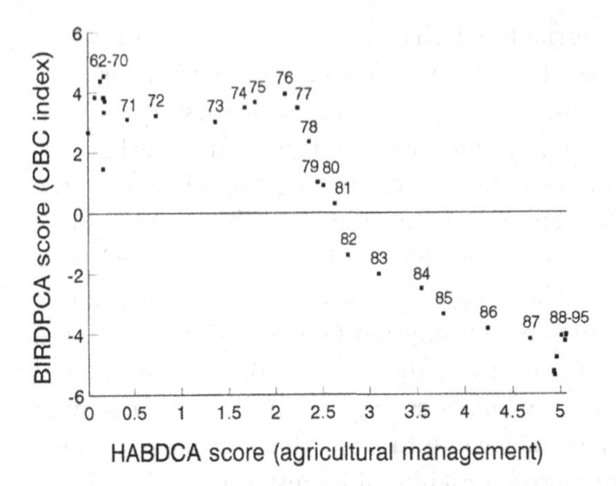

Fig. 6.2 The relationship between a composite measure of agricultural management intensity (increases to the right) and a composite measure of farmland bird abundance in England and Wales between 1962 and 1995. Reproduced, with kind permission, from Chamberlain *et al.* (2000), which see for details of derivation of the axis scores.

were increases. Interestingly, this coincidence was not perfect, with a strong indication of a time lag between the onset of rapid agricultural intensification in the early 1970s, and the start of widespread population declines about five years later (Fig 6.2). Overall, this comparison of bird population change and agricultural change at the national scale indicates the strong probability that agricultural intensification has been the main driver of farmland bird population declines, particularly during the late 1970s and 1980s.

Intensification of farming practice has of course not been confined to the UK but has happened in most countries where EU membership has provided access to the huge financial incentives for increased productivity offered by the CAP (Donald *et al.* 2002b). We might therefore expect to see a similar association between trends in breeding bird populations and trends in agriculture in those countries that have experienced the same intensification of farming practice that has been seen in Britain over the past 40 years. The prediction was tested more widely by Donald *et al.* (2001d). First, they showed that various indices of agricultural intensity (e.g. cereal yield, milk yield, cattle density, fertiliser use) collated from the United Nations Food and Agriculture Organization database were strongly correlated and tended to increase markedly over time from the 1960s to the 1990s. This was especially so in EU Member States subject to CAP subsidies, but less so in the (then) non-EU eastern European countries. Here, many of these measures reached a plateau in the 1980s and then fell sharply during the 1990s following the collapse of state support. They then compared these trends in agricultural intensity with trends in bird populations available from the BirdLife International/European Birds Census Council European Bird Database (EBD) for 52 species associated with agricultural habitats (Tucker & Heath 1994; Tucker & Evans 1997) across 30 European countries

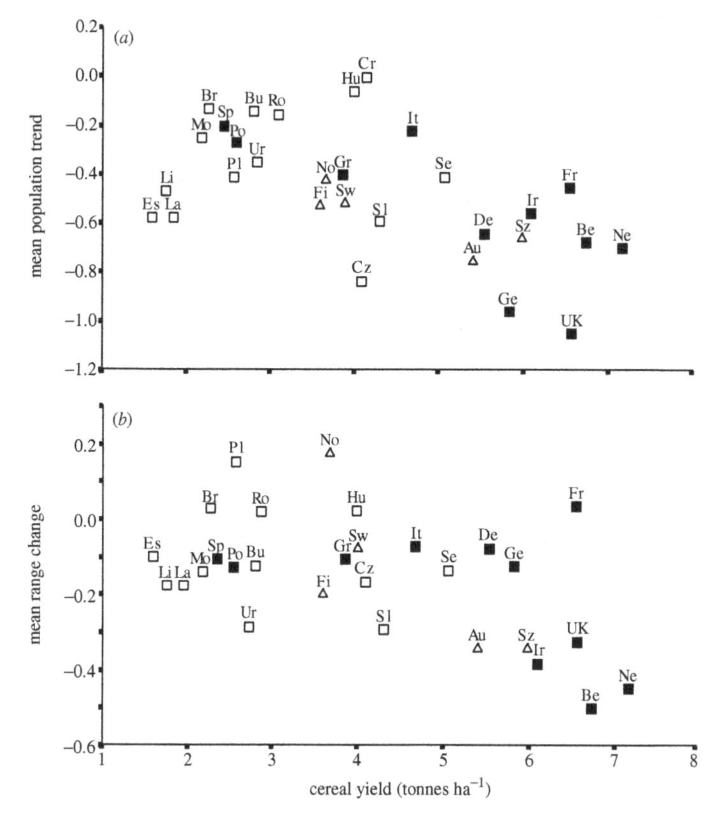

Fig. 6.3 The relationship between cereal yield (tonnes ha^{-1}, as measured in 1993) and population trend of farmland birds between 1970 and 1990, across 30 European countries. Filled squares indicate countries that have been EU members since before 1990, triangles indicate countries that joined the EU between 1990 and 2000, and open squares indicate eastern European countries that have joined the EU since 2000, or remain non-members. Au, Austria; Be, Belgium; Br, Belarus; Bu, Bulgaria; Cz, Czech Republic; De, Denmark; Es, Estonia; Fi, Finland; Fr, France; Ge, Germany; Gr, Greece; Hu, Hungary; Ir, Ireland; It, Italy; La, Latvia; Li, Lithuania; Mo, Moldova; Ne, Netherlands; No, Norway; Pl, Poland; Po, Portugal; Ro, Romania; Se, Slovenia; Sl, Slovak Republic; Sp, Spain; Sw, Sweden; Sz, Switzerland; Ur, Ukraine. Reproduced from Donald *et al.* (2001d) with kind permission.

over the period 1970–90. The analysis was also weighted to take account of the fact that data quality, and hence confidence in trend information, differs between species and countries. Cereal yield alone was found to explain over 30% of variation in the overall farmland bird trend between countries and, although the bird trend was negative in most countries (Fig. 6.3), the most marked declines were in those nations with higher cereal yields, notably those with long-term membership of the EU. The UK showed the most severe decline in farmland bird populations of all 30 countries considered. A recent update of these analyses covering the period 1990–2000 (Donald *et al.* 2006) found that these correlations have persisted more recently with continuing Europe-wide declines of farmland species (e.g. Herzon *et al.* 2008), a trend not shared by bird assemblages of other broad habitat types over the same period.

These analyses are of course only correlations, and when data for individual countries are examined in more detail, the picture is often more complex. For example, in Britain and Sweden respectively, Siriwardena *et al.* (1998a) and Wretenberg *et al.* (2006) found similar temporal patterns of decline amongst lowland farmland birds with a common tendency for declines to be most severe amongst species specialising in exploitation of agricultural habitats, and for generalist species to decline less severely or even increase. However, Wretenberg *et al.* (2006) also found that agricultural abandonment in some areas was also having detrimental effects by leading to development of tall, rank vegetation unsuitable for ground-nesting species such as lapwings and skylarks. In Denmark, Fox (2004) found that the populations of 27 farmland bird species were characterised more by stability or increase between 1976 and 2001 (5 declined, 10 were stable and 12 increased). This contrasts with the preponderance of declines (14 declines, 9 were stable and 4 increased) recorded by Chamberlain *et al.* (2000) in Britain between 1969 and 1995. In particular, species such as grey partridge, skylark, song thrush, tree sparrow, linnet, bullfinch, corn bunting and reed bunting, all of which have declined severely in Britain, had remained stable or increased in Denmark. Fox (2004) noted the same general intensification trends in Denmark as have occurred in Britain, though changes were often later in Denmark, but found that use of pesticides and inorganic fertilisers had declined substantially during the 1980s and 1990s as a consequence of the imposition of a pesticide tax, and legislative limits on fertiliser use. The same constraints probably also accounted for the more rapid growth of organic agriculture in Denmark, reaching 6% of the farmed land area by 2000. These trends certainly contrast with those in Britain, where agrochemical applications are consistently higher than in Denmark, and Fox suggests that more detailed studies of individual species responses may help us to understand whether these recent reversals of elements of agricultural intensification in Denmark might account for the better fortunes of Danish farmland birds. Such studies would certainly be worthwhile, but it is also worth bearing in mind that at least some measures of agricultural intensity in Denmark never reached those typical of intensive agriculture in Britain during the 1980s and 1990s. For example, levels of phosphate fertiliser application have always been lower in Denmark (Fig. 6.4). Also, even though Fox reports the switch from spring-sowing to autumn-sowing of cereals to be the greatest change in agricultural land use in Denmark, much as was the case in Britain, the proportion of cereals that are spring-sown (especially barley) remains three times higher in Denmark than in England (the part of the UK where recorded farmland bird declines have been most severe). Overall, Denmark thus retains a mix of autumn- and spring-sown cereals, whereas England is now dominated by autumn-sowing (Fig. 6.4). So, whether the contrasting fortunes of farmland birds in Denmark and Britain do reflect recent reversals of agricultural intensification in the former country, or

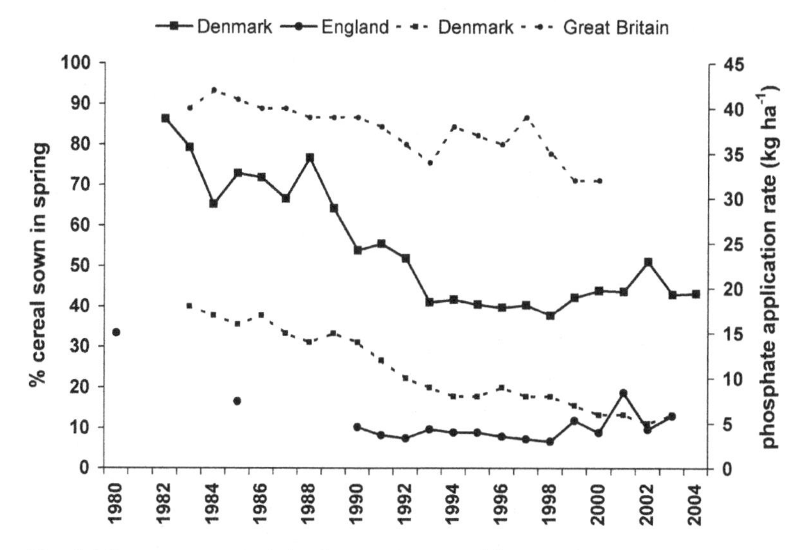

Fig. 6.4 Long-term trends in the percentage of the cereal crop sown in spring (solid lines) and the application rate of phosphate fertiliser (dashed lines) to farmland in Britain and Denmark. Data from the online databases of the UK Department for Environment, Food and Rural Affairs (www.defra.gov.uk) and StatBank Denmark (www.dst.dk).

whether they simply reflect the fact that Danish agriculture never reached levels of intensification sufficient to precipitate the scale of declines seen in Britain remains uncertain. Only more detailed studies of species' responses to different elements of agricultural intensification are likely to shed more light.

Back in Britain, BTO researchers pursued the relationship between farmland bird decline and agricultural intensification further by examining geographical patterns of decline using data from the two national breeding bird atlases (Chamberlain & Fuller 2000, 2001; Gates & Donald 2000; Donald & Greenwood 2001). They found that farmland species tended to show greater declines in areas where they were already sparsely or patchily distributed compared to areas where they were more continuously distributed. This implied a retreat of species to core ranges, rather than a fragmentation into many isolated populations, and is consistent with the effects of agricultural change being relatively even and synchronous across the range, such that peripheral areas with low and variable densities of birds tend to experience local extinctions (Channell & Lomolino 2000). For seven severely declining farmland species (grey partridge, lapwing, turtle dove, yellow wagtail, tree sparrow, reed bunting and corn bunting), 10-km squares from which these species were lost between the two atlases were grassland-dominated, and had experienced loss of tillage crops. These squares were also more similar (in terms of land-use, climate and topography) to squares where the species had not occurred in the *1968–1972 Atlas*, than to those where the species was still present in the *1988–1991 Atlas*. This led to the prediction that if range contraction continued in the same manner then all seven species would tend to show future losses from the northern, western and upland peripheries of their geographical ranges, and to contract to core populations in the lowland south and east. However, Chamberlain & Fuller (2001) also noted that for at least some species (e.g. grey partridge, lapwing, tree sparrow and corn bunting), there was a mismatch between geographical patterns of decline in population density and contraction in range, with abundance declining fastest in lowland arable and mixed farmland whilst range contractions were most apparent in the pastoral north and west. This last result is important because it raises further questions about the dynamics of decline in British farmland birds. First, can the drivers of decline in such agriculturally contrasting parts of Britain really be the same, or do extinctions in 10-km squares on the margins of the range have different causes to those in the core? The corn bunting provides an interesting test case for this question (see Chapter 8). Second, might it even be possible that extinctions at the peripheries of these species' ranges in Britain do not always demand a local ecological explanation? In some cases these peripheral populations may always have been population sinks (i.e. areas where local mortality is not balanced by reproduction so that the population relies on immigration for persistence), so that their loss is simply caused by decline in the rate of immigration from population sources (i.e. areas where local reproduction exceeds local mortality and therefore exports dispersing individuals) as populations decline in the core of the range (Brown 1984; Pulliam 1988; Maurer & Brown 1989; Gates & Donald 2000). Such a case has yet to be identified amongst farmland birds in Britain.

Trends and patterns in other wildlife groups

Plants and invertebrates

The New Atlas of the British and Irish Flora (Preston *et al.* 2002) summarises trends in plant populations between the 1960s and the 1990s. There was a preponderance of declines amongst the arable flora, amongst species with low nutrient requirements and amongst low-growing species of open habitats. Increases predominated amongst species that are shade-tolerant, tall-growing and with high nutrient requirements (Fig. 6.5).

This is of course exactly the response expected as a consequence of increasing fertiliser use, and a trend towards early sowing of vigorous crops supplemented by intensive fertiliser regimes in both crops and grassland. It is those species that are easily controlled by herbicides (e.g. charlock), germinate in spring (e.g. corn marigold) or are poorly competitive in cultivated, nutrient-rich environments (e.g. many species of unimproved grassland) which have suffered. Many other studies report similar trends, both in Britain (e.g. Rich & Woodruff 1996; Smart *et al.* 2000; Sutcliffe & Kay 2000; Stevens *et al.* 2004) and elsewhere in Europe (e.g. Andreasen *et al.* 1996). Nutrient enrichment, in particular, has been pervasive across a wide range of landscapes. A study of changing floral composition in almost 1500

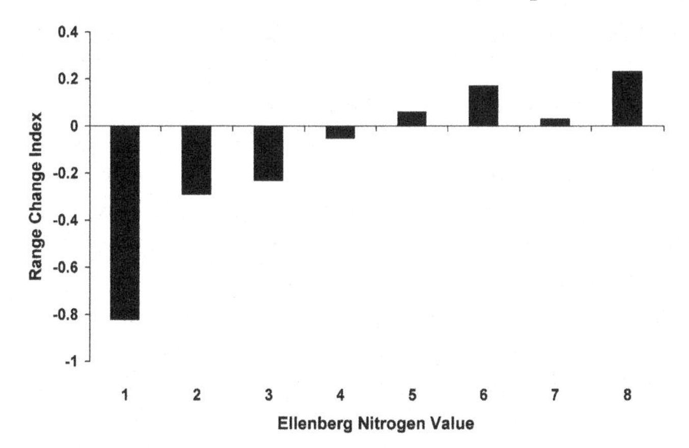

Fig. 6.5 An example of the impact of agricultural intensification on wild plant populations. The relationship between Ellenberg nitrogen value (a measure of the fertility of the substrate in which the species typically grows, with low values indicating low nitrogen levels) and range change in the British Isles between two broad time periods: 1930–69 and 1987–99. Species with positive values increased in range between these two time periods, whilst species with negative values contracted in range. Data from Table 7.8 of Preston *et al.* (2002), which see for details.

permanently marked Countryside Survey quadrats across Britain between 1978 and 1998 found that the pattern of species declines and increases was consistent with the effects of increasing nutrient availability on inherently low-fertility habitats such as semi-natural grassland, heathland and moorland (Smart *et al.* 2005). Preston *et al.* (2002) identify six factors that have had the most profound influence on plant population trends. Five of these are aspects of agricultural intensification: increasing use of fertilisers, loss of semi-natural heathland and grassland habitats, grassland intensification, loss of arable cultivation from marginal upland farmland, increasing herbicide use associated with more competitive crop varieties, and conifer afforestation. On the other side of the coin, three decades of benefit from stringent air-quality legislation introduced in the 1960s has allowed some lower plant groups to make a comeback as levels of particulate air pollution have declined; for example, lichens have tended to increase in Britain over the past 30 years (Gilbert 1992).

Long-term monitoring of invertebrate populations also suggests detrimental impacts of agricultural intensification. At the Rothamsted Experimental Station in Hertfordshire, light traps have been sampling moth populations in the local woodland and farmland since the 1930s, and a national network of light traps has been in operation since 1968. When Woiwod & Harrington (1994) examined long-term trends in species diversity and numbers of moths in the Rothamsted data, they found that both had decreased on farmland (with the annual catch since the 1960s being only a third of that before 1950), whilst the woodland catches remained unchanged in both species diversity and abundance. A more recent and comprehensive analysis of data from across Britain found that within a sample of 337 moth species, there were more than twice as many species showing significant long-term decreases (54%

of the total) as increases (21%). Seventy-one species (21% of the total) had declined by more than 30% per decade over the 35-year study period (Conrad *et al.* 2006). Strikingly, species with wide distributions across Britain tended to be decreasing, especially in the south, whilst species with initially southerly distributions tended to increase (Conrad *et al.* 2004). The extent to which agricultural and other environmental changes may be driving these trends remains to be established. However, Conrad *et al.* note that these declines have worrying implications for insectivorous species such as bats and many birds, and conclude starkly that 'the pattern of southern decline across so many species indicates a widespread deterioration of suitable conditions for many of the more abundant moths in the south of Britain'. Across Britain's butterfly fauna as a whole, J. A. Thomas *et al.* (2004) found that rates of decline were faster than for birds and plants, and certainly here agricultural intensification is implicated as an important driver of change. For example, the *Millennium Atlas of Butterflies of Britain and Ireland* (Asher *et al.* 2001) lists 23 species (39% of Britain's 59 resident butterfly species) where some aspect of agricultural intensification has been the main cause of decline, even though many species concurrently exhibit northwards range expansions (Warren *et al.* 2001). Loss and fragmentation of semi-natural grasslands coupled with changing grazing patterns on the remaining patches are the main agricultural culprits, but mechanical trimming of hedges and clearance of ditches on farmland, and changes in woodland management are also implicated for some species. Even amongst common and widespread butterfly species where the national range is stable, or in some cases increasing as the climate warms, the picture at a finer scale is much less rosy. For example, Cowley *et al.* (1999) found that for eight common species in 35 km^2 of North Wales, the proportion of the land surface over which these butterflies were flying had declined by 49–91% during the twentieth century even though their ranges are stable or increasing nationally.

One long-term study of invertebrate populations is much more intimately linked to intensively managed farmland. The Game and Wildlife Conservation Trust (GWCT) has collected invertebrate samples annually from a sample of over 100 cereal fields in Sussex since 1970. Although no significant changes in abundance have been observed for some groups (e.g. ladybirds, ground beetles and predatory flies), others (e.g. cereal aphids and their parasitoid wasps, spiders, sawfly larvae and rove beetles) declined in numbers substantially, in some cases by over 50% (Aebischer 1991; Ewald & Aebsicher 1999). In the cereal field environment, Ewald & Aebischer attributed most of these declines either to the direct toxic impacts of insecticides, especially those applied in summer to control aphid populations, or to the indirect effect of herbicides removing invertebrate food plants for phytophagous insects, and hence also removing the food of their predators. This latter mechanism is generally assumed, rather than proven, to operate but one classic study by Sotherton (1982a, b) showed that the knotgrass beetle, an herbivorous species that feeds solely on knotgrass and black bindweed, could be rendered locally extinct by application of herbicides designed to kill these arable weeds. Moreover, even if the weed populations reappeared quickly in subsequent years, the beetle populations took many more years to recover.

Vertebrates other than birds

There are no long-term studies of the distribution or abundance of British amphibian, reptile or mammal populations, although monitoring schemes are now being developed for some. Even reliable estimates of population size are difficult because of the problems in surveying many of the species directly (Harris *et al.* 1995). What evidence there is suggests that populations of some species (e.g. amphibians, reptiles, bats, brown hare, hedgehog, weasel, harvest mouse, field vole) have declined whilst those of others (e.g. stoat, rabbit, fox, wood mouse) have remained stable or increased (Robinson & Sutherland 2002). In some cases, understanding the impact of agricultural change on these species is confounded by changes in gamekeeping intensity over the same time period, but in others there is good evidence of the impact of changing agricultural practice. For example, bats are nocturnal consumers of the same invertebrate food resource base that is exploited by birds during daylight hours and, like birds, declines of bats associated with agricultural habitats are widespread across Britain and Europe (Harris *et al.* 1995; Hutson *et al.* 2001). On farms where recent intensification has been reversed (in this case organically managed farms – see below), measures of bat activity and foraging (mostly by pipistrelles) were higher (by 61% and 84%, respectively) on organic than on neighbouring farmland. These differences could be explained by higher abundance and species richness of nocturnally flying insects on the organic farms (Wickramsinghe *et al.* 2003, 2004). There is also evidence that brown hare populations have declined in direct response to agricultural intensification. Hares require a variety of vegetation covers on farmland (e.g. spring- and autumn-sown crops, fallows, grassland) to provide a suitable mix of feeding habitat and cover from predators through the year, and reduction in habitat variety both within and between fields may have contributed to their decline (e.g. Smith *et al.* 2004).

Links between declines in different wildlife groups

So far in this chapter we have reviewed the evidence that trends in bird and other wildlife populations on farmland have been associated with agricultural intensification in ways that suggest cause and effect. Trends in different groups may themselves be linked causally. For example, Biesmeijer *et al.* (2006) examined changes in bee and hoverfly assemblages over time in two countries that have experienced rapid agricultural intensification – Britain and the Netherlands. They found that declines were most prevalent amongst species that specialised in particular habitats or in feeding on particular plant species. In addition, outcrossing plant species that are

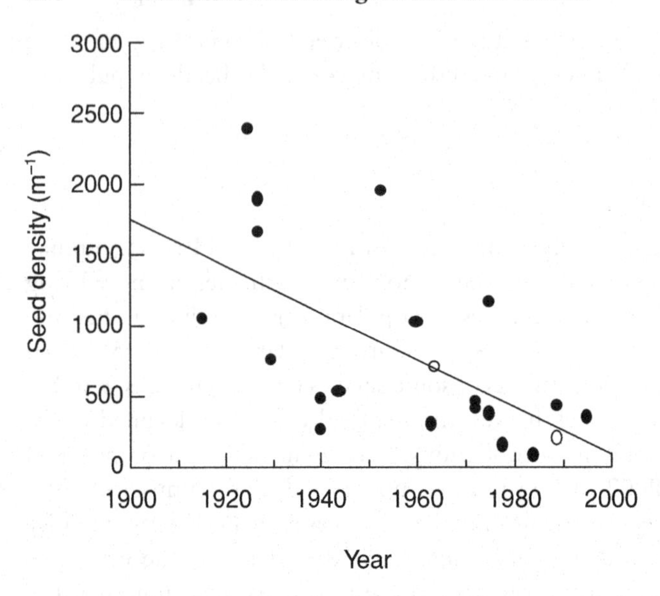

Fig. 6.6 Published estimates of seed density in the top 1 cm of arable soils. Filled symbols indicate data from Britain; open symbols indicate data from Denmark. Reproduced from Robinson & Sutherland (2002) with kind permission.

reliant on insect pollination for reproduction had tended to decline more severely than other plant species. Similarly, Gabriel & Tscharntke (2007) compared arable plant diversity at the centres and edges of 20 pairs of organic and conventional cereal fields and found that the diversity of insect-pollinated plant species benefitted most from the adoption of organic management.

These functional linkages between trends in different species groups are likely to involve birds too. Many of the invertebrates, and the seeds and fruits of many of the plants discussed above, are of course also food for birds exploiting agricultural habitats. Wilson *et al.* (1999) reviewed the impacts of agricultural change on those invertebrate and plant groups that are important components of the diet of 26 seed-eating farmland birds of western Europe (gamebirds, pigeons, larks, corvids, finches, sparrows and buntings). They did not find that declining bird species were associated with particular plant foods, but did find declines in the diversity and abundance of many of the most important seed-food species (notably knotgrass, fat-hen and chickweed), especially on arable land, where the combination of herbicide usage, seed-cleaning and the development of more competitive crop varieties has caused widespread reductions in weed populations. Robinson & Sutherland (2002) have collated published studies of broadleaved weed seed densities in the top 1 cm of arable soils. They found strong evidence of substantial declines in the densities of seeds likely to be available to birds (Fig. 6.6), with seed densities in excess of 1000 m^{-2} common before the 1950s but values below 500 seeds m^{-2} more typical since the 1960s.

On grassland, although species diversity of seeding weeds declines with intensification, a smaller number of competitive, grazing-tolerant species (e.g. docks, dandelions and chickweed) do well and continue to provide an important food source for birds where grassland management permits them to set seed. Many of the bird species concerned rely on invertebrates as a protein-rich food source for chicks, and here Wilson *et al.* (1999) found that grasshoppers, sawflies, spiders and leaf-beetles, four invertebrate groups that are highly sensitive to insecticide usage, all formed part of the diet of declining bird species. Overall, the review concluded that pesticide usage, intensive cultivations and loss of uncultivated field margin habitats on arable land, coupled with agricultural improvement of grassland through drainage, reseeding and fertilization, were likely to have combined to reduce the availability of key seed and invertebrate foods for birds on farmland.

Benton *et al.* (2002) compared population trends of 15 farmland bird species in lowland Scotland with agricultural change, climatic variation and changes in aerial arthropod populations, and found strong relationships between all of these sets of variables. Birds tended to be more abundant in years, or following years, of high arthropod abundance and lower indices of agricultural intensity. Though purely correlative, these results are consistent with an effect of intensity of agricultural management on bird populations via food supply. The partially time-lagged relationship between arthropod and bird populations is noteworthy (see also Chamberlain *et al.* 2000). Perhaps this is exactly what would be expected if the process at work is an effect of food supply on breeding success or survival. This could be the case, especially if the presence of non-breeding ('floating') individuals in populations creates a buffer effect in which total population size may begin to decline before any effect is detectable in the size of the surveyed territorial breeding population (e.g. Durell *et al.* 2004).

Wildlife population responses when agricultural intensification is reversed

The strong associations between wildlife losses and agricultural intensification over large spatial scales and long time periods provide compelling circumstantial evidence of a cause–effect relationship. However, before we move on to consider the mechanisms in more detail, it is worth considering two recent changes in agricultural management that buck the pervasive trend of intensification, and which have happened over sufficiently large areas of land in Britain and elsewhere in Europe that their impact on birds and other wildlife is measurable. These are, first, the conversion of increasing areas of lowland farmland to organic production standards and, second, the widespread adoption of set-aside as a production control measure following the CAP reforms of 1992. Neither was introduced to British agriculture specifically with the intention of delivering wildlife benefits, but both represent reversals of the general trend of agricultural intensification, which offered conservation scientists an opportunity to assess whether wildlife benefits followed.

Organic farming

Organic farming (also known as 'biological' or 'biodynamic' farming) has existed as a philosophy for sustainable food production in Britain since the 1940s, and has objectives that enshrine the principle of minimising the external inputs and externalised costs (e.g. agrochemical production and pollution) of the farming system. Organic farming is characterised by prohibition of the majority of synthetic chemicals in crop and livestock production, coupled with a requirement to manage non-crop habitats in ways that encourage control of crop pests by natural predators, and that are sympathetic to the interests of farmland wildlife (Lampkin 2002). Farms must reach and maintain rigorous quality assurance standards for their produce to be certified as organic. Organic farms rely on diverse crop rotations to maintain soil fertility and control crop pests and diseases. Coupled with the requirement for environmentally benign management of non-crop habitats, this means that conversion to organic standards entails a reversal of many of the management trends that characterise modern, industrial farming. In southern England, for example, an organic farm is more likely than its neighbours to be a mixed enterprise producing both crops and livestock, more likely to retain spring-sowing of some crops, and more likely to have these crops undersown with grass–clover leys as well, of course, as having no synthetic pesticide or fertilizer usage. Fuelled by growing concerns amongst European consumers over the environmental and human health impacts of intensive agriculture, the land area devoted to organic agriculture across Europe has increased from just 0.3 million hectares in 1990 to 7 million in 2006. In the UK, the land area either managed at organic standards or in conversion peaked at 740 000 hectares in 2002 but has since declined to around 620 000 hectares in 2006 (www.organic.aber.ac.uk). Consequently, throughout the 1990s there have been increasing opportunities for research to assess whether the reversal of agricultural intensification reflected by many organic farming practices has had beneficial consequences for farmland wildlife.

Hole *et al.* (2005) recently reviewed studies across Europe and North America that specifically set out to compare the abundance and diversity of wildlife on organic and 'conventional' farms. The problem of deciding exactly what the biodiversity of organic farms should be compared with has always been fraught with difficulty. Should one use the most intensive of 'conventional' farms for comparison, or those that in some sense typify local agricultural practice? Should the comparison seek to control for variations in some factors (e.g. availability and management of non-crop habitats) in order to focus on the withdrawal of agrochemical usage as the 'core' principle of organic farming, or should it allow for the full range of environmental contrasts that is brought about by organic management standards? The studies considered by Hole *et al.* (2005) undoubtedly cover this range of approaches but, broadly speaking, the 'conventional' farms used as comparators can be taken to be non-organic farms, characteristic of the local region where the study took place, and that rely on external inputs to achieve economically viable yields. In

Table 6.1 *Summary of the effects of the relative benefits of organic and conventional farming across a range of wildlife groups on farmland*

Group	Organic beneficial	Conventional beneficial	Mixed results or no significant difference
Birds	9	0	2
Mammals	3	0	0
Butterflies	3	0	2
Spiders	8	0	3
Earthworms	7	2	4
Beetles	13	5	5
Other arthropods	7	1	2
Plants	17	2	3
Fungi	0	0	1
Soil biota	9	0	9
Total	76	10	31

Source: Updated from Hole *et al.* (2005) to include later studies (2005–2007) sourced from Web of Knowledge. Numbers of different studies are shown. It should be noted that individual studies vary in design, duration and range of groups studied. A single study may therefore contribute to more than one cell.

summary, the review found that the abundance or species richness of a wide variety of wildlife groups, including birds, was typically higher on organic farms than their 'conventional' counterparts (Table 6.1), and that the reduced use of agrochemicals, sympathetic management of non-crop habitats and the retention of a diverse mixed farming enterprise on many organic farms were the three most beneficial aspects of organic farming for wildlife. Similar results were also obtained in an independent review of broadly the same literature by Bengtsson *et al.* (2005).

Table 6.1 suggests that, in general, the conversion of a modern intensive farming system to organic standards is likely to increase invertebrate and seed food sources for farmland birds. However, studies that compare the diversity and abundance of several wildlife groups on organic and non-organic farms often find markedly contrasting responses, with plants in particular usually showing greater and more consistent positive responses to organic management than other taxa (e.g. Fuller *et al.* 2005). This may reflect the fact that organic farms are often small and isolated in an intensively managed landscape so that the management of the surroundings dominates the extent to which the effects of organic management on wildlife are measurable. Effects on more mobile groups may be especially difficult to detect unless the contrast between the organic farms and management of the surrounding landscape is stark. For example in Denmark and Germany, respectively, Petersen *et al.* (2006) and Roschewitz *et al.* (2006) found strong, consistent, positive effects of organic management on plant diversity, but the latter study found that effects were greater when the surrounding landscape was homogeneous and simple as a consequence of intensive management. Similarly, impacts of organic management on bumble-bee and butterfly diversity and abundance depend heavily on landscape context, with beneficial effects only marked in intensively managed, homogeneous landscapes. In more varied agricultural landscapes, converting farmland to organic management had little effect (Rundlöf & Smith 2006; Rundlöf *et al.* 2008a, b).

Against this background, it is perhaps surprising that six studies that focused on bird communities – three in North America (Lokemoen & Beiser 1997; Freemark & Kirk 2001; Beecher *et al.* 2002), one in Denmark (Christensen *et al.* 1996) and two in Britain (Chamberlain *et al.* 1999; Fuller *et al.* 2005) – all found evidence of benefits to birds of organic management. However, most of these outcomes reflect the discovery of greater abundance of individual species on organic than neighbouring non-organic farms rather than effects on species diversity. This may be because individual species abundances are more likely to respond to the abundance of common invertebrate and plant food and suitable nesting habitat on farmland rather than diversity of prey species present. For example, the first British study (Chamberlain *et al.* 1999) was based on a comparison of 22 pairs of organic and conventional farms across England and Wales, with data collected over three breeding seasons and two winters between 1992 and 1994. Species richness differed little between the two farm types, but densities of individual species were consistently higher on organic than on conventional farms both in the breeding season and in winter. No species was found to be significantly more abundant on conventional than on organic farms at any season. Many of these differences were attributable to structural differences between organic and conventional farming such as hedgerow structure and crop rotation (Chamberlain & Wilson 2000). Organic farms, for example, tended to have greater lengths of tall, wide hedgerow, and smaller field sizes than conventional farms, and a larger, more recent study of organic farms across Britain reached similar conclusions (Fuller *et al.* 2005). It remains unclear to what extent the withdrawal of large-scale agrochemical applications in organic systems is directly beneficial to farmland bird populations relative to other aspects of organic management, although specific studies of the impacts of pesticide use on farmland birds to which we will return in Chapter 9, give some indication of likely responses. We should also remember that this study took place in the early 1990s and relied on data collection from a relatively small number of farms, many of which had only recently converted to organic standards, and most of which were small 'islands' in landscapes of more intensive agricultural practice. Today, there are many more, well-established organic farms, and ever-larger contiguous areas of land coming under organic management. A repeat study able to take advantage of these changes over the past 10–15 years could usefully determine whether the benefits of organic management to birds are more clearly measurable on larger areas of longer-established organic farmland, though comparisons may be complicated by the larger-scale implementation of agri-environment management on conventional farmland over the same time.

The remaining four avian studies reviewed by Hole *et al.* (2005) focused on individual bird species (skylark, yellowhammer and linnet) on organic and conventional farmland, and sought to gain a clearer picture of the mechanisms through which organic farming might prove beneficial. We will return to these and other studies as part of the more general review of the mechanisms linking agricultural change to bird population change that follows in later chapters.

Set-aside

In 1992 substantial reforms of the CAP sought to reduce the levels of surplus production of a range of agricultural products that had been brought about by price support and protection of European farmers from international markets. Subsidies for grain and protein crops were switched from a yield to an area basis. In order to receive these subsidy payments (termed Arable Area Payments), farmers were required to remove a variable proportion (up to 18%) of arable land growing grain or protein crops from production each year. As a consequence, land removed from agricultural production, or 'set-aside' as it was known, became a dominant feature of the agricultural landscape in Britain almost overnight. With over 600 000 ha in the 1992/93 season, set-aside instantly became the third largest land-use type in the lowlands after grassland and cereals. Across the EU as a whole, 6.4 million hectares were set-aside. The management regulations governing set-aside land were, and remain, somewhat complex (see Evans *et al.* 1997a for details in Britain) but, fundamentally, there are two forms of set-aside. Land may be set aside for a single cropping year and moved around the farm from year to year (rotational set-aside) or a fixed parcel of land may be set aside for several years (non-rotational set-aside). In either case, a green cover is established on the land either by natural regeneration or by sowing, to prevent nutrient leaching and soil erosion. Harvesting of this green cover is not permitted. Instead, it is managed on an annual basis by cutting, cultivation or use of non-residual herbicides, though initially the latter option was not available to farmers.

Despite the fact that the implementation of set-aside was driven by economic rather then environmental concerns, conservation scientists quickly appreciated that this large-scale withdrawal of arable land from production could offer benefits. For example, a rotational set-aside field left to regenerate naturally after harvest was effectively a stubble field left over winter that might offer valuable foraging habitat for seed-eating birds. Any form of set-aside, freed from agricultural operations, might provide suitable habitat for ground-nesting birds and a richer source of invertebrate food, as well as providing opportunities for conservation of declining arable flora (Clarke 1992; Wilson & Fuller 1992; Wilson *et al.* 1995; Evans *et al.* 1997a). At first, however, few of these benefits were realised as the management regulations governing set-aside did little to assist farmers in managing their set-aside sympathetically for wildlife interests. Indeed, set-aside may have proved directly damaging to breeding bird interests. For example, in 1992/93, farmers were required to cut set-aside

Table 6.2 *Summary of patterns of field type selection in winter by six Red-listed farmland birds on farmland in Devon and East Anglia*

	Set-aside	Non set-aside stubble	Winter cereal	Brassica	Grassland
Grey partridge	+	+	–	O	O
Skylark	+	+	–	–	–
Song thrush	–	O	–	+	–
Linnet	+	+	–	+	–
Cirl bunting	+	+	–	O	–
Yellowhammer	+	+	–	O	–

Notes: + indicates that selection was shown (i.e. more birds were counted than expected from random distributions generated by resampling); – indicates that avoidance was shown (i.e. fewer birds were counted than expected from random distributions generated by re-sampling); O, no significant selection or avoidance shown. Grey partridges were recorded only in East Anglia and cirl buntings only in Devon.
Source: Modified from Buckingham *et al.* (1999) and Henderson & Evans (2000).

vegetation before 1 July, or cultivate it during May. Many observers (e.g. Poulsen & Sotherton 1992) noted widespread destruction of nests of ground-nesting species such as skylarks, pheasants and partridges, and even those not directly destroyed by cutting or cultivation were rendered much more susceptible to predation once the vegetation cover had been removed. The attendant outcry provided the impetus for changes to the set-aside management regulations which, by the following year, allowed farmers the opportunity to turn set-aside into a form of land management with real wildlife benefit, and counter those who contended that the set-aside scheme was achieving no more than paying farmers to 'do nothing'. Specifically, farmers were now permitted to manage green cover through the use of non-residual herbicides, so reducing mechanical destruction of nests, and non-rotational set-aside was introduced, thus permitting farmers to sow seed mixtures specifically designed to create nesting cover and enhance food resources for birds (Sotherton 1998).

Two subsequent studies in Britain set out to assess the extent to which set-aside provided tangible benefits across a wide range of farmland bird species. One focused on the use of set-aside as a winter foraging habitat (Buckingham *et al.* 1999), whilst the second measured bird abundance on set-aside fields and neighbouring fields under crops or grassland during the summer (Henderson *et al.* 2000a, b). Buckingham *et al.* surveyed 40 farm plots (20 in Devon and 20 in East Anglia, totalling over 2500 hectares) over two winters, making three to four visits each winter. For a suite of six Red-listed, declining farmland bird species, they found strong evidence that both rotational and non-rotational forms of set-aside were selected as a foraging habitat relative to their availability by five of the species, and that alternative, commonly available habitats such as winter cereal crops and grassland were avoided (Table 6.2).

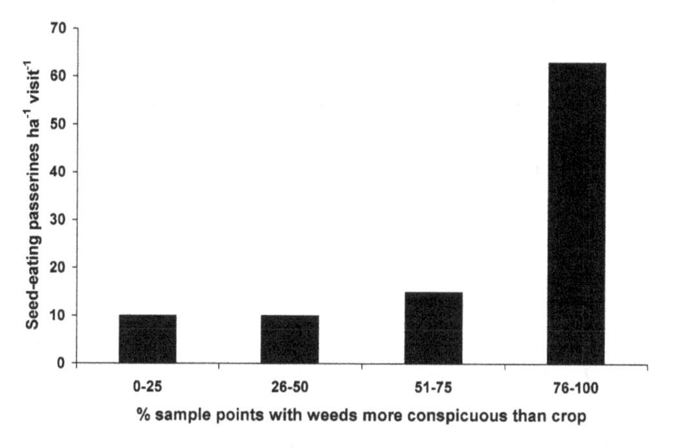

Fig. 6.7 The density of seed-eating passerines found in winter in fodder crops of varying weediness. Redrawn from Hancock & Wilson (2003) with kind permission.

All five of the species found to select set-aside rely on seed foods over winter, and the similar pattern of selection of non-set-aside stubbles suggests that the set-aside fields were probably being selected because they held a rich seed supply. Many other studies of set-aside and other stubble fields in winter have shown similar concentrations of seed-eating birds. Evans & Smith (1994) and Donald & Evans (1994) found that cirl buntings and corn buntings selected winter stubbles. Wilson *et al.* (1996b) found strong selection of stubbles by greenfinches, reed buntings, linnets, chaffinches, yellowhammers, goldfinches, pied wagtails, skylarks, grey partridges and meadow pipits on mixed farmland in Oxfordshire. For the first six of these species, more than 50% of all the individuals recorded were on cereal stubbles, even though this field type (present largely as set-aside) occupied no more than 10% of the surveyed area and, in total, 88% of all finches, sparrows and buntings were recorded on stubbles. Even more extreme selection was noted by Mason & Macdonald (1999a) who recorded 96% of these birds on stubbles in Essex farmland. In Scotland, Hancock & Wilson (2002, 2003) found higher densities of grey partridges, skylarks, house and tree sparrows, chaffinches, greenfinches, goldfinches, linnets and twite on cereal stubbles and fodder brassica crops than on other habitat types. In all of these studies, the importance of 'weediness' as a characteristic of stubbles should not be underestimated. In their study of winter habitat selection by corn buntings, Donald & Evans (1994) found that although the birds selected stubbles of all types, weed-rich stubble held roughly twice as many flocks and birds as weed-free stubble. Similarly, a study of the winter habitat associations of seed-eating birds in Scotland (Hancock & Wilson 2003) found that the most weed-infested fodder brassica crops and stubbles held more than six times as many seed-eating birds as fields that were weed-free or had very few weeds (Fig. 6.7).

Application of broad-spectrum herbicides such as glyphosate is one of the main reasons why winter stubbles may be weed- and seed-free. For example Robinson (2001) found that skylarks wintering in his East Anglian study area selected unsprayed stubbles, but avoided sprayed stubbles, a difference almost certainly accounted for by the fact that seed densities were approximately six times higher on the unsprayed stubbles. This difference is certainly the main reason why rotational set-aside has proved such a valuable winter foraging habitat for a wide range of seed-eating farmland birds, since farmers are not permitted to apply herbicides until the spring following establishment. Stubbles in the set-aside scheme are therefore much richer in regenerating vegetation and associated seed supply than those outside the scheme, which will usually be subjected to herbicide application shortly after the harvest of the previous crop. Indeed, a study by BTO found very few birds and few arable weeds on cereal stubbles subject to pre- or post-harvest spraying with the broad spectrum herbicide glyphosate (BTO 2002). Almost 80% of the variation in the number of granivorous birds using a field in midwinter (expressed in terms of their energy demand) in this study was explained by the density of seeds of Chenopodiaceae (e.g. fat-hen) and Polygonaceae (e.g. knotgrass), and the number of chemicals used on the preceding crop.

Henderson *et al.* (2000a) surveyed breeding season usage by birds on 92 farm plots in England in 1996, each of these including a preselected set-aside field, plus approximately 5–10 neighbouring fields. A subset of 63 of these plots was surveyed again in the following year, and each field season involved four visits to each plot to map bird distributions and collect habitat data. The results were summarised across five functional groups of farmland birds (gamebirds, pigeons, crows, thrushes, and seed-eating passerines), plus skylark – the one species nesting commonly in the set-aside itself. With only one exception (crows in the first year of the study), set-aside held higher densities of birds of every group than did grassland or arable crops. Densities were universally higher in set-aside than in winter cereal (the crop type that most set-aside replaces), and winter cereals held the lowest densities of birds in most cases (Fig. 6.8).

With the exception of skylark for which set-aside can provide excellent nesting habitat (see below), the majority of this use of set-aside reflects use by foraging birds. In this context the study found that rotational set-aside tended to hold higher densities of birds than non-rotational set-aside, and that the margins of set-aside fields were used more intensively than the field centres by birds making forays from nearby hedgerow and woodland nesting habitat (Henderson *et al.* 2000b). The general preference for set-aside by foraging birds is unsurprising given the fact that set-aside land is freed from most intensive agricultural operations and rapidly develops a diverse, early successional sward with a rich invertebrate fauna (Clarke 1992). However, the greater use of rotational than non-rotational set-aside across a wide range of farmland birds is noteworthy. The authors attribute the difference to the diverse, sparse and heterogeneous structure of set-aside vegetation in its first year offering a greater abundance and accessibility of invertebrate and seed food to foraging birds than the denser, perennial-dominated sward characteristic of older, non-rotational set-aside (Clarke 1992; Fisher *et al.* 1994). Similar benefits of younger swards have also been found in the Conservation Reserve Program (an equivalent of set-aside) in North America (Millenbah *et al.* 1996).

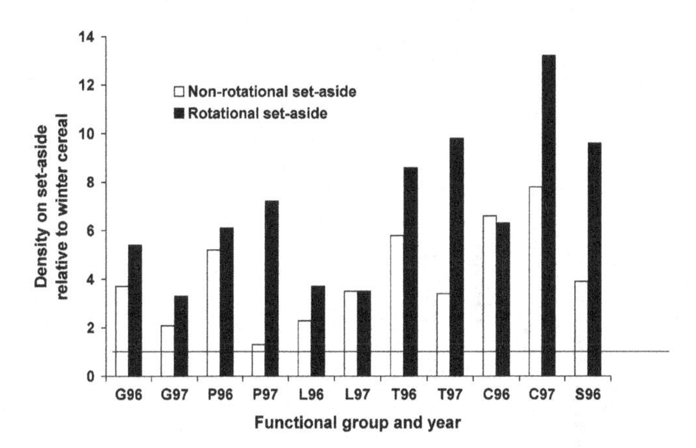

Fig. 6.8 Ratio of density of five functional groups of birds (G, gamebirds; P, pigeons; L, skylarks; T, thrushes; C, crows; S, seed-eating passerines) on set-aside relative to winter cereal in summer (April–July) in 1996 and 1997. The ratio value of 1 (equal bird density on winter cereal and set-aside) is marked with a horizontal line. All groups show significantly greater density on set-aside than winter cereal except crows in 1996. Adapted from Table 1 of Henderson *et al.* (2000a).

Skylarks make greater use of set-aside as a nesting habitat, as well as a foraging habitat, than any other species in Britain. Territory densities are higher in set-aside than any other lowland farmland habitat, and may be exceeded only by those on steppe grasslands and coastal saltmarshes (Donald 2004). Clutch sizes are higher on set-aside than other farmland habitat types (Donald *et al.* 2001b), set-aside is heavily selected by adult birds collecting food for chicks (Wilson 2001) and the breeding season is longer, resulting in higher breeding productivity even though nest predation rates may sometimes be high (Donald *et al.* 2002a; Eraud & Boutin 2002). Provision of fallow nesting plots on set-aside also played a key role in the early recovery of stone curlew populations, although agri-environment schemes are now the main means of providing such habitat (Chapter 8). In addition, a 2006 national survey showed that 7% of the Woodlark population bred on set-aside (see Fig. 4.2). We will return in more detail to the relationships between skylarks, stone curlews and set-aside in the case studies in Chapter 8. However, the evidence for set-aside having population-level impacts on widespread farmland birds is generally less clear. Many farmland birds have continued to decline since 1993, albeit at a slower rate in several cases. For example, Henderson *et al.* (2001) expected the skylark population to have increased by more than 10% during the mid 1990s, but this did not happen, and they cited the lack of suitably managed set-aside as the reason.

Summary

Across lowland farmland bird communities in Britain and more widely in Europe, there is a strong correlation between measures of agricultural intensification and periods of population decline across a broad range of species. A study in lowland

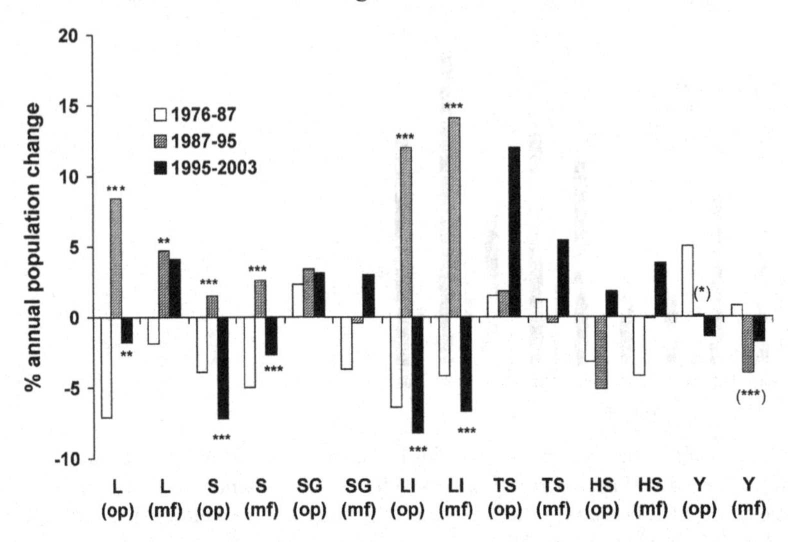

Fig. 6.9 Percentage annual population changes of seven farmland bird species in three time periods in two farmland types (op, open plains; mf, mosaic farmland). Asterisks (∗, $p < 0.05$; ∗∗, $p < 0.01$; ∗∗∗, $p < 0.001$) indicate cases where the population trend switched to a significant degree between the first and second (set-aside) period, or between the second and third period. Cases in brackets indicate where the switch was opposite in direction to that expected if set-aside had beneficial effects. L, lapwing; S, skylark; SG, starling; LI, linnet; TS, tree sparrow; HS, house sparrow; Y, yellowhammer. Data from Wretenberg *et al.* (2007).

Scotland was able to show correlations at a finer resolution with year-to-year changes in lowland farmland bird populations tending to be more positive after years with low indices of agricultural intensity and high measures of insect abundance. Reductions in abundance and diversity in association with agricultural intensification also characterise a wide range of other taxa including flowering plants, moths, butterflies, the wider invertebrate fauna of cereal fields and some mammals. Spatial patterns in declines of farmland birds are less easy to interpret. Many widespread species are characterised by local extinction of previously low-density populations in areas of northern and western Britain dominated by pastoral agriculture, but with numerically larger declines in areas of mixed and arable agriculture further south and east. Functional linkages between different declining species groups are common. For example, outcrossing plant species reliant on insect pollination have declined more severely than other species in the agricultural landscape of both Britain and the Netherlands. Similarly, many of the most important food plants of declining granivorous birds are themselves declining, and four taxonomic groups of arthropods that are of high importance in the nestling diets of declining farmland birds are also known to be highly sensitive to insecticide usage.

Two examples of localised reversal of agricultural instensification – organic farming and set-aside – both suggest that if the trend of agricultural change is reversed then the fortunes of bird and other wildlife populations can be reversed too.

Comparative studies of bird populations on organic and 'conventional' farms show evidence of species densities that are slightly but consistently higher on organic farms, though reliance on cultivations to control weeds in organic systems may detrimentally affect ground-nesting species. Some bat and bird species may respond similarly to higher invertebrate diversity and abundance on organic farms relative to their conventional neighbours. There is strong evidence that the withdrawal of individual fields from intensive agricultural production under EU set-aside regulations has provided important benefits for both nesting and foraging birds, especially when set-aside management rules have been modified to accommodate these potential benefits. However, set-aside fields and organic farms are typically small islands in wider landscapes of intensive agriculture, and the potential benefits they can bring at the population level may be limited by this for many bird species. Areas of land being converted to organic management have been increasing, and this may provide opportunities to test whether the greater wildlife benefits of organic management accrue when it is practised over larger, contiguous areas of farmland for longer periods. In Britain, no study has shown any large-scale population-level effect of either set-aside or organic farming on bird populations. Intriguingly however, in Sweden Wretenberg *et al.* (2007) have found evidence that population trends of several species were more positive during the period when set-aside and other agricultural extensification measures were most prevalent (1987–95) than before or after that period, when agricultural polices were more supportive of increased production (Fig. 6.9).

This study is purely correlational and the association of more positive trends with the set-aside period did not apply to all species. However, the positive association of lapwing, skylark and linnet population trends with set-aside, in particular, is very striking, and this study may provide some indication that large-scale extensification measures, whether by means of set-aside or agri-environment schemes, have the capacity to reverse decline in farmland bird populations.

Plate 6.1 Corn marigold in an arable strip on the South Uist machair. Once considered one of the most troublesome agricultural weeds, the combination of autumn sowing, liming, improved seed-cleaning and herbicides has made this species a much less frequent find in intensive agricultural systems. © Jeremy Wilson.

Plate 6.2 The common blue butterfly still merits the epithet 'Common' in comparison with other species of blue. However, widespread declines have resulted from loss of semi-natural habitats to agricultural improvement. © Jeremy Wilson.

Plate 6.3 Declines in brown hare numbers have been caused partly by the loss of seasonal variety in food sources and cover as agricultural landscapes have become simplified and more uniform. © RSPB.

Plate 6.4 Though introduced as a production control mechanism, fields set-aside from agricultural production, and sensitively managed, have become an oasis for wildlife, including birds, in intensive agricultural landscapes. Ground-nesting species such as grey partridges, lapwings, skylarks, yellow wagtails and corn buntings may nest and a much wider range of species will find good breeding season foraging opportunities. In winter, rotational set-aside can provide a rich seed source for larks, finches, buntings and sparrows. © RSPB.

Part III

The effects of agricultural change on birds

If we wish to understand the causes of population trends, whether they be increases or declines, it is helpful to keep in mind that external, environmental factors drive population trends through their influence on demographic rates such as breeding success and survival (Green 1995a; Newton 1998). To take a simple example, the proximate cause of the decline to extinction of the dodo was reduction in breeding success and survival rate as both nests and adult birds were plundered; however, the ultimate cause was the exploitation of Mauritius by humans, an invasive, non-native predator. If we wish to manage bird populations to influence trends then, as Newton (2004) put it, 'it is the external limiting factors that must be changed before any lasting change in population level can be achieved; the necessary demographic changes will follow naturally'.

Here our interest focuses on understanding to what extent aspects of agricultural intensification are the main external limiting factors on bird population trends on farmland. For example, several bird species nesting in agricultural crops have suffered declining breeding success. In some cases, the relationship with agricultural production may be simple and direct as earlier or more frequent harvesting destroys nests, whereas in others the relationship may be indirect, involving intervening processes such as predation risk and the birds' ability to find and gain access to food in intensively managed crops. Table III.1 lists examples of possible proximate and ultimate causes of farmland bird population trends, as well as the intervening processes that may be involved.

The long-term decline of farmland bird populations has stimulated a huge growth in research since the late 1980s that has been directed specifically at disentangling the varied causal relationships between agricultural intensification and bird population response (Fig. 6.1). These studies can be categorised roughly into four general types:

(i) Large-scale studies of associations between bird abundance or distribution and environmental measures using data sets available at the national scale.
(ii) Large-scale studies of the relationships between trends in demographic rates and trends in populations, and of spatial variation in demographic rates as a function of agricultural practice.
(iii) Experimental or comparative studies of the ecology of individual species.

Table III.1 *Examples of possible proximate, intervening and ultimate drivers of bird population trends*

Cause	Examples
Proximate	Nest success, number of nesting attempts, survival, density dependence
Intervening mechanism	Availability of suitable habitat for nesting, foraging and roosting, food supply, predation rates, disease
Ultimate	Agricultural change (e.g. agrochemicals, non-crop habitat management, cultivation cycles, drainage, grassland improvement, changes in crop rotation), other environmental drivers (e.g. climate change, pollution, human disturbance)

(iv) Experimental or comparative studies of the impacts on birds of specific agricultural practices or land-uses.

Each of these approaches has it strengths and weaknesses. In particular, large-scale approaches have the advantage of detecting relationships between birds and agriculture that are robust in time and space. However, they give no direct insight into the ecological causes of those patterns as they affect individual birds and, as we shall see, can mislead. More intensive field studies are much more likely to yield evidence of the mechanisms linking patterns of abundance or demography to their environmental causes, and may also offer the opportunity to follow up intriguing correlations with the rigour of experimental manipulation. Their limitation is that it may not be reasonable to extrapolate their findings from the specific study areas concerned to the wider agricultural environment in order to explain national population trends. For example, different mechanisms may operate at different times and in different places but with a similar overall impact on a bird population. The combination of both approaches has been critical to the dramatic improvement in our understanding of the impacts of agricultural intensification on birds that has accrued over the past 20 years.

Here we review the evidence offered by each of these approaches in turn, dealing with large-scale studies of distribution and demography in Chapter 7, and studies of individual bird species and specific aspects of agricultural management in Chapters 8 and 9, respectively. We then go on to draw some general conclusions concerning the main causes of farmland bird population declines in Britain. This sets the scene for the final chapter which considers how the growth in our understanding of the causes of farmland bird trends has allowed research to move on to designing, testing and implementing management solutions.

7 · *Large-scale studies of abundance, distribution and demography*

Large-scale distribution and abundance models

In Britain we are fortunate to have excellent, nationwide data on bird distributions and trends, and a variety of long-term, geographically referenced environmental data sets, describing variables such as weather, topography, soils and land-use. This makes it possible to conduct statistical modelling exercises to assess the extent to which change in bird distribution and abundance in space and time over large geographical scales can be explained by this environmental variation. Any such analyses must pay heed to the fact that relationships between organisms and their environment will more often than not be scale-dependent (Wiens 1989). For example, if we take a notional bird which nests in cereal crops but finds its richest food sources in grassland then, measured at the field scale, we would expect to see a strong positive association between the species' nesting distribution and that of cereal fields. At a larger scale, however, if we ask which agricultural landscapes hold the highest breeding densities of this species then it will probably be mixed landscapes of cereals and grassland, rather than landscapes dominated solely by cereal crops. In effect, the interpretation of large-scale distribution models needs to consider how the spatial grain at which data underpinning the model are collected (e.g. 1-km or 10-km squares) relates to the environment as perceived by the organism in question. For example, in Britain, a 10-km square will represent a landscape combining in various ways all of the (possibly many) environmental factors that will determine the density at which that square is occupied by multiple individuals of a given bird species. A 1-km square, however, comes closer to the scale at which, in a suitable landscape, individual birds are making decisions about the location of breeding territories and favoured foraging locations.

A good example of this approach at larger scales is provided by the work of Gates *et al.* (1994). They took occupancy data at the 10-km square scale from the 1988–91 *Breeding Atlas* (Gibbons *et al.* 1993). They then modelled these as a function of a wide range of environmental variables, including agricultural land-use, weather, topography and soils. Models explained between 31% and 76% of variation in the occurrence of eight farmland bird species across Britain (grey partridge, lapwing, turtle dove, skylark, tree sparrow, linnet, reed bunting and corn bunting), but did not show clear overarching correlations and were not generally straightforwardly associated with species ecology. Overall, of 46 environmental variables considered, weather variables were more often found in the best-fit models for each species (54% of possible inclusions) than variables describing agriculture (19%), other habitats (31%), topography (25%) or soils (20%). As the authors themselves recognised, at such a coarse spatial scale the variables included in models are likely to be a mix of those that are genuinely linked to factors limiting large-scale species distribution and those that are simply incidental correlates of processes operating at finer scales. Hence such models need to be interpreted with a good deal of common sense and caution. For example, the strong positive association found between turtle dove distribution and breeding season temperature may reflect genuine climatic limits on the species' current distribution in Britain. A similar association between linnet

Table 7.1 *Relationships between farmland bird occurrence at the 10-km square scale (1988–91* Breeding Atlas*) and soil type (from Gates* et al. *1994). The distributions of turtle dove and linnet showed no strong associations with soil type*

Soil type	Species showing positive association	Species showing negative association	Species showing peak occurrence at intermediate values
Brown earths and podzols (well drained)		Lapwing Tree sparrow Reed bunting Corn bunting	Skylark
Rendzinas and other calcareous soils (well drained)	Grey partridge Skylark Corn bunting		
Gleys and peaty gleys (poorly draining)	Lapwing Skylark Tree sparrow Reed bunting		
Deep peats	Reed bunting Corn bunting		
Skeletal upland soils			Skylark

distribution and the availability of oilseed rape was one of the drivers for research which has since shown this species' increasing reliance on this crop as a food source in arable farmland (Moorcroft *et al.* 2006). On the other hand, Gates *et al.* (1994) found a strong positive association between skylark distribution and wheat area, and this can only be interpreted reasonably as an indication that high skylark densities are found in open, arable landscapes where wheat is a dominant crop. This association is also supported by the analysis of correlates of skylark abundance on 1500 BBS 1-km squares by Chamberlain & Gregory (1999). It would be misleading, however, to extrapolate these correlations to the conclusion that simply growing wheat is automatically beneficial for skylarks. In fact, changes in cereal husbandry have probably been the main cause of skylark decline on British farmland (Donald 2004 and Chapter 8). Overall, it seems that modelling at the 10-km resolution may be most useful in identifying broad landscape types with which species are associated, or which they avoid. In retrospect, perhaps one of the more intriguing aspects of Gates *et al.*'s modelling analyses, not explored further until very recently, was their inclusion of soil variables in their models. Given that soils are a fundamental intervening factor relating climate and geology to the capability of land to sustain different forms and intensities of agriculture, the relative rarity of studies considering soil variables is perhaps surprising. Gates *et al.* found some strong correlations with soil type across their sample of species, especially amongst those with a ground-nesting habit (Table 7.1). For example, whilst skylarks occupied a wide range of soils, corn buntings were associated with easily cultivated peats and calcareous soils over which spring tillage is most commonly found, whilst lapwings, reed buntings and tree sparrows were associated more with wetter gley soils than free-draining brown earths. Whether these differences reflect any direct response to the soil characteristics themselves or an indirect response to the forms of agriculture associated with different soil types is impossible to conclude from these analyses. Nonetheless, the fact that correlations were strongest for ground-nesting species suggests that some form of relationship between soil conditions and suitability for nesting is possible, perhaps early in the season when conditions are wetter and colder. For example, Paul Donald (2004) has suggested that skylarks tend to reach higher densities on sandy soils than clay soils, perhaps because the former are more freely draining and dry out more quickly after rain, or are less prone to surface run-off (which could potentially flood nests). Donald also notes that drainage of raised bogs in Russia was associated with their colonisation by skylarks (Mal'chevskiy & Pukinskiy 1983).

These analyses were developed further by Siriwardena *et al.* (2000c). Their study focused on agricultural land-use variables as predictors of bird distribution in the 1988–91 *Breeding Atlas*, but refined the earlier analysis by considering the proportion of 2-km squares (tetrads) within each 10-km square that were occupied as a finer scale measure of distribution and a surrogate measure of abundance at the 10-km square scale. They considered 14 granivorous species (grey partridge, stock dove, turtle dove, skylark, house sparrow, tree sparrow, chaffinch, bullfinch, linnet, goldfinch, greenfinch, reed bunting, yellowhammer and corn bunting) and, although the detail of the models was as highly species-specific as found by Gates *et al.*'s study, some general correlations were clearly identifiable. Specifically, correlates of high-intensity arable farming in arable landscapes (e.g. large areas of sugar beet, wheat and oilseed rape) were associated with low tetrad occupancy for 11 of the 14 species, whereas correlates of high-intensity grassland farming (large areas of reseeded grassland and high sheep densities) were associated with low tetrad occupancy in pasture-dominated landscapes for 12 of the 14 species. The presence of large areas of fallow land (a good indicator of lower-intensity agriculture) was associated with high tetrad occupancy across 11 species.

Not all of these large-scale modelling exercises have relied upon Atlas distribution data; some have made use of bird abundance and habitat data from CBC plots or sample 1-km squares from the BBS. Fuller *et al.* (1997) used a 20-year run of habitat data from CBC plots as the basis for models assessing to what extent habitat variation between years and survey plots explained variation in bird abundance. They found that variables relating to broad structural and topographical characteristics of the survey plots (e.g. altitude, woodland areas and field size) were good predictors of the distribution and abundance of species more associated with the non-cropped habitats of survey plots such as robin, great tit, blue tit, willow warbler and chaffinch. This was not true for species making extensive use of cropped habitats for nesting or feeding (e.g. song thrush and whitethroat). In a further study focusing on non-crop habitats on farmland CBC plots, Gillings & Fuller (1998) modelled CBC population

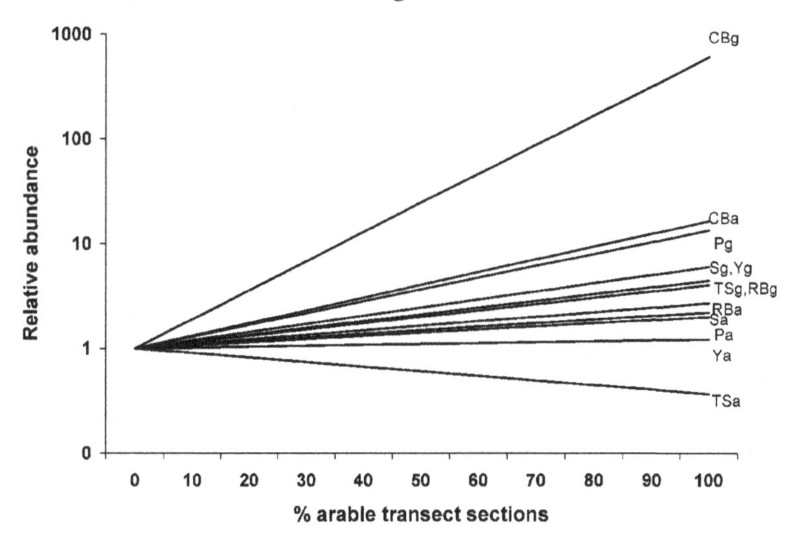

Fig. 7.1 Modelled relationships between breeding abundance of seed-eating birds, and extent of arable habitat on BBS squares in arable landscapes (arable : grass ratio >2.5 and grassland landscapes (arable : grass ratio <0.5). Modelled abundance is expressed relative to that for a BBS square containing only grass transect sections. Note logarithmic scale. a, arable landscape; g, grassland landscape. P, grey partridge; S, skylark; TS, tree sparrow; Y, yellowhammer; RB, reed bunting; CB, corn bunting. The slopes for TSa, Ya and Pa do not differ significantly from zero. Data extracted from Robinson *et al.* (2001).

trends as a function of loss or degradation of non-crop habitats. They found no strong relationships, implying that although loss and degradation of marginal habitats such as hedgerows, woodland edges, scrub and wetland have undoubtedly been a feature of agricultural intensification, they have, perhaps surprisingly, been of secondary importance in causing farmland bird population declines, at least over the relatively short timescales examined.

Most recently, Robinson *et al.* (2001) used a large-scale modelling approach with BBS data to take further previous analyses of *Atlas* data (Chamberlain & Fuller 2000, 2001; see Chapter 6). These analyses had suggested that although the numerical declines of many lowland farmland birds were most severe in lowland arable and mixed landscapes, range contraction and local extinction were more prevalent in grassland landscapes and at the upland fringe further to the north and west. Robinson *et al.* found that the abundance of seven species (grey partridge, skylark, whitethroat, tree sparrow, reed bunting, yellowhammer and corn bunting) tended to increase with the proportion of BBS 200-m transect sections in each square that were made up of arable habitat. The strength of this relationship was much stronger where the wider landscape was predominantly grassland than where it was predominantly arable (Fig. 7.1). Indeed for tree sparrow, dominance of arable habitat in a BBS square in arable landscapes tended to be correlated with lower abundance.

All of these species except whitethroat are granivorous. They are also amongst those showing the most substantial range contractions from northern and western areas of their breeding range between the two breeding atlases (Gibbons *et al.* 1993), including the virtual extinction of corn buntings in Wales, whilst their numerical declines were most severe in lowland arable and mixed farmland further to the south and east (Chamberlain & Fuller 2001). These results indicate the importance of retaining heterogeneity in agricultural land-use at the farm scale; some species (notably granivores) may depend on resources offered by tillage and the cultivation of arable crops, and may be unable to persist without at least some of these land-uses being retained in otherwise grassland landscapes. Conversely, other species may depend on resources offered by grassland and may be unable to persist without some of this land use being retained in tillage-dominated landscapes. Equally this study also suggests that loss of remaining pockets of arable agriculture in grassland landscapes at the periphery of the ranges of some lowland farmland birds may be the main driver of local extinction of these species (see the corn bunting case study in Chapter 8). Such species were already likely to have been present at relatively low densities due to the long-term specialisation in grassland enterprises in the west and north of Britain. The homogenisation of agricultural environments by specialisation in grassland agriculture is likely to be a very different driver of decline and local extinction than those that account for the substantial numerical declines of the same species in the core of their breeding ranges further south and east, where mixed or arable agriculture dominates.

Summary

Models relating large-scale distribution and abundance of birds to variation in their environment need to be interpreted cautiously. The larger the spatial grain at which data are collected, the more likely it is that the environmental measures found to explain variation in distribution and abundance of birds will be a mix of those that are genuinely linked to factors limiting large-scale species distribution (e.g. climate) and others that are incidental or indirect correlates of processes operating at finer scales (e.g. relationships between soils and agricultural land-use). Nonetheless, when these approaches have been applied to farmland birds in Britain, they have revealed that the most intensive grassland and tillage landscapes are associated with lower abundances of most species than landscapes supporting a mix of agricultural land-uses. They have also shown that the retention of at least 'pockets' of tillage in grassland landscapes (or, conversely, pockets of grassland in tillage landscapes) may be critical in preventing local extinction of some species. In addition, these approaches have suggested that although variation in the non-cropped 'skeleton' of Britain's agricultural landscapes (woodlands, hedgerows, field margins, ponds, farmyards and so on) is very important in explaining the local distribution and abundance of many species otherwise associated with woodland, changes in the physical management of such habitats are unlikely to have been a main driver of the recent declines of many species.

Demographic analyses of farmland bird populations

In some cases, the list of plausible candidate external drivers of decline for a given species is very short. For example, from the earliest observations of declines in corncrake populations in Britain, it was long suspected that the mechanisation of harvesting was likely to be causing problems for a species which both nests and finds most of its food in hay meadows (Alexander 1914; Norris 1947). In many other cases though, the list is sufficiently long that an initial focus on the proximate demographic changes has helped to narrow the search, for example by distinguishing whether a population trend has been driven by change in a vital rate such as breeding success or survival. As with our long-term knowledge of bird population trends, we are uniquely fortunate in Britain to have the long-term data sets amassed by the volunteer observers of the BTO that allow study of long-term demographic trends. First, the National Ringing Scheme's archive of recoveries of ringed birds allows survival rates to be calculated annually (Brownie *et al.* 1985) or for blocks of years of differing population trend (Fewster *et al.* 2000). Survival rates are usually estimated separately for first-year and older (adult) birds because it is usually found that survival rates are higher amongst adult individuals (e.g. Siriwardena *et al.* 1998b). Second, the Nest Record Scheme's database of individual nest histories allows population-wide estimates of components of individual nest success to be estimated (Crick & Baillie 1996; Aebischer 1999; Crick *et al.* 2003). Again, where data permit, these may be broken down in various ways, for example by nesting stage (egg versus nestling), by geographical region, by habitat type, or by periods of contrasting population trend.

Using these data sources, the most wide-ranging analyses of the relationships between demographic rates and population trends for a range of lowland farmland species have been carried out by the BTO and Oxford University (Baillie 1990; Baillie & Peach 1992; Siriwardena *et al.* 1998b, 1999, 2000a). These analyses are summarised in Siriwardena *et al.* (2000b), and are supplemented by more detailed studies of individual species, including sparrowhawk (Newton 1988), black grouse (Baines *et al.* 2007), grey partridge (Potts & Aebischer 1995), corncrake (Green *et al.* 1997), breeding waders, (e.g. Shrubb 1990; Peach *et al.* 1994), turtle dove (Browne *et al.* 2005), skylark (Chamberlain & Crick 1999), song thrush (Thomson *et al.* 1997; Robinson *et al.* 2004), spotted flycatcher (Freeman & Crick 2003), starling (Freeman & Crick 2002; Freeman *et al.* 2002, 2007), chough (Reid *et al.* 2004), house sparrow (Crick & Siriwardena 2002; Freeman & Crick 2002), bullfinch (Siriwardena *et al.* 2001b) and reed bunting (Peach *et al.* 1999). Taken together, these analyses allow us to assess whether, across species, population trends are more strongly associated with variation in survival rate or with variation in nesting success (Table 7.2).

Three general outcomes are immediately striking. First, there is much greater evidence that trends in populations of the smaller passerines have been driven by changes in survival rates than by changes in nesting success. Indeed for several species nest success tends to have increased as populations have declined, perhaps suggesting a density-dependent response. Moreover, further work by Siriwardena *et al.* (2000d) showed that across 10 seed-eating farmland passerines, there were no clear patterns

Table 7.2 *Associations between population trend and demographic rates in lowland farmland birds in Britain*

Positive association between survival rate and population trend	Negative* or no clear association between survival rate and population trend	Positive association between nesting success and population trend	Negative* or no clear association between nesting success and population trend
Barn swallow	Lapwing*	Sparrowhawk	Kestrel*
Pied wagtail	Turtle dove	Black grouse	Stock dove
Robin	Wren	Grey partridge	Turtle dove
Blackbird	Dunnock	Corncrake	Skylark*
Song thrush	Tree sparrow	Lapwing	Blackbird*
Mistle thrush	Linnet	Snipe	Song thrush*
Sedge warbler	Bullfinch	Magpie	Spotted flycatcher
Lesser whitethroat	Yellowhammer	Jackdaw	Starling
Whitethroat		Carrion crow	Dunnock*
Spotted flycatcher		House sparrow	Tree sparrow*
Blue tit		Linnet	Bullfinch
Great tit			Greenfinch
Starling			Yellowhammer*
Chough			Corn bunting*
House sparrow			
Chaffinch			
Goldfinch			
Redpoll			
Reed bunting			

of association between nest success and agricultural land-use either at the territory or landscape scale. Second, and in contrast, population trends of ground-nesting waders and gamebirds, increasing corvid populations and species recovering from the direct impact of organochlorine pesticides (e.g. sparrowhawk) do show evidence of being driven by changes in nesting success. Third, for a substantial number of species, including several of pressing conservation concern (e.g. turtle dove, tree sparrow, bullfinch, yellowhammer and corn bunting), these analyses have revealed no clear picture of the main demographic drivers of population change.

The presence of several long-distance (trans-Sahelian) migrant species in column 1 of Table 7.2 is unsurprising since it is well established that annual variation in conditions on the wintering grounds and migration routes and, in particular, the extent of drought in the Sahel region of Africa have played a dominant role in driving recent population trends of species such as sedge warbler and whitethroat (Baillie & Peach 1992). For these species at least, declines have been caused by catastrophic losses of birds between breeding seasons and it seems that any impact of agricultural intensification in Britain will be to limit population recovery to former levels. The remaining species span a range of ecologies. Given that we know that variation in winter weather has driven many of the observed fluctuations in resident songbird populations in Britain (e.g. Dobinson & Richards 1964; Marchant *et al.* 1990; Greenwood & Baillie 1991), it is easy to conclude that the population trends of these species are also being driven by changes in survival rate outside the breeding season (e.g. Peach *et al.* 1999). We should be cautious in doing so; limitations on resources at any time of the year may account for an association between survival rate and population trend. Correlations between population trend and survival rate estimates are not usually sufficiently sensitive to show whether the mortality changes that are driving population trends are concentrated at a particular season. The few exceptions include a study of the demography of song thrush decline undertaken Robinson *et al.* (2004). This study went further than others of its kind in three ways. First, it corrected for the possibility that reporting rates of ringed birds may have changed and thus be confounding attempts to estimate survival rates from recoveries of ringed birds (Baillie & Green 1987). Second, it estimated survival during the immediate post-fledging period, as well as for the remainder of the first year and for adult birds, for time periods of contrasting population trend. Third, it combined these survival estimates with estimates of breeding productivity to assess their relative contribution to driving the species' recent population decline. This study was able to confirm that survival during the first year of life, and specifically during the first winter and the immediate post-fledging period was sufficient to have driven the observed population decline. An intensive study of the demography of an individually marked population of choughs on Islay similarly suggests that variation in pre-breeding survival may have been the most important demographic rate driving population change in recent years (Reid *et al.* 2004). Further circumstantial evidence that many of the population trends being driven by changes in survival rate do indeed reflect constraints outside the breeding season comes from inspection of column 4 in Table 7.2, in which a partially overlapping range of species of similar ecology to those in column 1 show evidence of *increasing* nesting success as populations have declined. This seems unlikely if deteriorating breeding season conditions were causing increased mortality rates at that time of year.

In column 3 of Table 7.2, the presence of grey partridge, corncrake, lapwing and snipe suggests that the population trends of ground-nesting, precocial species on farmland have been driven largely by constraints on breeding success. As we shall see in Chapter 8, this finding is supported by more detailed studies of these and other individual species (e.g. stone curlew, skylark and corn bunting), which confirm that ground-nesters in agricultural fields are especially sensitive to the impacts of agricultural intensification on breeding success. The presence of sparrowhawk and three corvid species in this column is perhaps unsurprising since all have undergone strong population increases (Gregory & Marchant 1996) and correlations between population trend and breeding success are expected in situations where populations are not close to carrying capacity. The exceptions to the rule in this list are linnet and house sparrow. Linnet was the only one of 12 seed-eating passerines for which Siriwardena *et al.* (2000a, b) found that correlations between nesting success and population trend were consistent with the possibility that a decline in nesting success had driven population declines. Further work examining the relationship between measures of nesting success and local agricultural land-use (Siriwardena *et al.* 2001a) also found that low nesting success during periods of decline was driven mainly by greater loss of nests during incubation. The case of house sparrow is of interest because there is evidence both that changes in survival rate and nest success may have driven population trends. The most recent analysis of house sparrow data combined both ring-recovery data and nest record data into a single integrated demographic model (Freeman & Crick 2002). They concluded that variation in survival (specifically of first-year birds) may have driven population trends prior to 1975 as earlier concluded by Siriwardena *et al.* (1999). However, the main driver of trends since then (a steep decline in the late 1970s and early 1980s followed by a gradual slowing of decline to stability by 2000) had been variation in the proportion of nests surviving the clutch and brood stages. Added to the fact that we already know from Chapter 5 that recent house sparrow population trends contrast sharply in different parts of Britain, this demonstrates neatly that the demographic factors driving population trends in a single species may change over space and time. Although it is important to understand the cause of population decline, it may not be necessary to reverse that same cause to generate a recovery (see Green 1995a).

The third conclusion to be drawn from Table 7.2, that for many species demographic models have yielded no clear picture of the main drivers of population change, is a very important one. The general explanation is simply that limitations and possible biases in the ring-recovery and nest record data sets limit the power of analyses designed to detect correlations between population changes and demographic rates. The first and most important of these limitations relates to the fact that the demographic rates which drive population trends very often do so through the effects of density-dependent processes such as resource competition, predation or parasitism (Newton 1998). This means that the magnitude and duration of the change in breeding success or survival that drive a period of decline may be small, ephemeral and not detectable once the population has stabilised at a new equilibrium level (Green 1999). For example, consider a species occupying a new, favourable, seasonal habitat in which the birds depend upon different resources in the breeding season and non-breeding (winter) season. Initially, numbers increase rapidly and demographic rates vary independently of population size. As the population approaches the maximum that the area will support (the carrying capacity) either in winter or during the breeding season, then competition for resources intensifies, and demographic rates are likely to become density dependent.[1] If carrying capacity for breeding resources (e.g. food, nest sites) is approached first then density-dependent variation in breeding success will become the primary driver of between-year

[1] For simplicity, we ignore for the time being the possibility that natural enemies such as predators, parasites or pathogens might limit the population at a level lower than that which resources would permit.

population changes. Conversely, if the winter carrying capacity (e.g. food, shelter) is approached first, then density-dependent variation in survival will become the main driver of population change. If we now turn this scenario on its head to consider an environment in which carrying capacities are being reduced progressively by agricultural intensification then if carrying capacity is reduced by declining availability of food then we might expect to see survival rates or breeding success fall and drive a population decline. The crux of the problem is that the magnitude of the change in demographic rate may only be small and only detectable during the phase of decline, since it will subsequently recover in density-dependent fashion as the population settles at the new, lower carrying capacity. Another case where measurements of trends in demographic rates may not help us is where breeding season carrying capacity is reduced very directly by, for example, loss of suitable nest sites. For example, consider removing nest boxes from a wood where a population of great tits or pied flycatchers was largely dependent on these nest sites or, to take an agricultural example, replacing a large area of spring cereal sowing in which lapwings nest with an equivalent area of winter cereal in which they are unable to nest (see Chapter 8). If the main population response in these cases is simply for a lower number of birds to nest, then breeding output may either not change, or may even increase as a density-dependent response to increased per capita availability of food for the remaining lower-density population.

In summary, by no means all of the demographic changes that are responsible for driving population declines may be detectable as measurable trends in nest success or survival rates, especially given the inevitable limitations and biases in the available data. Some of these are outlined in Table 7.3.

Key amongst these are, first, the fact that we have no knowledge of trends in survival rates of some species of conservation concern (e.g. skylark, corn bunting) simply because the sample size of ringed and recovered birds is too small. Second, with the exception of Robinson *et al.*'s (2004) study of song thrushes, the potential effect of changes in post-fledging survival as a driver of population decline remains largely unknown. Third, nest record data offer no insights into the possibility that an effect of breeding success limitation on population size is operating via impacts on the number of breeding attempts made by individual females. For example, Table 7.2 shows that nest record data indicated either no relationship or a positive relationship between nest success and population trend for turtle dove, skylark, song thrush and corn bunting. However, more detailed data on seasonal productivity from field studies of all these species have revealed strong evidence of a reduction in season-long breeding productivity during population declines, driven by a reduction in the opportunity for repeat nesting attempts (Table 7.4).

Summary

The general conclusion from these demographic studies is that declines of altricial species are more likely to have been driven by changes in survival rates outside

Table 7.3 *Limitations of ring-recovery and nest record card data sets relevant to the detection of demographic drivers of population trends at the national scale*

Constraint	Example
Sample size	Ring-recovery data sets are too small for analyses for species that are rare or caught in only small numbers (e.g. skylark, corn bunting), and nest record data for scarcer species may also be limited and/or heavily geographically biased by large numbers of records from very small numbers of observers specialising in that species. Some attempts have been made to measure current survival by recording the disappearance of uniquely marked colour-ringed individuals. Few studies adequately address the need for large sample sizes and intensive, standardised monitoring of the marked population (but see Hole *et al.* 2002 for one exception).
Lack of specificity to farmland habitats	Although nest record card samples specific to farmland habitats can be sub-selected for analysis, analyses of survival rates based on ring-recovery data are not able to consider the habitats occupied by birds prior to recovery.
Post-fledging survival	With the exception of the study of song thrushes by Robinson *et al.* (2004), based on new methods developed by Thomson *et al.* (1999), ring-recovery analyses have not been able to quantify immediate post-fledging survival rates of birds ringed in the nest.
Variation in reporting rates of ringed birds	Trends in the probability that ringed birds will be reported are known to exist and may bias estimation of survival rate trends (Baillie & Green 1987; Catchpole *et al.* 1999). Again, until recently (e.g. Robinson *et al.* 2004; Freeman *et al.* 2007), analyses have not specifically taken account of this problem.
Failure to measure seasonal breeding productivity	Although nest record data permit estimation of a variety of measures of breeding success per nesting attempt, they cannot estimate the number of nesting attempts being made and, hence, total production of young over the whole breeding season (Siriwardena *et al.* 2000a). For these data, we remain reliant on intensive field studies for insights into the extent to which impacts of agricultural change on breeding success may have operated via effects on the number of breeding attempts made (see Table 7.4).
Seasonal effort in nest recording	The early part of the nesting season is more accurately monitored by nest records cards than later, partly because vegetation growth may make nest finding more difficult later in the season, and partly because observer effort wanes later in the season (Crick & Baillie 1996). Either effect may generate bias in nest record-based estimates of nest success if nest success varies systematically through the season as is known to occur for several species (e.g. Evans *et al.* 1997b; Newton 1999; Bradbury *et al.* 2000; Moorcroft & Wilson 2000; Proffitt *et al.* 2004).

Table 7.4 *Contrasting perspectives provided by nest-record-based analyses and field studies, demonstrating the potential importance of limitation of breeding season length in driving declines of lowland farmland birds*

Species	Findings of nest record studies	Findings of field studies[a]
Turtle dove	Number of fledglings per nest greater during a period of decline (1979–95) than previously (1962–78) when the population was stable (Siriwardena *et al.* 2000a). A later analysis using a longer run of data (Browne *et al.* 2005) found no correlation between nest success and population trend.	Comparison of breeding success at East Anglian study areas during the 1960s (Murton 1968) and 1990s (Browne & Aebischer 2004) found no differences in per nest measures of breeding success but found that the number of clutches laid per season had declined by 45% from 2.9 to 1.6. The resultant decline in season-long breeding output was more than sufficient to drive the observed population decline.
Skylark	Clutch and brood sizes increased progressively over the period 1950–94, and partial losses of chicks between hatching and fledging declined (Chamberlain & Crick 1999).	Abundant direct and indirect evidence from field studies that the number of nesting attempts possible during a single season has become progressively more limited on intensively managed farmland, and that this change has been sufficient to drive observed population declines (reviewed by Donald 2004, and see Chapter 8).
Song thrush	A greater number of fledglings per nesting attempt during population decline driven mainly by increasing nest survival rates at both clutch and brood stages (Baillie 1990).	In a rapidly declining farmland population in Essex, females made too few nesting attempts (2.5) in a season to sustain local numbers. In a stable farmland population (Sussex), females made a sufficiently greater number of nesting attempts per season (4) to account for the stability of the population (Thomson & Cotton 2000).
Corn bunting	The number of fledglings per nesting attempt during a period of population decline (1974–95) was more than double that during a period of population increase (1962–73), driven both by increases in nest success rate and brood size (Siriwardena *et al.* 2000a).	The frequency with which females are able to make repeat or second nesting attempts in this late-nesting species is diminished in intensive agricultural systems, and this is likely to have been a key driver of population decline (Brickle & Harper 2002).

[a] Field studies referred to in this table are discussed in more detail in Chapter 8.

the breeding season, whilst declines of precocial, ground-nesting species are more likely to have been driven by reduction in nesting success (see Sæther *et al.* 1996). However, although increases in breeding success per nesting attempt amongst many farmland passerines may have been driven largely by density-dependent increases in brood sizes and survival rates of individual nests, large-scale demographic analyses do not tell us anything about season-long breeding success for species making multiple nesting attempts. Where such studies have been undertaken, they frequently reveal that a limitation on the length of the breeding season may be a very important factor in driving population declines. It is difficult to disentangle the effects of food shortages, nest predation, weather conditions and the interactions between these factors in determining what may drive trends in individual nest success (see below). However, it is probable that the main impact of agricultural intensification on the breeding season for many altricial species has been to reduce the duration of the season over which adequate nesting or food resources are available. The reduction in the absolute size of the food resource peak for these species has perhaps been less important. In contrast, we might expect a given reduction in the size of the seasonal food resource peak to have greater impacts on precocial species since here the chicks depend on feeding themselves very soon after hatching, rather than benefitting from their parents' ability to seek out profitable foraging sites in a wide area around the nest.

Survival rate analyses give poor resolution of the seasonal timing of any increase in mortality that may be driving population declines, and the potential impact of mortality during the immediate post-fledging period is especially poorly understood. In species where analyses have resolved the seasonal timing of trends in mortality (notably song thrush, chough and starling), it seems that pre-breeding mortality, especially during the post-fledging period and first winter, is the key to understanding population trends. In addition, the studies of both song thrushes and house sparrows have revealed that the demographic causes of population trends may vary between different locations and at different times, such that trends at the national scale may need us to consider several demographic drivers. For example, if we wish to understand the causes of song thrush decline, we probably need to know both what has driven reductions in survival during the first of year of life, but also what is limiting the length of the breeding season in intensively managed arable habitats (Thomson & Cotton 2000; Peach *et al.* 2004b; Robinson *et al.* 2004; and see Chapter 8).

Plate 7.1 During winter, the yellowhammer specialises in feeding on large seeds – especially cereal grain. Its breeding range has contracted from many northern and western areas of Britain, as arable agriculture has been lost. Analyses of Breeding Bird Survey data show that small areas of arable cultivation are very strongly correlated with the persistence of this species in otherwise grass-dominated agricultural landscapes. © RSPB.

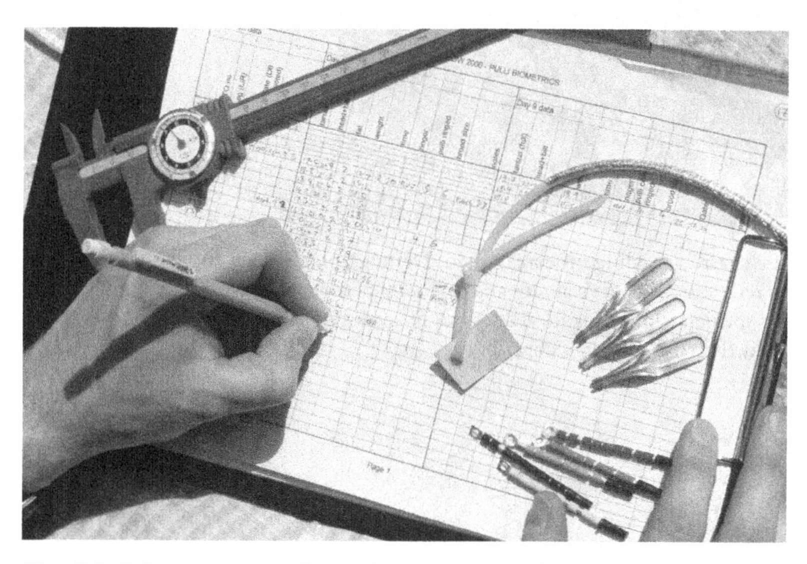

Plate 7.2 Colour-ringing and recording biometrics of tree sparrows in a population study at Rutland Water. Ringing studies such as these have played a critical role in understanding changing survival rates and understanding how individual birds exploit the habitats and food resources available to them through the seasons. © RSPB.

Plate 7.3 Male house sparrow feeding on cereal grain. Demographic studies suggest that at different times both reductions in survival rate and breeding performance may have contributed to the decline of this species. On farmland, reductions in seed supply, including the bird-proof storage of grain, are likely to have been a contributory factor. © RSPB.

Plate 7.4 The song thrush has recently been the subject of one of the most intensive demographic studies of any British farmland bird. Results suggest that both reduced survival during the post-fledging period and the inability of adults to make successful breeding attempts later in the summer on intensive, arable farmland have driven population declines. Lack of damp habitats, rich in earthworms, in later summer in the intensively drained agricultural landscapes may be the driver of both of these demographic trends. © RSPB.

8 · *Species case studies*

Introduction

Just five breeding species currently or formerly characteristic of agricultural habitats were identified as 'Red Data' species on the basis of severe population decline in the first analysis of priority bird species for conservation action in Britain (Batten *et al.* 1990). These were grey partridge, corncrake, stone curlew, red-backed shrike and cirl bunting. By the mid 1990s, the first revision of this Red List (Gibbons *et al.* 1996a) and the identification of priority species under the UK Biodiversity Action Plan (UKBAP) had increased this list to 15, and the second revision of the Red List (Gregory *et al.* 2002) increased it to 18. In 2007, revision of the UKBAP priority list added a further nine species. Table 8.1 summarises this progressive addition of species meriting urgent conservation action as population declines have taken hold, and also shows the tie-in to the start of research dedicated to diagnosing the causes of species' declines and identifying potential solutions. The clear identification of species priorities for conservation action initiated by the International Union for the Conservation of Nature (IUCN) and pioneered for British birds by Leo Batten and colleagues at NCC and RSPB undoubtedly galvanised the direction of scarce resources to conservation research and action for declining farmland birds. This focus has stimulated an explosion of research during the 1990s that has dramatically improved our understanding of the impacts of recent agricultural change on bird populations in Britain. It is perhaps worth reflecting that had this process of prioritisation been in place even a decade earlier we would have been able to respond far more quickly to this unfolding conservation challenge. Despite the excellence of bird monitoring in Britain, the declines of many species continued for 20 years before they were recognised as a generic issue.

Below we present 16 case studies of birds that have been the subject of detailed research effort and that, taken together, illustrate the diverse mechanisms through which agricultural intensification has affected their populations. These comprise nine of the Red-listed species in Table 8.1, two further declining species (lapwing and chough), two species characteristic of upland, grazed systems (black grouse and golden plover), one species whose decline in response to changing agricultural practice was early and precipitous and which has since recovered in both numbers and range (sparrowhawk), and case studies of wintering geese and woodpigeons as examples of species which have largely benefited from agricultural intensification.

Case study 1: Wintering geese (pink-footed, greylag, white-fronted, barnacle and brent)

The sight and sound of skeins of geese flighting in to their feeding grounds on a winter morning is a stirring wildlife spectacle, and one that can now be enjoyed the length and breadth of Britain from the south coast of England to the islands of Scotland. Dramatic increases in the populations of several species that have taken place over the past 40 years (Table 8.2) reflect the combined effects of site and habitat protection, reduced hunting pressure and increasing behavioural adaptation by the

Table 8.1 *Admission of breeding birds of agricultural habitats to the UK Red List of species of high conservation concern (1990, 1996, 2002), and UK Biodiversity Action Plan priority list (1995, 1998–99, 2007). No species, once added, has yet recovered sufficiently in numbers to be removed from this list*

Species[a]	**Included from**	**Start of diagnostic research**
Red grouse	2007	1960s
Black grouse	1996, 1999	1980s
Grey partridge	1990, 1995	1960s
Quail	1996	None
Corncrake	1990, 1995	1980s
Stone curlew	1990, 1995	1980s
Lapwing	2007	1980s
Black-tailed godwit	2007	1980s
Curlew	2007	1990s
Turtle dove	1996, 1998	1990s
Cuckoo	2007	None
Nightjar	1996, 1998	1980s
Woodlark	1996, 1998	1990s
Skylark	1995, 1996	1990s
Tree pipit	2007	None
Yellow wagtail	2007	Since 2000
Dunnock	2007	None
Ring ouzel	2002, 2007	Since 2000
Song thrush	1995, 1996	1990s
Grasshopper warbler	2002, 2007	Since 2000
Spotted flycatcher	1996, 1998	Since 2000
Red-backed shrike	1990, 1998	None in Britain
Starling	2002, 2007	Since 2000
House sparrow	2002, 2007	1990s
Tree sparrow	1996, 1998	1990s
Linnet	1996, 1998	1990s
Twite	1996, 2007	1990s
Lesser redpoll	2007	None
Bullfinch	1996, 1998	1990s
Yellowhammer	2002, 2007	1990s
Cirl bunting	1990, 1998	1980s
Reed bunting	1996, 1998	1990s
Corn bunting	1996, 1998	1990s

[a] Species in bold are the subject of detailed case study accounts in the main text.
Sources: Red List: Batten *et al.* (1990); Gibbons *et al.* (1996); Gregory *et al.* (2002). Biodiversity Action Plan priority list: Anon. (1995, 1998–1999); BRIG (2007).

Table 8.2 *Approximate British wintering population sizes (1960–2000) of wild, migratory geese that may cause agricultural conflicts*

	Estimated size of the British wintering population					
Species	1960	1970	1980	1990	1995	2000
Pink-footed goose	57 000	72 000	95 000	195 000	200 000	245 000
Icelandic greylag goose	26 000	65 000	90 000	115 000	83 000	79 000
Greenland white-fronted goose	3 000	2 000	4 000	15 000	22 000	21 000
Barnacle goose	5 000	18 000	21 000	34 000	47 000	64 000
Dark-bellied brent goose	15 000	24 000	67 000	116 000	101 000	93 000
Total	106 000	181 000	277 000	475 000	453 000	502 000

Source: Modified from Vickery & Gill (1999) and with data for 2000 from Pollitt *et al.* (2003).

birds to exploit the increasing abundance and nutritional quality of agricultural crops as a food source. Only a few populations, such as those of the European race of the white-fronted goose, and the bean goose, have declined. As a consequence of goose-grazing of crops, celebration of a conservation success story is in many areas tempered by the reality of managing conflicts between geese and agricultural production.

Geese are grazers that, before the advent of human agriculture, would have been restricted to natural habitats allowing this form of feeding (e.g. mudflats, coastal marshes and tundra) and open areas of ground created by herds of large mammalian grazers such as buffalo and horses (Ogilvie 1978). Agriculture, however, provides a huge range of protein-rich vegetation and grains, growing cereals and other grasses, roots and tubers all of which may be taken by geese. When migratory geese first arrive in autumn, the level of conflict with agricultural interests may be low, as the geese feed on stubbles, waste crop material (e.g. grain and potato or beet fragments), grassland (which is little used by farm animals at this time) or, in the case of brent geese, remain on mudflats and saltmarshes (Vickery & Gill 1999). Later in the winter, however, geese may move on to growing cereals, brassicas and grass, and damage may then become economically significant (Table 8.3). Even relatively low rates of offtake such as those recorded by Paterson (1991) for native greylag geese may be a significant loss to economically marginal agricultural systems such as the crofting land of the Hebridean machair.

Although some goose species have retained relatively restricted habitat and dietary niches, all of those whose populations have increased in Britain in recent decades have proved rapidly adaptable in the exploitation of farmland. Brent geese traditionally fed, and still do, on the leaves and rhizomes of eel-grass which grows in beds on estuarine mudflats. Later, when the supplies of eel-grass are exhausted, the geese move on to mudflat algae such as *Ulva* and *Enteromorpha*, and ultimately to saltmarsh succulents such as glassworts and grasses (Ogilvie 1978). During the 1930s,

Table 8.3 *Examples of yield loss caused by winter goose grazing*

Species	Yield loss
Brent goose	Up to 10% on winter wheat and up to 27.5% on oilseed rape
Pink-footed goose	Up to 7% on barley and 15% on wheat
Barnacle goose	Up to 82% of spring standing crop of grass and 38% of silage
Greylag goose	Up to 5% on machair cereals

Source: Data from Patterson *et al.* (1989); Summers (1990); Paterson (1991); Percival & Houston (1992); McKay *et al.* (1993); Vickery & Gill (1999).

eel-grass beds were attacked and wiped out in many areas by an amoeboid protozoan *Labyrinthula zosterae*. The immediate response was a rapid and severe decline in brent goose numbers, but some of the remaining birds did explore alternative food sources, including a move over the sea wall on to coastal pastures and cereal fields. Whether the primary cause of the development of this new feeding habitat was the loss of eel-grass or the gradual reduction in hunting pressure remains uncertain. Either way, however, the habit has spread alongside the subsequent revival of brent goose populations to the extent that resolution of conflicts between brent geese and the management of pastures in southern England became the subject of intensive research effort (Vickery *et al.* 1994a).

Though drastic loss of a traditional food may have been necessary to propel brent geese into the exploration of agricultural habitats, other species have been quicker to adopt new food sources. Pink-footed geese were first recorded feeding on waste potatoes in Lancashire in the late nineteenth century, but the habit had become widespread by the 1920s and now occurs wherever the distribution of potatoes and geese coincide (Ogilvie 1978). More recently, pink-footed geese have also discovered the merits of feeding on the harvested remains of sugar beet. The recent redistribution of the wintering population towards areas of north Norfolk where sugar beet is commonly grown may reflect cultural learning of the benefits of feeding on fields where there is no conflict with the farmer, and the birds are left relatively undisturbed (Gill *et al.* 1997). However, the possible collapse of the sugar beet industry in Britain as a result of withdrawal of lucrative subsidies has prompted fears for the future status of pink-footed geese in England and may substantially increase conflict with farmers.

Although culling remains a possible option for reducing the level of conflict between wintering geese and agricultural interests, it is nowadays considered an option of last resort, and two alternative approaches are usually adopted. The first, exemplified by several statutory Goose Management Schemes in Scotland, simply involves compensating farmers for allowing geese to forage undisturbed in areas where crop damage is heavy. On the Solway Firth, for example, the entire Svalbard population of barnacle geese winters and has increased from just 400 individuals in the 1940s to over 27 000 today, centred on reserve areas at Caerlaverock (Wildfowl and Wetlands Trust) and Mersehead (RSPB). As the population has increased, birds have moved from traditional feeding grounds on the 'merse' (saltmarsh) to feed on fertilised pastures inland where they have caused significant damage (Owen 1990). Traditionally, farmers scared foraging geese from their land, but the introduction of the Barnacle Goose Management Scheme (BGMS) in 1994 introduced payments to farmers to cease scaring in a zone around the main reserve areas (Cope *et al.* 2003). The BGMS has been successful to the extent that it has concentrated feeding geese within a defined area and permitted population increase beyond that which nature reserve areas alone could sustain. However, as the barnacle goose population continues to increase, further conflicts with agricultural interests beyond the current BGMS boundary seem inevitable and, as the reserve and BGMS areas cannot expand indefinitely, efforts to increase the quality of some of the feeding fields within the BGMS may be preferable.

The second approach – the design and management of Alternative Feeding Areas (AFAs) specifically to attract and hold foraging geese in order to keep them off nearby agricultural land – has been the subject of much recent research, particularly in relation to the creation of pastures ideal for brent geese (Vickery *et al.* 1994a, b). In general terms, the ideal field for brent geese is large (>5 hectares), undisturbed, located within a few kilometres of a favoured roosting area, sown with favoured grasses (e.g. ryegrass, timothy, fescue) and white clover, fertilised, and cut or grazed to around 5–10 cm in height (e.g. Vickery & Gill 1999; McKay *et al.* 2001). Clearly, the precise specification may differ for other species. Nonetheless, the incorporation of sacrificial AFAs into goose management schemes or agri-environment schemes at hotspots of goose–agriculture conflict may provide one of the best short- to medium-term means of mitigating goose impacts. As an example, the current Higher Level Environmental Stewardship scheme in England supports management of wet grassland for wintering geese of all indigenous species and, similarly, the Essex Coast Environmentally Sensitive Area (ESA) has offered a 'wildfowl pasture' supplement as payment to manage for brent geese. In the longer term, if goose populations continue to increase, the challenge of limiting the agricultural impact of burgeoning goose numbers may remain. Recently, an analysis of the economic costs (from grazing damage to crops) and benefits (from government compensation payments) to farmers of goose conservation was undertaken in north-east Scotland (pink-footed and greylag geese) and on Islay (barnacle and white-fronted geese). Encouragingly, this found that the benefits markedly outweigh the costs, and that the per capita costs of goose damage fall as goose density rises (MacMillan *et al.* 2004).

Case study 2: Sparrowhawk

The sparrowhawk is once again one of Britain's most common and widespread birds of prey, with a current population of approximately 40 000 breeding pairs (Table 5.2). However, its historical decline due to the toxic effects of the early generations of organochlorine insecticides led to the sparrowhawk becoming the subject of one of the seminal studies of the impacts of environmental pollution on wildlife by Ian Newton (1986), and his account is central to the material presented here.

Sparrowhawks hunt over all manner of habitats that contain a plentiful supply of small to medium-sized birds, and hence make extensive use of farmland. Prey is usually captured near to some form of cover through a combination of stealth and manoeuvrability, though sparrowhawks can fly faster than most songbirds. Extensive woodlands of young, closely spaced trees support the highest breeding densities, although sparrowhawks also nest readily in small woods, shelterbelts and patches of scrub. Whilst conifers are preferred for nesting, broadleaved trees are also commonly used. The nest tree is typically placed near an opening in the tree canopy such as a ride or stream. Suitable breeding territories are typically occupied year after year though a new nest is built each season, usually in a tree close to one used in a previous year. A single clutch of three to six eggs is laid sometime between late April and early June. Laying is timed to coincide with the peak in food availability, when newly fledged songbirds are most abundant. Young sparrowhawks vacate the natal territory when they become independent at about eight weeks old and usually take up residence within 20 km of where they were born. Most birds do not breed until they are in their second, third or even fourth year, and having bred within an area, they usually remain there for the rest of their lives. A large variety of avian prey is taken by sparrowhawks: the smaller males typically take smaller, sparrow-sized songbirds whereas the larger females usually take thrush-sized birds but can tackle prey up to the size of a full-grown woodpigeon. Studies suggest that sparrowhawk predation has no lasting effect on the abundance of songbird populations (Perrins & Geer 1980; Newton *et al.* 1997; Thomson *et al.* 1998), despite claims to the contrary. Indeed, there is evidence to suggest that sparrowhawk numbers are controlled by availability of their prey (Petty *et al.* 1995). A long-term study of sparrowhawk populations in southern Scotland found that the loss of birds between fledging and recruitment into the breeding population was the key factor determining the numbers of breeding pairs, and was probably related to food availability (Newton 1986, 1988).

The changing fortunes of the sparrowhawk over the last 50 years provide a compelling case study of the impact of environmental pollutants on birds that was instrumental in alerting society to the dangers to wildlife resulting from the widespread use of highly toxic and persistent pesticides on agricultural land (Newton 1986). Prior to the late 1950s, and despite decades of persecution in the interests of game preservation, the sparrowhawk was a common inhabitant of Britain's woodlands and farmlands. However, in the space of just a few years, the species had become scarce over much of the country and had almost disappeared from some eastern areas. This population crash coincided with the introduction of organochlorine pesticides on to British farmland, especially DDT, aldrin and dieldrin. DDT was introduced in the mid 1940s and soon became widely used as an insecticide spray on many arable crops and on grasslands. The so-called cyclodiene compounds, most notably aldrin and dieldrin, were introduced in 1956 and were most commonly used as cereal seed dressings to protect against attack by the larvae of the wheat bulbfly. These chemicals were persistent and fat soluble, and so quickly became concentrated in the tissues of sparrowhawks and other predatory animals that preyed upon birds that had previously fed on contaminated seeds and invertebrates. When aldrin and dieldrin were first used they were sprayed over the whole crop at a rate of roughly 1 kg ha^{-1}. Ironically, it was later decided to concentrate the insecticide on the cereal grain where the target pest attacked in order to minimise wider environmental effects of its use. This of course turned out to be the best way possible to maximise the ingestion of the chemical by seed-eating birds, and thence raptors (Mellanby 1981).

Thanks to the observations of birdwatchers and to the work of a dedicated group of government scientists, the impact of such contamination on sparrowhawks and other birds of prey, such as peregrine falcons, soon became apparent. Whilst DDT is only mildly toxic to birds, it was discovered that sub-lethal doses caused embryo deaths and the thinning of eggshells, leading to egg breakage, and hence reduced breeding success (Ratcliffe 1970; Cooke 1973, 1979; Newton & Bogan 1978). In contrast, the cyclodiene compounds were found to be 150 times more toxic than DDT and killed many birds outright (Cramp *et al.* 1962). Eggshell-thinning first occurred in the late 1940s but it wasn't until the late 1950s, after the introduction of the cyclodienes, that the sparrowhawk population crashed (Prestt 1965). This suggests that the decline was driven by the increased mortality caused by aldrin and dieldrin contamination, though most of the surviving birds also suffered reduced breeding success. Birds of prey like sparrowhawks were more affected than other birds because, as predators, they accumulated higher levels of pollutants and because they were especially sensitive to the chemicals and so experienced more eggshell-thinning. Such impacts were noted wherever these chemicals were in wide use across the world (Newton 1979).

Pesticide contamination had dire consequences for the sparrowhawk population. Whilst the first signs of decline became apparent from around 1950, the real crash began from 1957 and, by 1960, was evident to most birdwatchers. Virtually the whole country was affected to some degree, but the greatest declines were seen in eastern England where the amount of tilled land and hence pesticide use was greatest. The bird even became extinct in arable areas of East Anglia (Newton & Haas 1984). Numbers were least affected in south-west England, Wales and northern and western Scotland, and reasonable populations persisted in small pockets of southern and eastern England, such as the New Forest, where there was less cultivated land. The plight of the sparrowhawk and other predatory animals was quickly brought to the attention of Government and society as a whole, leading to successive restrictions on the use of the offending pesticides on farmland between the early 1960s and mid 1980s, and culminating in eventual bans. Indeed, there was public outrage at the impacts of these chemicals, prompted by such landmark publications as Rachel Carson's *Silent Spring* (1963), and widespread concern over the potential effects on human health.

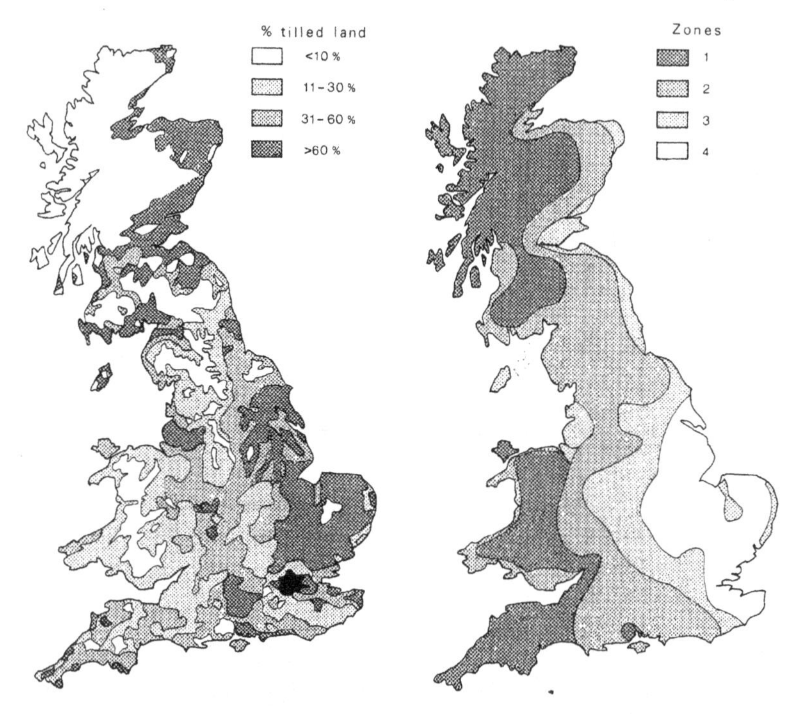

Fig. 8.1 Changes in the status of sparrowhawks in relation to agricultural land use and organochlorine use. The agricultural map (left) shows the proportion of tilled land, where almost all organochlorines were applied. The sparrowhawk map (right) shows the status of the species in different regions and time periods (see text). Reproduced with kind permission from Newton & Haas (1984), Newton (1986).

Key milestones were the withdrawal of the use aldrin and dieldrin on spring-sown cereal grains in 1962 and on autumn-sown grains in 1975. As a consequence, the area treated with these two chemicals in the late 1970s was a mere fraction of that previously treated. The use of DDT declined during the 1960s and 1970s but it was still used in considerable quantities into the 1980s (e.g. on fruit and brassicas) until, finally, its use was banned in 1986. As early as the mid 1960s, there was evidence of partial recovery in sparrowhawk numbers in parts of the north and west of Britain, and in the following years, the recovery occurred in a wave-like pattern, spreading eastwards across Britain. Newton & Haas (1984) identified four zones of recovery, stretching from west to east across the country, which largely reflected the proportion of tilled land (Fig. 8.1): Zone 1 was in the west and north of Britain where the sparrowhawk survived in greatest numbers (suffering less than a halving of numbers) and recovery was complete before 1970; Zone 2 included much of western Scotland and northern, western and south-central England where the population had recovered to more than 50% of its former level by 1970; Zone 3 was just to the east of Zone 2 and included much of central England where the population had recovered to more than 50% of its former level by 1980;

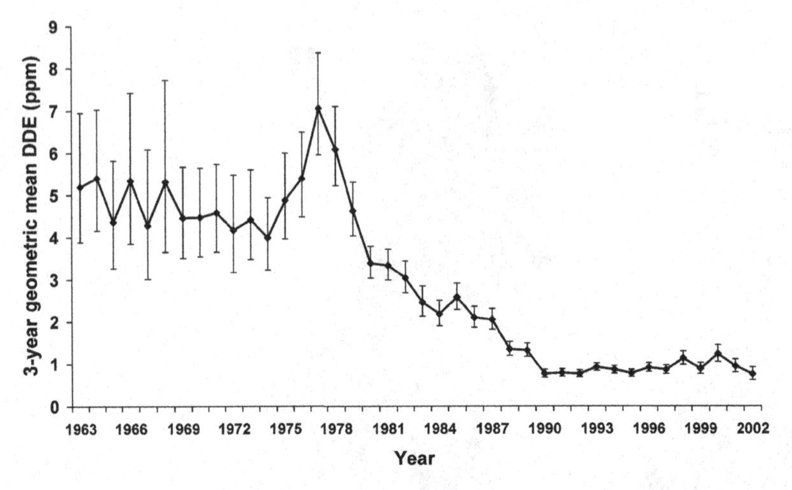

Fig. 8.2 Changes in DDE concentrations (ppm ± 1 geometric SE) in sparrowhawk eggs, 1963–2002. Data provided with kind permission of the Centre of Ecology and Hydrology's Predatory Bird Monitoring Scheme (Shore *et al.* 2006).

and Zone 4 consisted of much of eastern England, where the population had become extinct and little or no recovery was apparent by 1980.

Since 1963, scientists at Monk's Wood Research Station have examined a range of predatory birds found dead in Britain and established the cause of mortality and pollutant content, as well as analysing addled eggs. This work has shown that there were definite declines in the amount of HEOD (a chemical derived from aldrin and dieldrin) found in sparrowhawks in the periods following the two main restrictions in their use. Similarly, the levels of DDE (the main metabolite of DDT present in avian tissues) found in sparrowhawks declined from the 1980s (Fig. 8.2) which had a corresponding effect on eggshell thickness and breeding success. Information from the BTO's Nest Record Scheme suggests that nest failure rates at the egg stage fell by 92% between the late 1960s and 2005 (Baillie *et al.* 2007).

Since the 1980s, sparrowhawks have recolonised eastern England and have probably reached carrying capacity over much of the country, even colonising many towns and cities. The pattern of recovery is also illustrated by data from many local areas, no better exemplified than by a study in the Rockingham Forest area of Northamptonshire. The county was probably recolonised by breeding birds in 1976, and in Rockingham Forest alone the number of sparrowhawk nests increased from three in 1979 to 84 in 1989 (Newton & Wyllie 1992). This coincided with a significant reduction in HEOD and DDE levels in sparrowhawks collected across eastern England and an improvement in average breeding success in the Rockingham Forest pairs from 1.8 young per clutch in 1980 to 2.9 in 1989. Numbers subsequently stabilised at a slightly lower level probably reflecting the availability of avian prey in the area (Wyllie 1995). This dramatic recovery was probably typical of many parts of eastern England and, albeit in a less extreme fashion, of much of Britain. Along with the recovery of the peregrine falcon and several other predatory birds that were similarly affected, the return of the sparrowhawk is a true conservation success story. It serves as a testament to those dedicated scientists and lobbyists who championed such a worthy cause and as a salutary lesson of the dangers to wildlife of introducing untested chemicals into the environment.

Case study 3: Black grouse

The black grouse is one of just three breeding species in Britain whose mating system is based upon lekking – the others being the capercaillie and ruff. Black grouse leks are typically found on open, sparsely vegetated ground such as forest clearings and rides, grassy or boggy patches on heather moorland or on hill pastures. Cock birds gather at lek sites early on spring mornings where their reverberating 'roo-koo' calls and 'crowing-hiss' displays may be audible over several kilometres, and are a spectacular birdwatching experience. Males compete for central areas of the lek and for the attentions of the promiscuous females, with the males defending the central stances within the lek (usually older birds) securing the majority of the copulations. Females then move away to nest in tall vegetation, usually mature heather or rushes, but occasionally in hay meadows (e.g. Starling-Westerberg 2001). For the first three weeks after hatching, the chicks rely for food on invertebrates collected from vegetation such as soft rush, bog myrtle and bilberry that support high densities of prey items. Although broods are capable of travelling several hundred metres from the nest within days of hatching, their prospects are better when they are able to find adequate food sources without moving far (Picozzi & Hepburn 1986).

Quite apart from the charismatic appeal of its lekking display, the black grouse has also been a prized quarry species throughout its range. For this reason its population trends have been well documented and its catastrophic decline in Britain during the twentieth century has stimulated intensive research to understand the causes and identify solutions. To describe the ecological needs of black grouse in Britain is a difficult task. They are often described as being birds of 'edges', 'fringes' and 'early successional stages' (Baines 2003), indicating a need for a mosaic of habitats within a small area. Over much of their Scandinavian range and in the Scottish highlands, this habitat mix is provided by conifer forest and birch wood edges next to moorland or open, boggy areas within wooded land; in the Alps by herb-rich alpine grasslands at the upper forest edge; and in some parts of central Europe, and their former range in England, by lowland heathland. This diverse habitat requirement reflects the variety of foods needed to support the species' nutritional needs, in marked contrast to the heather-dominated diet of the red grouse. Table 8.4 illustrates the range of habitat and food requirements characteristic of black grouse in Britain.

The remnant mosaics of Scots pine forest, birch woodland and heather moorland occupied by black grouse in northern Scotland are an echo of natural habitats that are now rare due to intensification in human land-use that has created coarser-grained

Table 8.4 *Food and habitat requirements of black grouse in Britain*

Season	Typical foods	Habitat
Spring	Cotton-grass,[a] bog myrtle,[a] bilberry, conifer buds and shoots, heather shoots	Blanket bog, rough pasture, heather moorland, pine, birch and conifer woodland
Summer (adults)	Seeds of grasses, rushes and sedges, green parts of bilberry, heather and birch	Rough pasture, wet flushes and bog on heather moorland or in woodland
Summer (chicks)	Invertebrates[b] (especially ants and caterpillars of moths and sawflies, but also spiders, plant bugs, beetles and flies)	Wet flushes with stands of rush or bog myrtle, hay meadows
Autumn/Winter	Heather, birch, green parts and fruit of bilberry and crowberry, berries of rowan and hawthorn	Heather moorland, rough pasture, birch and conifer woodland

[a] Foods rich in protein that are of particular importance in spring when hens are attaining breeding condition.
[b] Invertebrate foods are taken particularly by young chicks, up to the age of three weeks, by which time they have adopted the herbivorous diet typical of full-grown birds.
Source: Modified from Hughes *et al.* (1998) and based on the studies of Parr & Watson (1988), Baines (1994), Baines *et al.* (1996) and Starling-Westerberg (2001).

land-use mosaics and 'harder' edges between habitats. In northern England black grouse are now found in largely treeless landscapes, occupying the moorland fringe where heather, rough grassland and hay meadows are found in close proximity. In Wales and many areas of Scotland, they are also associated with upland conifer plantations adjoining open moorland and blanket bog. Former populations on lowland heathland from Norfolk to Cornwall, and throughout much of Wales, have been extinct since the turn of the twentieth century, probably as a consequence of the widespread loss and fragmentation of heathland habitat mosaics.

The historical decline of black grouse populations in Britain has been spectacular. Shooting bags suggest a decline in excess of 90% over the hundred years between 1890 and 1990 (Baines & Hudson 1995), and the two estates with, respectively, the highest annual bag (Drumlanrig, Dumfries) and the highest day bag (Cannock Chase, Staffordshire) no longer support any black grouse (Baines 2003). By 1995/96, the population of displaying males was estimated at only 6500 (Hancock *et al.* 1999) and under 5100 by 2005 (Sim *et al.* 2008), with further local extinctions in the Peak District, the Lake District and parts of southern and western Scotland. In contrast, local increases may be seen where black grouse occupy new conifer plantations or native woodland plantings. For example, for the first 10–15 years after planting conifers on moorland, ground vegetation such as heather and bilberry proliferates due to the exclusion of sheep and deer, and the resulting increase in both cover and food availability may attract black grouse at high densities. Later, canopy closure excludes field layer vegetation, and the habitat gradually becomes wholly unsuitable (Cayford & Hope-Jones 1989; Pearce-Higgins *et al.* 2007). In Scotland, any short-term, beneficial impacts of commercial afforestation may be outweighed by the erection of fences to protect forestry from deer browsing. Such fences, typically 1.8m high and made of wire mesh or rylock, are difficult for flying grouse to see and avoid, and cause substantial numbers of fatal collisions from both black grouse and capercaillie (Baines & Summers 1997), although marking fences to make them more visible, for example with wooden droppers or fluorescent tape, does help to reduce mortality (Baines & Andrew 2003).

The growth of commercial conifer forestry during the twentieth century has certainly had a big effect on black grouse populations, with local increases after initial planting being followed by extirpation of populations after canopy closure. However, the ultimate cause of the decline of black grouse in the British uplands, especially in England, Wales and southern Scotland, has probably been the progressive loss, by increasingly intensive land-use, of the fine-scale habitat mosaics of forest edge, heathland and grassland upon which the species depends. The felling and grazing out of native woodlands and scrub with their open stand structure and understorey of heathers and bilberry has been going on for several hundred years but, more recently, increased grazing pressure from sheep on upland heathland and grassland has exacerbated the problem. Heavy grazing pressure has the immediate effect of reducing sward height and cover, potentially exposing black grouse nests and broods to higher risk of predation. Over the longer term, however, if mosaics of heather, other dwarf shrubs and grassland are lost in favour of a dominant cover of coarse grasses (e.g. Anderson & Yalden 1981; Hester & Sydes 1992; Thompson *et al.* 1995), then the diversity and abundance of foods for both adults and chicks is likely to decline. Particular damage may be caused where sheep are over-wintered on heather moorland (this is the season when their grazing impact on heather is greatest: Armstrong *et al.* 1997) and when accompanied by burning extensive areas of heather to improve grass growth for grazing (e.g. Yallop *et al.* 2006). The effects of higher grazing intensities are likely to have been further compounded by the drainage, fertilising and reseeding of existing rough pastures at the moorland edge to create agriculturally improved but species-poor swards that are lacking in both the plant and invertebrate foods favoured by black grouse (Hughes *et al.* 1998). For example, Pearce-Higgins *et al.* (2007) found that lek size and change in lek size in Perthshire were both inversely correlated to cover of improved agricultural pasture in the surrounding area. Ludwig *et al.* (2008) have found that drainage of bogs in Finland is associated with reduced breeding success of black grouse. Reductions in invertebrate populations, increased predation risk in drier habitats and drowning of chicks in large drainage ditches may all play a part in causing these reductions. In a study of 20 moors in northern England and Scotland, Baines (1996) found that moors with higher grazing intensities had vegetation height and nesting cover reduced by a third or more. Heavily grazed moors also supported about 40% fewer

Table 8.5 *Measures of black grouse density and breeding success on 20 moors in northern England and Scotland as a function of grazing intensity*

	Light grazing	Heavy grazing	Significance
Male density (birds km^{-2})	2.1	1.3	<0.05
Female density (birds km^{-2})	3.2	1.8	<0.02
Percent females with broods	56	45	<0.05
Brood size	3.5	2.9	<0.01
Juveniles per female	2.3	1.4	<0.01

Source: From Baines (1996).

invertebrates, including threefold fewer caterpillars, and 50% fewer spiders and plant bugs. Black grouse achieved significantly higher densities and breeding success (Table 8.5) on lightly grazed moors, irrespective of whether or not a gamekeeper was present, and even though gamekeepers achieved on average a threefold reduction in carrion crow density.

Further monitoring work by Calladine *et al.* (2002) in the same area of northern England showed 4.6% per annum increases in male black grouse numbers on 10 sites which had been lightly grazed under agri-environment agreements, compared with a 1.7% per annum reduction at 10 reference sites where no grazing restrictions had been imposed. Grazing restrictions had a less clear impact on densities of black grouse hens although their effect tended to be positive where grazing restrictions were applied at a fine grain (i.e. with a high edge: area ratio) over a small area, and neutral or negative where they covered larger areas in larger patches. The reason for this effect is unclear though it may be that some areas of shorter vegetation are beneficial, perhaps to reduce risk from mammalian ground predators that rely on stalking and olfactory detection, and perhaps to allow chicks to dry out during wet weather. These ideas remain untested at present, but it is clear that the requirement of black grouse for habitat mosaics exists even at the finest scale.

The implications of this research are that grazing, burning and drainage regimes might all be managed to provide better mixes of nesting, brood-rearing and year-round foraging habitat for black grouse than exist at present at the moorland edge. Where these habitats exist in association with forestry, then management of forest structure can be adapted to complement agricultural management on the surrounding land (Pearce-Higgins *et al.* 2007). Forest management should include provision of open areas rich in dwarf shrubs and wet flushes rich in cotton-grass or bog myrtle, manipulation of tree species composition (to provide species such as birch, rowan, hawthorn and larch whose fruits and seeds are taken by black grouse as food) and reduction of deer densities by culling (rather than use of fences). Control of generalist predators such as corvids and foxes may assist black grouse populations where densities are now very low, but in the longer term improving the quality of the habitat mosaic so that nests and broods are more secure may reduce the need for this.

Detailed management recommendations for black grouse are now available (e.g. Baines 2003), and are being applied through recovery schemes that are under way in Wales, the north Pennines, Dumfries & Galloway and in northern Scotland. Perhaps one of the greatest successes thus far has been in Wales where, in conjunction with control of generalist predators such as foxes and corvids, forest edges have been 'softened' by creating a graded change from dense plantation to open ground, dense stands of heather on adjacent moorland have been broken by cutting, and mires and wet flushes have been restored by drain blocking (Lindley & Smith 2002). In the project's core area in mid and north Wales, lek counts of displaying males have increased by a minimum of 41% over five years. The Welsh population was estimated at 213 lekking males in 2005, an encouraging recovery from the low point of 131 males reached in 1997. Research is currently under way to try and disentangle the relative contributions of habitat management and predator control to this population response. Similarly, in the north Pennines targeted management advice has allowed the population to increase from 773 males in 1998 to 1029 in 2006 (Warren & Baines 2008). More generally, however, Britain's black grouse population remains in a parlous state, at least outside the core areas of the range in highland Scotland, and much remains to be done if the future of this spectacular bird is to be secured. Nonetheless, recent experience in Wales and in the north Pennines suggests that the knowledge and management techniques do now exist to succeed.

Case study 4: Grey partridge

The grey partridge is arguably Britain's most studied farmland bird, with a long history of field-based research dating back to the 1930s. Most notably, this includes a long-term study of the causes of population decline which began in 1968 and continues to this day. Instigated by G. R. Potts, this has been one of the longest-running studies of a farmland bird anywhere in the world. This exceptional degree of research attention stems from the species' historical importance as a gamebird, and because its population has been in steady decline since the 1950s (Potts 1986). The bird is one of Britain's few true steppe species which evolved for a life in open, dry, grassy habitats. In Britain, grey partridges are closely associated with large, open, arable fields bordered by hedgerows or other uncultivated margins, although they also occupy grasslands, lowland heaths, low-level moorlands and coastal sand dunes. Birds have usually paired by the end of February and the nesting season extends from late April until August. Later nests are likely to be replacement clutches, as there are high rates of nest predation. Predators often also take incubating females. The nest is a shallow scrape concealed in dense cover, usually in grass or herbaceous vegetation at the base of a hedge or on a bank, although autumn-sown crops may also be used if suitable marginal cover is not available. Breeding densities are highly variable, partly due to supplementation of wild populations with releases of captive-bred birds. Potts (1986) found densities of between <0.8 and 53 pairs km^{-2} in 12 English studies conducted between 1903 and 1985. Hen grey partridges typically lay clutches of 12–18 eggs, larger than any other British bird. The hen quickly leads the hatchlings into some form of invertebrate-rich cover, usually within the margins of cereal fields (Green 1984) where the chicks forage for themselves. The chicks feed almost wholly on invertebrates in the first two weeks or so of life, taking a wide range of groups but especially caterpillars (mainly those of sawflies and moths), beetles and plant bugs. Beyond two weeks of age, the young are able to fly and feed on the seeds and fresh green plant material that is the staple diet of the adults. Grey partridges are highly sedentary and remain within the breeding area throughout the year. By late summer, the birds have gathered in small groups or coveys, which persist through the winter. They forage on spilt grain and the seeds of broadleaved plants and grasses (the seeds of black bindweed and other members of the Polygonaceae family being particularly important), and on growing crops and other green material as seed supplies become depleted (Potts 1986). The coveys break up in late winter as their members seek mates and, unlike most other birds, it is the young females that remain within the wintering area while the young males disperse, although only short-distance movements are usually made (Potts 2002). It has been estimated that around 100 000 captive-bred grey partridges are released each year in Britain by shooting interests (Aebischer 1997).

By the start of the First World War, the grey partridge population had probably reached its zenith in Britain; driven partridge shooting had become a highly fashionable pastime and large shooting bags were common on many country estates. Prior to this, three key changes had occurred in the British countryside, each of which had been of great benefit to the species (Tapper 1992). First, there had been a large increase in the extent of cereal growing; second, progressive land enclosure since the seventeenth century had provided many miles of new hedgerows; and third, the actions of farmers, game preservers and egg-collectors had greatly reduced the populations of both avian and mammalian predators. As a consequence, the grey partridge was provided with a large quantity of suitable habitat and greatly reduced levels of predation. Based on the shooting bags, Potts (1986) estimated that the pre-First World War breeding population in Britain was around 1 million pairs. Although, not surprisingly, the shooting bag declined during the Second World War, high population levels were probably maintained until the 1950s during which numbers fell steeply. Apart from a brief reprieve in 1959 and 1960, the decline in the national population has continued to this day, as indexed by both the declining size of the shooting bag on English estates (Tapper 1992), and by the results of the BTO's CBC. The results from the CBC's successor, the BBS, suggest that the population has continued to decline since in the mid 1990s (Table 5.2). Despite these declines, the grey partridge has remained a reasonably widespread bird, being present in 58% of 10-km squares in Britain at the time of the 1988–91 *Atlas* (Gibbons *et al.* 1993), although this did represent a 19% reduction of breeding range since 1968–72. The main losses were in the west of the range, which is probably linked to a decline in arable cultivation and its associated seed resources in these areas (Robinson *et al.* 2001). Indeed, whilst large numbers of birds still persist on a few estates in eastern and southern England, breeding densities across much of Britain are now extremely low and local extinctions are becoming increasingly common.

After more then 70 years of study, almost exclusively carried out by the Game and Wildlife Conservation Trust and its predecessor bodies, we have an excellent knowledge of the bird's ecological requirements, the causes of its decline and the measures necessary for populations to recover. This research included long-running demographic studies (notably on the South Downs) which allowed the development of population models to evaluate the implications of past trends in breeding productivity (Potts 1980, 1986; Potts & Aebischer 1995), and a series of field experiments designed to test the effectiveness of management measures (Sotherton *et al.* 1989; Tapper *et al.* 1996). The results of all this work led to the conclusion that there are three key factors which control the size of the grey partridge population in Britain: chick survival rates, the quantity of nesting habitat, and the proportion of nests and incubating females lost to predators.

Potts (1980) discovered that the survival of chicks up to two weeks old was closely related to the availability of key insect groups in crop margins and in a later analysis (Potts 1986), he found that there had been a fall in chick survival rates from over 40% prior to 1952 to around 30% after 1962. This coincided with a period of great increase in the use of herbicides that markedly reduced the quantity of broadleaved plants within cereal crops, the host plants for the chick food. Indeed, Southwood & Cross (1969) estimated that the widespread use of herbicides may have halved the abundance of invertebrates within cereals during the 1950s and 1960s. Chick survival rates remained broadly stable until the late 1980s when a further fall to 22% was linked to the introduction of broad-spectrum insecticides (Aebischer & Potts 1998). This work remains the most compelling evidence of an *indirect* effect (i.e. by reducing food availability) of pesticides on a farmland bird population (Campbell *et al.* 1997). These studies also highlighted the importance of increasing predator populations following a decline in gamekeeping since the 1950s, and the loss of hedgerows and other uncultivated field margins (Potts 1980, 1986). Rands (1986) highlighted the importance of the length of hedgerow and the amount of dead grass present at the hedge base as key factors determining nesting densities.

The results of these studies encouraged the GWCT to trial various management measures. Rands (1985) found that leaving the outer 6 m of cereal fields unsprayed by pesticides resulted in a near doubling of brood sizes. Sotherton (1991) confirmed that these 'conservation headlands' could support four times as many broadleaved plants and up to three times as many insects of the groups consumed by partridge chicks, with mean partridge brood sizes increasing by 50%. Moreover, diversity of insects also seems to be important; experimental studies by Borg & Toft (1999, 2000) found that grey partridge chicks whose natural diverse insect diet was substituted for aphids experienced relatively poor growth and flight feather development.

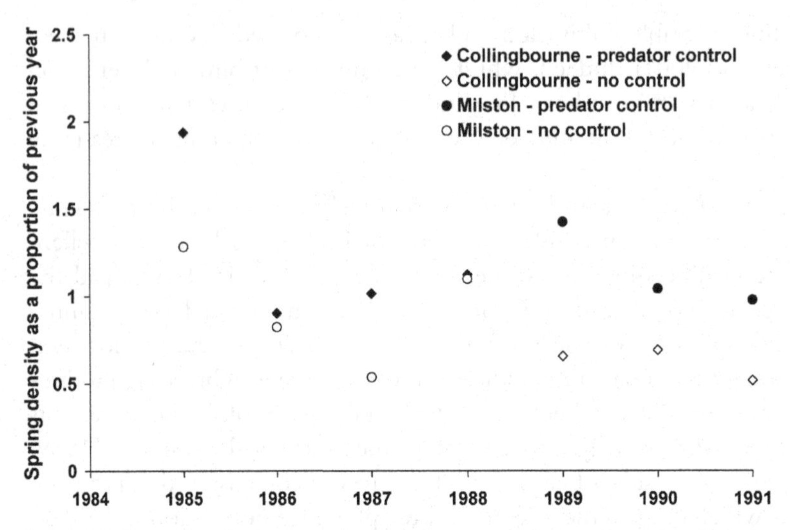

Fig. 8.3 Effect of predator control on between-year changes in spring densities of grey partridges on two study areas on Salisbury Plain. From data presented in Tapper *et al.* (1996).

Tapper *et al.* (1996) conducted a formal field experiment designed to test the impact of changed management on grey partridge numbers. They took two areas of Salisbury Plain with similar grey partridge densities and a history of habitat and predator management to favour this species, and subjected one to legal control of predators (mainly foxes, stoats, rats and corvids) over a three-year period, whilst the other remained as a control. Predator control was concentrated in the breeding season since its aim was to reduce predation rates on grey partridges rather than to exert any overall limitation on predator populations. After three years, the experimental treatments were reversed. Overall, the experiment significantly increased grey partridge breeding success and brood size, and increased breeding density by a factor of 2.6 after three years on sites where predator control was exerted (Fig. 8.3). This is a convincing demonstration that predation during the breeding season does play a role in limiting grey partridge breeding success and densities (Tapper *et al.* 1996).

Other work showed how additional nesting and foraging habitat could be provided by creating 'beetle banks' (grass strips raised by ploughing) across large fields (Rands 1987; Thomas *et al.* 1991), and how areas of set-aside could be managed to provide invertebrate-rich brood-rearing habitat (Moreby & Aebischer 1992). A study in north-east Scotland found significantly higher densities of breeding partridges in first summer set-aside fields than in adjacent crops (Watson & Rae 1997). Other studies have highlighted the importance of stubble fields and fodder brassica crops as winter feeding habitats, especially those that are weedy and seed-rich (Wilson *et al.* 1996b; Hancock & Wilson 2002; Moorcroft *et al.* 2002). Based on this rich history of research, the GWCT has been able to provide farmers, gamekeepers and other land managers with high-quality management advice. This involves providing a combination of nesting habitat, chick-rearing habitat and winter and spring cover and food in close juxtaposition, and adopting appropriate shooting practices (the Trust recommend not shooting wild grey partridges if fewer than 20 birds per 100 hectares are present in the autumn). Evidence that such measures can work at a local scale can be seen on a number of estates, notably in Norfolk where spring densities increased fourfold between 1992 and 2003 on five estates practising sympathetic management (Aebischer & Ewald 2004). In addition, at the Grey Partridge Restoration Project demonstration area at Royston in Hertfordshire, spring densities increased by two-thirds after just two years of management between 2002 and 2004. By 2002 some 21 479 km of sympathetically managed cereal field margins has been created by a range of agri-environment schemes in Britain (of which 18 309 km were in England), some with associated conservation headlands. Interestingly, these successes contrast with recent findings in France where grey partridge populations are also declining in association with intensification of agricultural practice (Bro *et al.* 2000, 2001). Here, a recent experiment compared grey partridge population densities across four areas, for three years before and three years after strips of wild bird cover (strips of mixed maize, sorghum and millet, or kale–cereal mixes or grass and lucerne) were planted in two of the areas (Bro *et al.* 2004). The experiment found no beneficial effect of the wild bird cover on grey partridge breeding densities and the authors speculated that increased predation risk associated with these strips might account for this. Perhaps a more likely explanation, however, is simply that this experiment did not offer management to improve nesting habitat and chick food supplies in conjunction with the provision of cover.

Despite local success on individual estates, there is no evidence of a recovery in the national grey partridge population in Britain. Agri-environment schemes may prove to be a turning point for the species since they now offer the full mix of management options that are likely to be needed to secure population recovery. Indeed, we predict that the widespread, and combined adoption of such measures as grassy margins, headlands, low-input spring cropping, summer fallows, over-winter stubbles and planted wild bird cover will be needed throughout Britain's lowland farmland before we can expect to see a meaningful recovery of populations of this species. In addition to this, it will be essential that partridge shooting practices do not have unintended detrimental impacts on grey partridge populations, especially given recent increases in commercial shooting of red-legged partridges in Britain. For example, even though considerable effort is invested in encouraging voluntary restrictions on shooting of grey partridges in low-density populations (Tapper 2001), a recent study by Watson *et al.* (2007) in Sussex has shown that accidental overshooting of grey partridge stocks during commercial shooting of red-legged partridges removed 35–39% of the autumn grey partridge population, with a predicted reduction of 68–85% in the equilibrium density of spring pairs. In local areas where inadvertent autumn shooting mortality exceeded 50% of the local grey partridge stock, extinction followed. Clearly such levels of mortality are capable of reversing any of the benefits brought by sympathetic habitat management and, as Watson *et al.* conclude, it will be critical to implement training measures to improve shooters' ability to distinguish the two partridge species and warning systems to alert shooters to the approach of flying grey partridges.

Case study 5: Corncrake

The skulking, secretive habits of the corncrake conceal the fact that it lives fast and dies young. The vast majority of birds reaching adulthood live less than two years, but in that time they undertake several 10 000-km journeys between their breeding and wintering grounds. Females may lay two clutches of up to a dozen or so eggs, usually fertilised by two different breeding partners. Adult survival rates are only approximately 20% (Green 1999), and this means that corncrake populations can fall very rapidly if reproductive success is poor – indeed, females must rear an average of five young each season if populations are to remain stable (Green *et al.* 1997). This requirement for high productivity, along with a preference for breeding in agricultural grasslands cut for hay and silage, underlies the chronic problems faced by the species for over a hundred years.

Corncrakes spend the winter on the grasslands and savannahs of central and southern Africa. They are mainly nocturnal migrants and begin to arrive back in Britain from late April, with the males often singing from suitable cover whilst on migration. Whilst the characteristic 'crek-crek' song can be heard on the breeding grounds at any time of the day between late April and August, the most prolific period is between 11 pm and 3 am during June and early July. Densities of over 10 singing males per square kilometre can be found in the most suitable habitats. However, males sing much less often once they have attracted a female and pairs remain close together while they mate and the female selects one of the nests prepared by the male, hidden away in dense vegetation near his singing place. These first nests are usually sited in herbaceous vegetation rather than in the grass fields, which may not yet provide sufficient cover. The males leave their mates halfway through egg-laying and then resume singing some distance away in the hope of attracting another female. The females incubate the eggs for 17–18 days. The young leave the nest soon after hatching and develop rapidly, gradually learning to forage for themselves in the food-rich cover. Both the chicks and adults feed on a wide range of invertebrates, with earthworms, beetles, slugs and snails being particularly important. The first broods are abandoned by the females when they are only 12 days old and before they can fly, so their mothers can go in search of other singing males and have a second breeding attempt – a practice known as successive polyandry. The second nests are usually placed in ungrazed grass fields that have been left for hay or silage. The second broods may not leave the nest until late July or even August and the female stays with them for 15–20 days after hatching. Most birds have left Britain by the end of October.

Prior to the clearance of Britain's primaeval forests, corncrakes would probably have inhabited floodplain meadows and coastal grasslands kept naturally open by regular flooding and grazing, but the species would have been quick to colonise the many hand-cut hay meadows that formed an important part of many traditional farming systems (Shrubb 2003). Despite its name, the bird was always far more common in grasslands than in cereal fields, and in the nineteenth century bred across Scotland and in virtually every county in England and Wales (Holloway 1996). However, declines were noted from the second half of the nineteenth century, and the bird was very scarce or absent from much of central and southern England by 1911 (Alexander 1914). The results of a BTO 'inquiry' in 1938 and 1939 found that it had virtually disappeared to the south of a line from the Gower Peninsula to the Tees estuary and was numerous only in the western and northern isles of Scotland, with only localised populations on mainland Scotland and in northern England (Norris 1945). Norris (1947) also found that the decline of the species, over space and time, was closely linked to the introduction of horse-drawn mowing, which was able to cut much larger areas and so the hay harvest could be completed much quicker. He concluded that 'no species can stand up to the annual destruction of its eggs and especially its young and remain unchanged in numbers'. The replacement of horse-power with tractors later in the twentieth century exacerbated the problem, as did the widespread intensification of grassland management permitted by the drainage of floodplain meadows and the development of fast-growing grass varieties, inorganic fertilisers and methods of silage production which quickly replaced hay-making on many farms (Chapter 1, and see later). All this meant that grasslands could be cut much earlier and, in some cases, several times, during the corncrake's breeding season, and that the grass crop could be taken in a much shorter time window.

As a consequence of these changes, the corncrake population became progressively confined to the islands of Scotland and by the time of a 1988 survey, the Inner and Outer Hebrides supported around 90% of the British total (Hudson *et al.* 1990). The first national survey of singing male corncrakes took place in 1978 and 1979 when 700–746 singing birds were recorded (Cadbury 1980). The 1988 survey found 551–596 singing birds and a 1993 survey found just 480, a decline of around a third since 1978, and possibly more as several key islands were not completely covered in the earlier survey (Green 1995b). Over the same period the number of 10-km squares in which singing birds were recorded fell from 160 to 93, a loss of 42%. The largest numbers were found on the islands of Tiree and Coll, the Western Isles, Skye, and Colonsay and Oronsay. Corncrakes had survived here because of the combination of climatic conditions, soils and crofting agriculture meant that changes in agricultural practice were less widespread and occurred at a slower pace. However, it was predicted that the population would halve within 20 years if current trends continued (Green 1995b); a likely prospect as mowing was becoming earlier and faster, and the extent of hay and silage was being reduced as fields were being turned over to year-round sheep grazing or abandoned altogether. This presented conservationists with a challenge – to integrate the needs of the corncrake with those of the crofting and farming communities on the islands.

Table 8.6 *Developing a suite of management prescriptions for the corncrake*

Research finding	Management prescription
Need for vegetation at least 20 cm tall throughout the time on the breeding grounds (Stowe *et al.* 1993; Green & Stowe 1993; Green 1996)	• Create and/or maintain areas of nettles, yellow flag iris, cow parsley and other tall herbaceous vegetation in field corners and along field margins to provide early and late cover (corncrake 'corridors' and 'corners') • Maintain extensive areas of late-cut meadows
Mowing date of meadows has a huge impact on breeding success (Green *et al.* 1997)	• Cut hay and silage fields after 31 July
Mowing fields from the inside outwards increases productivity by up to 20% (Green *et al.* 1997)	• Adopt corncrake-friendly mowing practices
Corncrake-friendly mowing is most successful if there is an adjacent strip of tall vegetation (Tyler *et al.* 1998)	• Place corncrake corridors adjacent to hay/silage fields
Nesting and brood rearing occurs within 200–300 m of the position of the singing male (Tyler & Green 1996)	• Target late-cutting and corncrake-friendly mowing on fields within a 250-m radius of singing males

Intensive studies of the habitat requirements and demography of the corncrake began in the mid 1980s, led by Rhys Green of the RSPB, with the aim of identifying measures that would reverse the downward trend in the population. The work involved ringing and retrapping, radiotelemetry, intensive field observations and mathematical modelling which increased greatly our understanding of the bird's ecology and the impact of habitat availability and farming operations on the population. For example, intensive observation during mowing operations revealed how flightless chicks were trapped by the usual method of mowing from the outside of the field to the centre; those unwilling to break cover were killed by the last swaths and those which did were often taken by waiting predators such as gulls. Overall, the work highlighted the importance of tall vegetation throughout breeding season, the fact that nesting and chick-rearing occurred within 300 m of the position of singing males, and the huge effect of mowing date and mowing method on breeding success (Green & Stowe 1993; Stowe *et al.* 1993; Green 1996; Green *et al.* 1997; Tyler *et al.* 1998). This knowledge-base enabled conservationists to develop a 'recipe' for the recovery of corncrake numbers, based on increasing the area of suitable habitat, especially early and late vegetation cover, and boosting breeding success through a combination of late hay or silage cutting and corncrake-friendly mowing techniques (Table 8.6). This formed the basis of a species action plan developed initially by RSPB and statutory conservation agencies in 1989. Action was mainly focused on providing management advice to crofters and farmers in the key areas by deploying fieldworkers, influencing agricultural policies, and the acquisition and management of nature reserves (Williams *et al.* 1997).

In the summer of 1989, the first fieldworkers were employed to provide advice on the islands of Tiree and North and South Uist. In 1991 RSPB purchased the freehold on an area of Coll and created a 1221-ha reserve. Management here significantly increased the area of late-cut meadows and early/late cover. This coincided with an increase in the number of singing males on the reserve from six in 1991 to 31 in 1995 while those on the rest of the island remained stable, providing confidence that the recipe was right. During the 1990s, corncrake-friendly mowing was applied to a number of existing and new reserves on the Western Isles and on the island of Egilsay, Orkney in an attempt to create a network of well-managed sites. Following initial success with a grant scheme in Northern Ireland, the Corncrake Initiative was established by RSPB, Scottish Natural Heritage (SNH) and the Scottish Crofters' Union in 1992, and received EU funding from 1994 to 1996. This provided an area-based payment to farmers who cut their hay and silage fields after 31 July, with an additional payment for adopting corncrake-friendly mowing techniques. All fields within a 250-m radius of a singing male corncrake were eligible for entry to the scheme. In the period 1992–96, the scheme secured the late cutting of a total of 4072 hectares of hay/silage fields and 4398 hectares of corncrake-friendly mowing at a cost of around £250 000 (Williams *et al.* 1997). In addition, a pilot scheme to provide early and late cover was introduced by RSPB to several islands in 1994/95 which created 35 corncrake 'corners' and corridors'.

The success of these early initiatives prompted the then Scottish Office Agriculture and Fisheries Department to introduce prescriptions for late cutting and cover provision into its existing Machair of Uists and Benbecula, Barra and Vatersay ESA, into a new Argyll Islands ESA in 1993, and into the new Countryside Premium Scheme in 1997. Many crofters and farmers now receive payments for corncrake-friendly management under these schemes. Both schemes were incorporated into a new Scotland-wide Rural Stewardship Scheme in 2001 and into Rural Development Contracts in 2008, though there remains concern to ensure that the transition of corncrake management between successive schemes runs smoothly. In addition, nine Scottish Special Protection Areas (SPAs) were identified for corncrake under the EU Birds Directive. These nine SPAs collectively support around 40% of the breeding population (Stroud *et al.* 2001). Scottish Natural Heritage introduced a dedicated management scheme for land within SPAs, and by 2003 over 5000 hectares of land were under direct management for corncrakes every year (O'Brien *et al.* 2006) (Fig. 8.4).

The response of the corncrake to these initiatives has been encouraging with the population being on an overall upward trend since 1993 (Fig. 8.5). Annual surveys of the core areas have been undertaken since 1993 with the last full surveys in 1998 when 589 singing birds were recorded, an increase of 23% since 1993 (Green & Gibbons 2000), and in 2003 when 832 singing birds were found. In 2004, there was a further spectacular increase to 1040 calling males in the core of the range alone, and

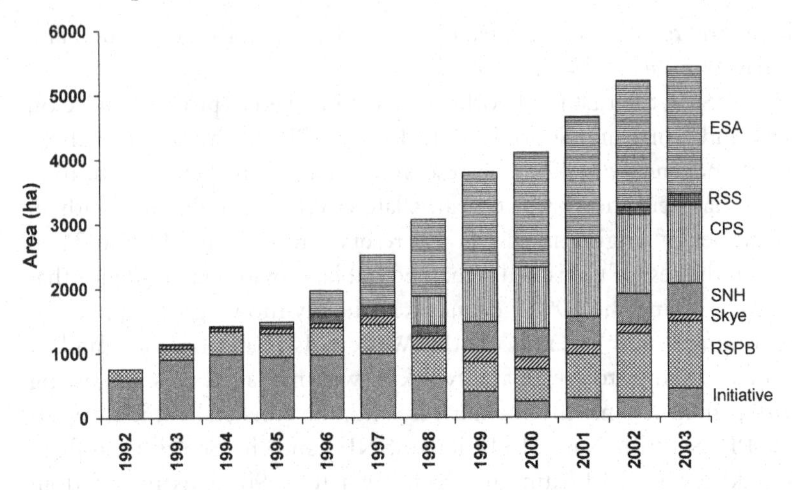

Fig. 8.4 Annual increase in area of land under management for corncrakes, 1992–2003. ESA, Environmentally Sensitive Area; RSS, Rural Stewardship Scheme; CPS, Countryside Premium Scheme (predecessor of RSS); SNH, management by SNH on Special Protection Areas; Skye, Skye Grassland Scheme; RSPB, RSPB management on nature reserves and under management agreements; Initiative, RSPB/SNH Corncrake Initiative. Data from O'Brien *et al.* (2006).

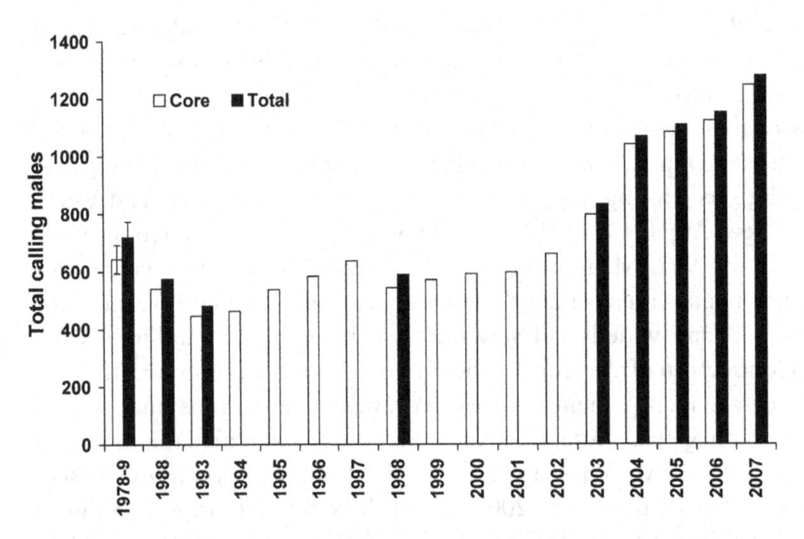

Fig. 8.5 Number of singing male corncrakes in Britain and the Isle of Man, 1978–2007. Open bars show totals from annual surveys in the core range (see O'Brien *et al.* 2006); filled bars show complete totals from periodic full surveys. Data for 1978/79 show mean, minimum and maximum estimates.

the increase has continued steadily since then (O'Brien *et al.* 2006). One challenge to the future of this core population in western Scotland is that current management depends on the maintenance of crofting agriculture and production of hay as a cattle feed. On islands where cattle husbandry has declined, as on Lewis, abandonment of hay meadows is taking place and current corncrake population levels are barely being maintained (M. Scott pers. comm.). Here management initiatives based on periodic mowing of otherwise abandoned meadows may need to be instigated as a conservation measure to maintain long-term suitability of habitat no longer under active crofting management, and to prevent succession to unsuitable habitat types such as scrub or heather moorland. Encouragingly, recent research in Sweden (Berg & Gustafson 2007) suggests that mowing every few years may be sufficient to achieve this.

If current overall trends continue, we can but hope that the species' characteristic call might become a regular feature at suitably managed sites on the Scottish mainland and in England and Wales. A potential barrier to this is that the birds are extremely site-faithful, returning to within a few kilometres of where they born and sometimes to the same field (Green 1999). Indeed the latest full survey found that although the corncrake population had increased since 1998 and now occupied a greater number of 1-km squares than previously, its range when measured by the number of 10-km squares occupied had actually declined slightly (O'Brien *et al.* 2006). In other words, corncrake populations are infilling their current breeding range at ever-higher densities, but this range is not expanding. This may mean that conservationists need to give the corncrake a helping hand, either through appropriate habitat management at the fringes of the current range on the western and northern coasts of the Scottish mainland, or perhaps through translocation of birds to new areas. In the latter context, RSPB, Natural England and the Zoological Society of London, with the assistance of Pensthorpe Conservation Trust, are currently trialling the release of captive-bred corncrakes to the Nene Washes in Cambridgeshire. Here, grassland management for breeding waders provides large areas of cover and late-cut grassland, but the surrounding agricultural land is almost entirely arable, thus helping to ensure that birds are not lost to 'sink' areas of intensively managed grassland. Only time will tell whether these techniques prove feasible as a means of restoring corncrake populations to parts of their former range from which they have long been absent. In the meantime, the reversal of the fortunes of Scotland's corncrake population through a combination of basic research, identification of practical management solutions and incorporation of this management into agri-environment schemes represents one of the first and most important successes of the modern marriage between nature conservation and agriculture policy in Britain.

Case study 6: Stone curlew

The stone curlew is unique among Britain's birds – it is a mainly nocturnal, migratory 'wader' that lives in dry lowland habitats. The bird both nests and forages on bare or sparsely vegetated ground, and is most active between dusk and dawn when its haunting wails and cries can be heard. It is these calls that have led to it being known as 'curlew' though it is in fact unrelated to true curlews, and is the only European representative of the thick-knees. In Britain, the stone curlew's habitat requirements are mostly provided on chalk downland and grass heaths with a short sward produced by intense rabbit and sheep grazing, and on sandy, stony soils planted with relatively slow-growing spring-sown arable crops such as sugar beet, carrots, onions, kale and maize. Britain's breeding birds winter mainly in Iberia, France and north Africa, and possibly also in west Africa (Green 2002). The first birds arrive on their breeding grounds from mid March and clutches of two eggs are laid in a bare earth scrape during a protracted breeding season which stretches from April to August. Some pairs raise two broods in a season. The chicks are tended by both parents who fly up to 3 km from the nest site, mainly at night, to collect food in a range of invertebrate-rich habitats. These can include short improved grasslands, pig fields and even manure heaps within a home range of around 30 hectares (Green *et al.* 2000). A wide range of prey items are taken, but the diet consists mainly of surface-active invertebrates such as earthworms, millipedes, beetles and spiders. Small mammals and birds are taken on occasion. Due to the extended breeding season, some pairs may still have dependent young in August and September. At this time, both the adults and juveniles gather together to form flocks, generally of a few tens of birds, prior to their autumn migration. Most birds have left the country by the end of October.

The stone curlew was formerly widely distributed across the free-draining soils of southern and eastern England, and was even considered locally numerous in some areas as recently as the nineteenth century (Brown & Grice 2005). However, a major decline in the nineteenth century and through much of the twentieth century was associated with either a reduction in the quality or the complete loss of their favoured breeding haunts. The most important factors were land enclosure, the conversion of semi-natural grasslands to arable farmland and forestry, a general reduction in mixed farming and the demand for low-intensity grazing lands. This was compounded by a decline in rabbit grazing due to the cessation of warrening and the crash in rabbit populations following the first myxomatosis outbreak of the mid 1950s which led to 'scrubbing-up' of much remaining open downland (see Chapter 1). The national population of stone curlews was estimated at 1000–2000 breeding pairs in the 1930s but this had fallen to just 300–500 pairs by the late 1960s (Sharrock 1976). Annual monitoring of the population began in the mid 1980s following concern over the bird's decline, and this revealed that there were only around 150–160 breeding pairs remaining and that birds had been lost from significant parts of the breeding range. Indeed, during fieldwork for the 1988–91 *Breeding Atlas* the species was found in 40 fewer 10-km squares than in the 1968–72 *Breeding Atlas*, representing a decline of 42% in the breeding distribution in just 20 years (Gibbons *et al.* 1993). In addition, the population had become largely confined to two areas: the Breckland of East Anglia, where the birds inhabited tightly grazed grass heaths and fields of spring-sown arable crops, and the extensive chalk grasslands of Salisbury Plain including Porton Down, and spring tillage on the adjacent farmland of Wiltshire, Hampshire and Dorset (often referred to as the 'Wessex' population). A long-term study on Porton Down from 1987 to 1996 showed the fluctuations of the population (between nine and 19 pairs) were strongly correlated with current and recent rabbit abundance, suggesting that rabbit grazing was critical in maintaining areas of sparsely vegetated, stony ground on which the species nests (Bealey *et al.* 1999). Very small remnant populations also still persisted in south Cambridgeshire, Oxfordshire, Berkshire, north Norfolk and east Suffolk. Not surprisingly, the stone curlew became recognised as one of Britain's most threatened birds (Batten *et al.* 1990).

The reduction of the breeding population brought added problems as a large proportion of the birds now nested in arable fields where their nests and young were subject to an increasingly high risk of destruction by ever-more mechanised crop husbandry. Indeed, operations such as spraying, irrigation and harrowing are regularly practised in the sparsely vegetated, broadleaved crops favoured for breeding. This prompted the RSPB to begin employing dedicated wardens in the mid 1980s to find nests and liaise with farmers who were usually more than willing to adjust their practices to avoid destroying active nests and chicks. Such interventions can boost annual breeding productivity by over 20% and this enabled the population to remain more or less stable over the next 10 years. Meanwhile, detailed research by the RSPB into the bird's ecological requirements used observations of colour-marked individuals, radio-tracking and dietary analysis (Green & Griffiths 1994; Green & Taylor 1995; Green *et al.* 2000), and allowed conservationists to draw up a more ambitious plan for the recovery of the species. This involved a combination of increasing the suitability of the remaining semi-natural habitats whilst safeguarding the birds nesting on arable land by providing breeding pairs with safe nesting areas, and continuing to protect those birds which nest in growing crops. This two-pronged approach was considered important as it provided insurance against detrimental changes affecting one habitat having a disproportionate impact on the population as whole, and because breeding birds do switch between habitats at a local scale, both within and between breeding seasons, in response to habitat suitability (Green & Griffiths 1994).

A key theme of the recovery plan was to take advantage of any new opportunities to secure stone curlew-friendly land management as they arose. The first opportunity came with the launch of the Breckland ESA in 1989 which provided a means to reinstate the necessary grazing management to the grass heaths in the bird's main stronghold. Ten years later, nearly a third of the Breckland pairs made their first breeding attempt on heathland managed under ESA agreement (Swash *et al.* 2000), and many more of the arable-nesting birds were able to find suitable short-sward feeding habitats nearby. The next opportunity came with the introduction of compulsory set-aside land in 1992. Although designed as a production control measure, on the advice of the stone curlew wardens, farmers were able to seek special 'derogations' from the then Ministry of Agriculture, Fisheries and Food (MAFF) to create special bare-ground nesting plots for the birds that would then be free from potentially destructive farming operations. Such plots were also created by the Ministry of Defence on their Salisbury Plain Training Area in places where a

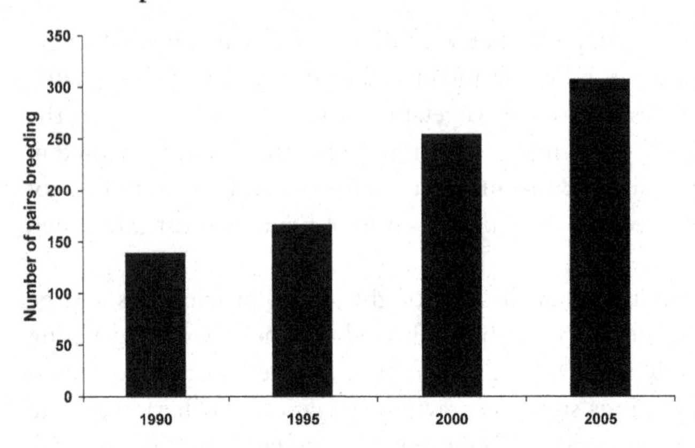

Fig. 8.6 Stone curlew breeding population in England, 1990–2005.

short downland sward could not be guaranteed due to insufficient levels of grazing. The success of these plots prompted RSPB and English Nature to ask MAFF to introduce a special project within its Countryside Stewardship Scheme which would create bare-ground nesting plots. This was introduced in the Wessex area in 1998. Crucially, the bird also became part of English Nature's (now Natural England) Species Recovery Programme in 1994, and funds became available to increase the amount of wardening effort substantially. Much of this extra effort went into encouraging farmers to take up the various schemes and advising them on suitable placement and management of the resulting nest plots; protecting nests and chicks from sprays and mechanical hoes thus became only part of their work. In 2005, over 150 nesting plots had been created by the Countryside Stewardship Scheme special project in Wessex: birds were recorded on 48 of these plots of which 36 were used for breeding, fledging 19 juveniles (over a quarter of the total production in the area). Plots on set-aside also still play an important role in the area: there were 48 plots in 2005, of which 12 were used for breeding, fledging seven juveniles.

The response of the stone curlew population to these initiatives has been most encouraging. Fuelled by a run of good breeding seasons in 1994, 1996, 1997, 1998 and 1999, the number of known breeding pairs increased from 139 in 1990 to 254 in 2000 and 307 in 2005 (Fig. 8.6).

However, much of the increase has occurred on arable land where around 60% of the population now nests; the numbers breeding in peripheral populations have increased only slowly, and no birds now nest in south Cambridgeshire. Clearly, increasing the numbers breeding on semi-natural habitats and restoring the past range of the stone curlew will entail even greater conservation efforts involving much broader-scale changes in land management, including linking the remaining isolated fragments of downland and grass heaths through ambitious, large-scale habitat recreation schemes. Such initiatives will also benefit a wider range of threatened flora and fauna associated with extensively managed grasslands and heathlands.

Case study 7: Lapwing

Breeding lapwings occupy open terrestrial habitats, especially where sparsely vegetated ground with a short sward is found close to damp substrates rich in soil invertebrates – the species' favoured prey. A wide variety of habitats may be occupied from muir-burn patches on heather moorland to saltmarsh, flooded gravel pits, airfields and waste ground. However, the great majority of the population nests on agricultural grasslands, especially the less agriculturally improved and damper pastures, in fields of spring-sown arable crops and, in recent years, set-aside. This broad habitat occupancy makes the lapwing Britain's most numerous and widely distributed breeding wader species. It is an early nester with the first eggs laid from mid March, though replacement clutches can be laid until early June. Very high nest losses to avian and mammalian predators can occur, as well as trampling by livestock and destruction by agricultural operations, so the lapwing's ability to lay up to four replacement clutches in a season is a crucial aspect of its breeding biology. Clutches are usually of four eggs. The chicks are precocial and are brooded and guarded by both parents. Chick mortality can be high in the first five to six days after hatching, especially on improved pastures and arable land – chicks hatched on arable land are typically moved to adjacent, more invertebrate-rich grass fields during this period (e.g. Galbraith 1988a). The surviving chicks fledge after 35–40 days and are soon independent.

After the breeding season, lapwings are highly gregarious and gather in flocks which range widely over Britain's lowland farmland and coastal habitats. A large proportion of Europe's breeding birds may pass through or over-winter in Britain, joining our resident birds and their offspring, although some of these migrate to Ireland, France and Iberia, especially during severe weather (Appleton 2002). Indeed, Britain supports over two million lapwings outside the breeding season (Rehfisch *et al.* 2003), representing over half of the total wintering in the whole of Europe, Asia Minor and North Africa. On farmland outside the breeding season, lapwings often occur in mixed feeding flocks with golden plovers and, in daylight hours, in the company of black-headed gulls which steal food from both species (Barnard & Thompson 1985). Earthworms are a key food item for lapwings outside the breeding season, and this underlies their preference for feeding on permanent pastures (Milsom *et al.* 1985; Tucker 1992). Nonetheless, they also forage on recently harvested and ploughed fields when these are available, as well as stubbles and autumn-sown cereals (e.g. Shrubb 1988; Mason & Macdonald 1999b). Recent studies have also highlighted the importance of nocturnal foraging for large earthworms on arable fields during winter, at all phases of the moon (Milsom *et al.* 1990; Gillings *et al.* 2005a, 2007). Aside from earthworms, Lapwings feed on a variety of invertebrates that live on or just below the soil surface including insects, spiders, molluscs, millipedes and woodlice, and even take a small amount of plant material (Wilson *et al.* 1996a). Like other plovers, lapwings locate their prey largely by sight, and this probably underlies the bird's close association with short vegetation.

Given the lapwing's close association with farmed habitats, and much anecdotal evidence testifying to its former abundance, it is perhaps surprising that it is only comparatively recently that large declines in breeding lapwing numbers, both locally and nationally, have been properly quantified. The 1968–72 *Breeding Atlas* found that the bird was still very widely distributed, being present in 92% of Britain's 10-km squares (Sharrock 1976). A dedicated survey of breeding lapwing in England and Wales, conducted by the BTO in 1987, estimated the population at just over 123 000 pairs (Shrubb & Lack 1991). These authors also compared their data with counts made in 27 areas of England and Wales during the period 1956–65 to discover that numbers had declined overall by 61%. Furthermore, the CBC index fell significantly during the second half of the 1980s (Marchant *et al.* 1990), with an overall decline of 53% between 1980 and 2005 (Table 5.2). In the 1988–91 *Breeding Atlas*, the species was found in 233 fewer British 10-km squares than in 1968–72, a decline in distribution of 9%, and a concern for such a ubiquitous species (Gibbons *et al.* 1993). Most of the losses occurred in north-west Scotland, south-west Wales and south-west England. The most compelling evidence for a decline in abundance in England and Wales emerged from a second lapwing survey in 1998 using the same methods as in 1987. The revised population estimate for these two countries was just over 62 900 breeding pairs, 49% lower than the first survey only 11 years earlier (Wilson *et al.* 2001). Declines were recorded in all regions but were greatest in Wales (77%) where the total population was estimated at fewer than 1700 pairs. The declines in English regions ranged between 64% (south-west) and 28% (Yorkshire/Humberside). Similarly, a survey of lowland wet grasslands in England and Wales in 1982 found 6721 nesting pairs (Smith 1983) but a random resample of these sites in 1989 suggested a decline of 38% (O'Brien & Smith 1992), and this was confirmed by a full resurvey in 2002 (A. M. Wilson *et al.* 2005). In Scotland, which now holds well over half of Britain's breeding lapwings, long-term declines had hitherto been less severe, and the species was still found in over 40% of 1-km squares surveyed in 1997 (O'Brien *et al.* 2002). However, trends very similar to those observed further south are now being observed, and should be a cause for immediate concern. Thus, the BBS shows a 38% decline in Scotland between 1994 and 2007 (Table 5.2), and a 20-year study on marginal upland farmland in southern Scotland revealed a 77% decline between 1980 and 2000, with further declines by 2002, associated with loss of spring-cultivated fields and unimproved grassland, in favour of drained, improved grassland (Taylor & Grant 2004). In Northern Ireland, Henderson *et al.* (2002) recorded a decline of 66% in the breeding population of lapwings between 1987 and 1999, and ascribed this to similar causes – the loss of arable cultivation and drainage and intensification of management of agricultural grasslands.

The growing concern over the fate of Britain's lapwings has generated a considerable body of research over last 20 years that has studied birds breeding in different farming systems and in different regions of Britain. The results of the four most detailed published studies are summarised in Table 8.7.

Table 8.7 *Selected studies of lapwings breeding on British farmland*

Study area	Key findings
Unimproved rough grazing land and mixed farmland in the Carse of Stirling, Scotland	• Nest and chick survival heavily affected by predation (75% of all nests were lost to predators) (Galbraith 1988a) • Arable-nesting pairs were often disrupted by cultivations (which led to smaller clutches) and crop growth (which shortened laying season) (Galbraith 1988a) • Arable-nesting pairs selected spring cereals adjacent to pasture (Galbraith 1988b) and lay larger eggs which produce larger/heavier chicks than those without pasture adjacent (Galbraith 1988c) • High chick mortality occurred when arable-nesting pairs moved their chicks to pasture – chicks which had a shorter distance to move to suitable pasture survived better (Galbraith 1988a) • Productivity of arable-nesting pairs was insufficient to maintain the breeding population (Galbraith 1988a)
Enclosed upland farmland in the Eden Valley, Cumbria and Teesdale, Co. Durham	• Nesting densities on improved grass fields were 69% lower than on adjacent unimproved fields (Baines 1988) • Birds nesting on unimproved fields reared 0.86 chicks per pair whereas those on improved fields reared only 0.25 chicks per pair (Baines 1989) • Differences in productivity were largely accounted for by differences in clutch predation rates (76% on improved versus 47% on unimproved fields) and clutch destruction by agricultural operations (Baines 1990) • Productivity on improved fields was insufficient to maintain the breeding population (Baines 1990) • Adult birds return to breed in the same or an adjacent field year after year and around half of young birds breed in the same or an adjacent field to where they were hatched (Thompson *et al.* 1994)
Lowland mixed farmland in Shropshire and lowland arable in south Cambridgeshire	All from Sheldon (2002a, b; Sheldon *et al.* 2007) • Nest survival rates were high (68% of nests fledged at least one chick) but chick survival was poor (<11%) • Nest losses were due mainly to predation (52%) and agricultural operations (26%), and most chick losses were to predation • Nests over 50 m from the field boundary were twice as likely to be successful than those within 50 m • Chicks that ate more earthworms were in better condition • Summer fallows, created by rotational set-aside and through a pilot agri-environment scheme, provided important nesting and chick-rearing habitat • Overall productivity was insufficient to maintain the breeding population
Arable farmland in Hampshire	• Breeding population on spring tillage declined to extinction, probably due to a combination of nest destruction by agricultural operations, predation and loss of nearby source populations to provide recruits (Milsom 2005)

These studies and others (e.g. Beintema & Müskens 1987; Shrubb 1990; Berg *et al.* 1992; Johannson & Blomqvist 1996; Hart *et al.* 2002; Kragten & de Snoo 2007) suggest that both reduction in the availability of suitable, safe nesting habitat, and reduced reproductive output from those pairs that do nest are jointly driving population declines. Indeed, productivity was considered to be high enough to maintain the local breeding population in only three out of 11 British, habitat-specific situations reviewed by Sheldon (2002a). By contrast, lapwing survival rates have increased in Britain since the 1960s, probably because of the generally mild winters experienced over this period (Peach *et al.* 1994). The reduction in the area of suitable nesting habitat has several causes. The key changes are the widespread switch from spring to autumn sowing of cereals resulting in crop vegetation that is too tall and dense to permit nesting, and the drainage and subsequent conversion of unimproved grassland (both in lowland floodplains and marginal, upland situations) in favour of autumn-sown arable crops, or dry, improved grassland. The causes of reduced breeding success are more complex and involve decline in the quality of the remaining habitat. Probably of greatest importance has been the loss of mixed livestock and arable farming (see Chapter 1) which means that many of the remaining fields of spring tillage are not sited close to pastures suitable for chick-rearing. In arable Cambridgeshire, chicks have been seen being led for many hundreds of metres by parent birds in search of foraging habitat (Glen Tyler pers. comm.). Also, a general increase in the number of agricultural operations during the nesting season on arable fields is likely to have increased nest destruction rates and reduced the abundance of soil invertebrates (Edwards & Lofty 1975; Tucker 1992). Faster growth of spring crops, promoted by the use of inorganic fertilisers and improved crop varieties, has truncated the length of the potential laying season and reduced the number of nesting attempts that are possible. Spring tillage is thus both a diminishing resource and one that is declining in quality. Also important has been the agricultural improvement of the vast bulk of Britain's permanent grasslands and rough grazing lands (Chapter 1). This has greatly reduced the suitability of agricultural grassland since rates of nest loss to predation, trampling and cutting tend to be higher, and invertebrate food availability and accessibility tend to be lower in intensively managed grasslands (Vickery *et al.* 2001).

Overall, the picture seems bleak; lapwings have been faced with declining habitat quantity and quality in all of their major agricultural habitats for several decades. 'Everything that can go wrong has gone wrong' (Marren 2002). However, considerable efforts have been expended on improving the quality of lowland wet grasslands for lapwings along with other breeding and wintering waders and waterfowl. Most of the remaining lowland wet grasslands are nature reserves or designated nature conservation sites, often within ESAs, where it has been possible to instigate potentially beneficial land-management practices. Careful manipulation of the grazing management and hydrological regime can produce the ideal nesting and foraging conditions for lapwings. A balance must be struck between having sufficient grazing pressure to create the required short sward conditions but minimising the risk of nest

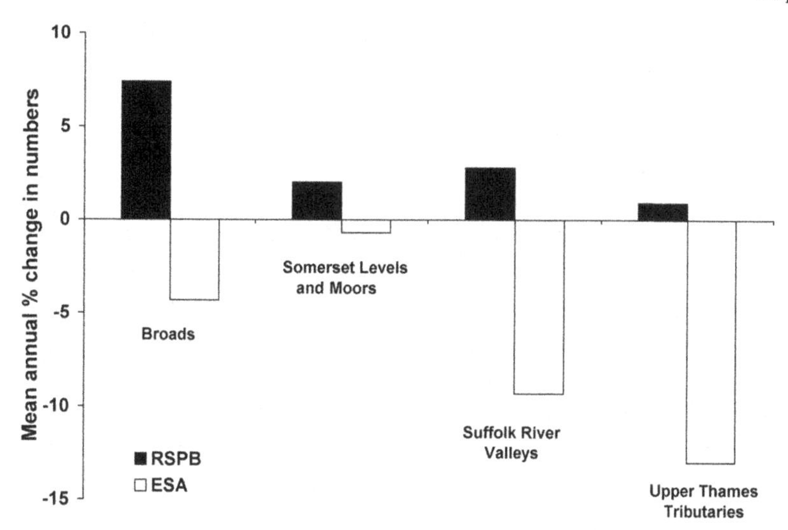

Fig. 8.7 Population trends of breeding lapwings on lowland wet grassland on RSPB reserves and on land under ESA management. From data presented in Ausden & Hirons (2002).

trampling; this can include restricting stocking rates in the spring to a maximum of two animals per hectare or even keeping some areas free from livestock between late March and late June (Hart *et al.* 2002). Ideally, there should be a mosaic of unflooded and winter-flooded areas with some shallow surface water and damp soil areas maintained into late spring to provide foraging areas for both adults and chicks (Milsom *et al.* 2000; Ausden *et al.* 2001; O'Brien 2002; McKeever 2003). The flooding of rills and footdrains in April and May can provide an alternative feeding habitat on clay/silt soils in places where creating more extensive wet areas is not practicable (Milsom *et al.* 2002). Eglington *et al.* (2008) found that high densities of flooded footdrains in wet grasslands attracted significantly higher densities of lapwings than fields without these features and that more nests were found within 50 m of flooded footdrains than further away. Later in the season, foraging chicks were significantly more likely to use fields with these features than those without. On the Somerset Levels, monitoring studies have revealed that management agreements which create wetter conditions in return for higher payment rates support more pairs of breeding waders (including lapwings) per unit cost than agreements that produce lower water levels (Ausden & Hirons 2002). Moreover, RSPB reserves within ESAs held higher densities of breeding lapwings than surrounding ESA agreement land, and numbers were increasing on RSPB reserves but still decreasing on the surrounding ESAs (Fig. 8.7). Clearly, there is still more to do to if agri-environment measures are to help lapwing populations recover more fully.

Where such measures have been implemented on nature reserves, the results can be spectacular. On RSPB's reserve in the Nene Washes, Cambridgeshire, the number of lapwing pairs increased from 22 in 1984 to 60 in 1988 following the introduction of a sympathetic grazing regime and the artificial retention of floodwaters. Similarly, at Holkham National Nature Reserve on the north Norfolk coast, the average number of pairs increased from 79 in 1986–88 to 229 in 1993–95 with a marked increase in productivity per pair following the instigation of improved grazing and hydrological management. At RSPB's reserve at Loch Gruinart on Islay, vegetation growth has been controlled by later application of fertiliser (necessary to produce the rich sward favoured by the barnacle geese) and, together with raising water levels, this led to an increase in the number of lapwing pairs, from 111 in 1984 to 250 in 1993. At RSPB's West Sedgemoor reserve on the Somerset Levels, not only did the number of breeding lapwings more than triple following improved hydrological management, but the number of wintering individuals also markedly increased, from a peak mean of 7500 birds in 1985/86–1989/90 to 23 591 in 1990/91–1994/95. Here, the creation of a disturbance-free refuge was also an important factor.

In the uplands, it is essential that the remaining rough grazings and unimproved pastures are safeguarded by offering farmers incentives that prevent agricultural intensification, especially drainage. The suitability of semi-improved pastures might also be enhanced by fine-tuning stocking regimes to produce a more varied sward without rush infestation, to reduce trampling and to prevent rolling during the nesting season (O'Brien 2001). Reverting back to the use of farmyard manures as opposed to inorganic fertilisers might also be beneficial to soil invertebrate populations (Vickery *et al.* 2001), though we need to learn more about the best ways to re-establish healthy, productive breeding populations on improved grasslands, given the opportunities now offered by management options within conservation designations and agri-environment schemes.

On arable land, conservationists noticed that lapwings quickly took advantage of the sparsely vegetated plots created for Stone Curlews on set-aside land in southern England. As a consequence, a summer fallow option was included in the Arable Stewardship Pilot scheme which ran in Shropshire and south Cambridgeshire from 1998 to 2000. This option involved light spring cultivation on a plot at least 2 hectares in size which was then left undisturbed. A detailed study in these areas found that this management provided suitable nesting conditions throughout the season and supported higher nesting densities and higher breeding success than any other field type, with no nests or chicks lost to agricultural operations which were forbidden between 21 March and 14 July inclusive (Sheldon 2002a; Sheldon *et al.* 2007). Following this successful trial, a summer fallow option was rolled out across England as part of the Countryside Stewardship Scheme from 2002, and by the end of 2005 this management covered over 15 000 hectares.[1]

To conclude, agricultural land-use and timing of field management have a huge impact on the productivity and therefore population trends of lapwings. High nest and chick predation rates have always been a fact of life for lapwings but, until recently, this was offset by their ability to undertake multiple nesting attempts.

[1] Recent monitoring studies show that up to 40% of these plots are used by nesting or foraging lapwings (Chamberlain *et al.* 2008b).

However, this ability is dependent upon having high-quality nesting and brood-rearing habitat present throughout the breeding season. Habitat mosaics are the key, whether this is the juxtaposition of spring tillage and pasture, short grass and longer vegetation, or dry areas and wet areas. Nature reserves and designated sites, where sympathetic management can be more easily instigated, may have a key role to play if the national population is to recover. However, a balance must be struck between management here and in the countryside as a whole where targeted management at the field, farm and landscape scale is essential to secure this species' long-term future.

Case study 8: Golden plover

The golden plover is included amongst the case studies because its annual cycle in Britain takes the same individual birds to a remarkably wide range of agricultural habitats. The species nests on upland heathlands and grasslands, blanket bogs and montane plateaux from Dartmoor to Shetland. Highest densities (typically five to eight pairs per square kilometre, but historically reaching 16 pairs per square kilometre on upland limestone grassland in the Pennines: Ratcliffe 1976) are found in the Northern Pennines, Grampians and on the extensive blanket bogs of Lewis and Harris, the Caithness and Sutherland Flows and in the Northern Isles (Gibbons *et al.* 1993). The breeding grounds are occupied from March to mid July, and during the main period of territory defence between March and May, the song flights and plaintive, haunting calls of the males capture the character of windswept moorlands.

Female golden plovers typically lay a clutch of four eggs in a nest usually placed on flat ground amongst short vegetation. Clutches lost to predators may be replaced, often with clutches of just three eggs, but loss of a brood usually results in abandonment of breeding for that year. Strips of rotationally burned heather may be favoured as nest sites for a few years after burning, whilst long, dense heather and tall swards of cotton-grass or purple moor-grass are generally avoided (Parr 1980; Ratcliffe 1990; Whittingham *et al.* 2002). On blanket bog, areas of cotton-grass are favoured for nesting, reflecting the subsequent abundance of adult tipulids (craneflies) for young chicks (Pearce-Higgins & Yalden 2004). Whilst grazing or burning interventions may be important to maintaining suitable nesting habitat for the birds on moors managed for grouse, sheep or red deer, montane heaths or blanket bogs dominated by *Sphagnum* mosses may more naturally retain a suitable vegetation structure in the long term (Ratcliffe 1990). A recent study sought to understand why golden plovers, like many other upland-nesting waders, select the flat upland plateaux, rather than more steeply sloping ground towards the moorland edge. In a north Pennines study area, 59% of nests were lost, mostly to egg predation by stoats, but nests on flat ground survived markedly better than those on slopes (Whittingham *et al.* 2002) (Fig. 8.8).

Although this study was unable to identify the reason for the relatively high vulnerability of nests on slopes, one possible explanation is that adult golden plover

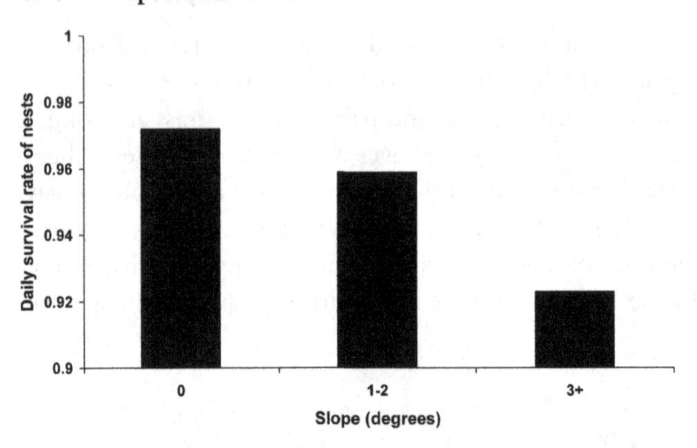

Fig. 8.8 Daily survival rates of golden plover nests in relation to the slope of the ground on which the nest was sited. Data from Whittingham *et al.* (2002).

rely on detection of potential nest predators at a distance (Byrkjedal & Thompson 1998), and the convex slopes of the moorland edge may make this difficult. It is perhaps no coincidence that although golden plover do tend to nest in short vegetation, in suitable habitat they often place their nests on top of tussocks of vegetation (e.g. Ratcliffe 1976; Whittingham 1996). Equally, it may simply be that stoats prefer to hunt on the lower, steeper slopes where rabbits are more abundant and are less often found on the higher, flatter tops where golden plover nest in areas of cotton-grass with its high densities of craneflies as food for chicks (Pearce-Higgins & Yalden 2004).

During incubation, both sexes incubate, but off-duty birds may fly several kilometres both during the day and at night to feed in pastures rich in earthworms and cranefly larvae (leatherjackets) at the moorland edge (Whittingham *et al.* 2000; Pearce-Higgins & Yalden 2003a). Intriguingly, in the Peak District, females and males used different fields, with females feeding by day up to 6–7 km from the nest, whilst males, off duty at night, fed only 2–3 km away in the closest available pastures (Pearce-Higgins & Yalden 2003a). Here, relatively intensive grazing may be beneficial as the birds tend to select fields with short swards which presumably provide for easier access to prey and detection of predators. Once the young hatch, foraging activity switches almost wholly to the moorland area around the nest. Radio-tagging studies of broods in the Pennines have found that parents with broods tend to select heterogeneous areas comprising short grass, bare peat, wet flushes or patches of blanket bog cotton-grass for foraging, interspersed with longer heather or soft rush which may have some value in providing cover (Whittingham *et al.* 1999, 2000, 2001b; Pearce-Higgins & Yalden 2004) (Fig 8.9). These wetter habitat patches tend to hold the highest densities of key prey items such as craneflies, leatherjackets and beetles. Pearce-Higgins' study found that craneflies and leatherjackets dominated the diet of golden plover broods, although older chicks also exploited areas of bilberry and crowberry where caterpillars were the likely prey. Areas of bare peat were

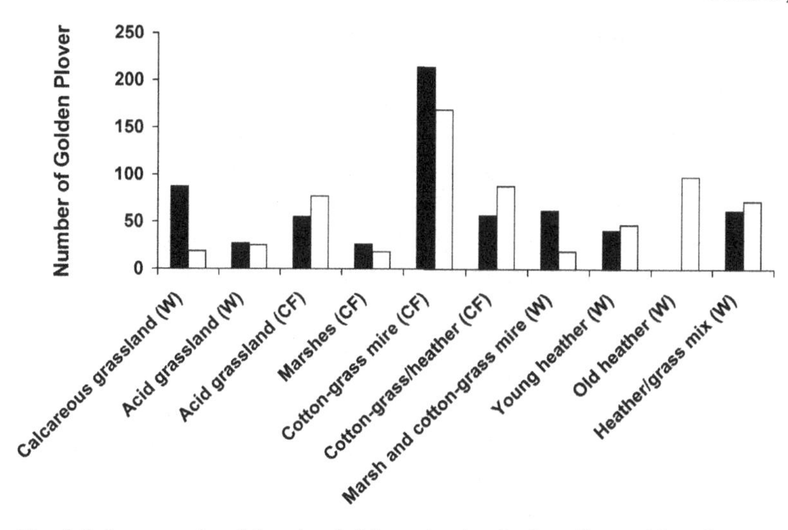

Fig. 8.9 An example of foraging habitat selection by breeding golden plover. Actual counts of adult golden plover (filled bars) from transects in June on two study areas in County Durham (Widdybank, W, and Chapel Fell, CF) relative to the counts expected if birds were distributed randomly across the available habitat areas (open bars). Data from Whittingham *et al.* (2000).

important for allowing easy access to leatherjackets, which dominated the diet of older chicks. Accordingly, chick weight gain and survival rates were strongly associated with the availability of bare peat and the abundance of craneflies, confirming that the summer flush of mature leatherjackets and hatching craneflies is likely to be a key determinant of breeding success (Pearce-Higgins & Yalden 2003b, 2004).

By mid July, the breeding grounds have been vacated and the British golden plover population joins with birds from Iceland, Scandinavia and as far east as Siberia to winter in mixed flocks (usually with lapwings) on lowland arable fields and pastures, as well as coastal grazing marshes, saltmarsh and mudflats. Approaching 300 000 birds may winter in Britain, some 17% of the combined populations that winter in Europe and north Africa (Delany & Scott 2002; Rehfisch *et al.* 2003). On agricultural land, earlier studies suggested that golden plovers preferred old, earthworm-rich pastures (e.g. Fuller & Youngman 1979), with a tendency for arable fields to be used more where suitable pasture is scarce (e.g. Mason & Macdonald 1999b). However, more recent studies using image intensifiers to allow nocturnal surveys have found that both golden plovers and lapwings feed actively in small, single-species flocks at night, exploiting a much wider range of fields than are seen to be used during the day. Gillings *et al.* (2005a) found that golden plovers tended to select bare, ploughed fields during the day but made more use of growing cereal and oilseed rape crops at night, whilst sugar beet stubbles were used throughout. In arable, eastern England Gillings *et al.* (2007) found that this combination of diurnal and nocturnal foraging on a variety of stubbles, ploughed fields and growing crops allowed both lapwings and golden plovers to winter on farmland in the absence of large areas of pasture. Intriguingly, Gillings (2003) suggests that nocturnal feeding may be especially beneficial in arable landscapes because the nocturnal diet consists of larger, deeper-burrowing earthworm species that are not available during the day and whose abundance is less affected by cultivations. Whilst there is no evidence that the British wintering population has changed in recent decades, there has been a marked increase in the numbers recorded at coastal sites by the Wetland Bird Survey in recent years, with the maximum count surpassing 200 000 individuals for the first time in January 2005 (Banks *et al.* 2006). Much of this increase has been at sites on the English east coast and probably results from a redistribution of birds within the country (Gillings *et al.* 2006).

The size of the British breeding population is poorly known with estimates varying between 22 600 and 59 400 pairs (Baker *et al.* 2006; Thorup 2006), some 3–13% of the estimated European total. Golden plovers have certainly been lost from marginal sites, and range contractions have taken place in the Welsh uplands, parts of the Southern Uplands and southern and eastern highlands of Scotland, and the Caithness and Sutherland peatlands, attributed variously to afforestation, disturbance and highly intensive sheep grazing (e.g. Thompson *et al.* 1988; Lovegrove *et al.* 1994; Ratcliffe 2007). There is no indication, however, of the substantial population declines experienced by lapwings and other breeding waders. One detailed study followed a local golden plover population to extinction in north-east Scotland (Parr 1992). Decline was found to be associated more with low survival rates in severe winters and not with loss of breeding habitat or poor breeding success. Only in the final stages of decline did breeding success fall sufficiently to drive the population to extinction, and this was associated with high rates of nest predation by increasing populations of egg predators whose appearance coincided with local afforestation with conifers. Similarly, although golden plover populations in the Peak District have undergone significant shifts in habitat association (from heather–grass mosaics to cotton-grass), long-term monitoring of the Snake Summit population since the early 1970s shows no evidence of any long-term trend, even though substantial fluctuations have occurred, driven partly by variation in winter survival (Yalden & Pearce-Higgins 1997).

The apparent robustness of the golden plover population to agricultural change perhaps reflects the very wide range of modern agricultural habitats that the species appears able to exploit to its advantage. Improved, upland edge pastures provide a rich soil invertebrate food source for adults in the early phases of breeding, whilst a structurally heterogeneous mix of dwarf shrub, grass and marsh vegetation provided on moorland managed for either grouse or sheep provides good foraging conditions for broods. The species is also certainly one of those whose breeding success benefits from predator control on grouse moors (Tharme *et al.* 2001). Golden plover may indeed be a good indicator of moorland management that is beneficial for a wider range of species since neither exceedingly intensive drainage and sheep grazing to produce 'billiard table' grasslands, nor abandonment of grazing and burning practices on moorland are likely to favour the species. Rather, a judicious mix of grazing, burning and hydrological management to create structural mosaics of vegetation cover (e.g. dwarf shrub, grasses, sedges and rushes) with areas of invertebrate-rich wet flush or blanket bog appear to provide ideal conditions for this species and are likely to favour others too. The exception may be the species' apparent predilection for areas of eroded blanket bog where extensive areas of exposed peat appear to provide good access to cranefly larvae for foraging birds (Pearce-Higgins & Yalden 2004). In these situations, management designed to restore blanket bog condition and function – for example by reducing grazing intensity or more directly revegetating exposed peat – may have locally detrimental effects on golden plover numbers (J. W. Pearce-Higgins pers. comm.). Outside the breeding season, golden plovers appear to find adequate food resources on both arable and pasture fields in modern, agricultural landscapes, perhaps aided by being able to feed at night. In future, golden plover is one of the species for which the potential impacts of climate change will be worthy of exploration. For example, a recent study by James Pearce-Higgins *et al.* (2005) suggests that under current climate change predictions, golden plover may suffer from an increasing mismatch between the timing of their laying and the emergence of craneflies as key food for chicks. Mason & Macdonald (1999b) also note that rapid autumn growth of winter cereal crops in arable landscapes may render increasing numbers of fields unsuitable for occupation by feeding plovers in southern England.

Case study 9: Woodpigeon

The woodpigeon has thrived during the post-war period of agricultural intensification. It is an abundant bird throughout Britain, excepting only exposed uplands and the highlands and islands of Scotland where it is rather more sparsely distributed. The British breeding population is estimated at 2.5–3.0 million breeding pairs (Table 5.2), more than that of any other European nation (BirdLife International 2004). Throughout their wide European breeding range, woodpigeons are associated with woodland and forest edge, alongside well-vegetated open ground. Here they exploit a huge range of plant food, ranging from grazing on green material to taking seeds, berries and flowers, both from the ground and whilst clambering with some agility in bushes and trees. Lowland agricultural land in which protein-rich tillage crops and grassland are interspersed with copses, spinneys and hedgerows provides ideal habitat for woodpigeon, and their success made them one of the major agricultural bird pests in Britain (Murton 1965).

Woodpigeons have an unusually long breeding season, partly reflecting differences in the timing of breeding in different habitats. Birds nesting in urban areas may lay as early as February, with most clutches laid in April and May, whilst on farmland, where cereal grain is the staple nestling diet, laying typically begins in June and July with some young being fledged as late as October and November. Nests are usually in forks or branches of trees and shrubs, and the majority of clutches contain two eggs. Several breeding attempts may be made during the course of a season, but high levels of nest predation usually mean that only one or two broods are successfully fledged. Young are fed on roughly the same diet of green plant material and grain as adults, but smaller weed seeds and animal food – especially snails – are more frequent. During the remainder of the year, cereal grain remains the favoured food supply when available, but brassica leaves, peas and beans, clover leaves, beech mast, acorns, hawthorn and ivy berries and the seeds and fruits of agricultural weeds such as buttercup and chickweed are amongst the huge variety of plant material eaten.

The dubious honour of being the 'major bird pest in the United Kingdom' (Inglis *et al.* 1990) means that the ecology and behaviour of woodpigeons were studied in detail long before those of many species whose declines have become the source of increasing conservation concern. Indeed the studies of woodpigeons begun by R. K. Murton, A. J. Isaacson and N. J. Westwood of the then Ministry of Agriculture in 1960 constitute one of the classic long-term studies of avian ecology (Inglis *et al.* 1990). These studies reveal the degree of dependence of this species on food made available through local agricultural practice. The 10-km^2 study area is centred on the Cambridgeshire village of Carlton. It is an area of arable farming, with woodlands and hedgerows providing up to 35 hectares of suitable nesting habitat for woodpigeons. In the 1960s, agricultural practice was dominated by spring-sown barley and winter-sown wheat, some of it undersown with clover, and with permanent pasture also providing a clover food source. At that time, woodpigeon population densities were largely determined by the autumn area of cereal sowings and stubbles which controlled the number of birds remaining in the area until the end of the year, and subsequently by the availability of clover ley which determined over-winter survival rates. Winter 'battue' shoots of woodpigeons probably then had little long-term impact on populations since they took place before the main season of limitation of population density by depletion of clover (Murton *et al.* 1974).

During the late 1960s and early 1970s, however, agricultural practice changed substantially, as it did across Britain; winter-sown wheat increased in area, autumn-sown barley replaced spring-sown varieties, the practice of undersowing with clover declined and pastures were ploughed to allow for greater areas of cereal. At this time, the woodpigeon population at Carlton crashed, probably because of the increasingly limited supply of clover as a late-winter food source, and perhaps partly because of increasing nest predation caused by the spread of introduced grey squirrels into the area at the time (Inglis *et al.* 1994). This downturn in fortunes is also reflected nationally with the CBC index for this species showing a shallow population decline during the early 1970s (Marchant *et al.* 1990). From the mid 1970s, however, the factors limiting the Carlton woodpigeon population changed once again. Oilseed rape was introduced under subsidy following the UK's accession to the European Economic Community (as it was then known), and rapidly became an important autumn- and spring-sown break crop in cereal rotations. Oilseed rape leaves are just as nutritious as clover for woodpigeons, and more efficiently consumed. The limitation on over-winter food supply was thus lifted, and populations increased

(Inglis *et al.* 1997). They have continued to grow ever since (Table 5.2), in parallel with increasing oilseed rape acreages. During this period, damage to both yield and oil content of oilseed rape crops caused by woodpigeon grazing became a major cause of abandonment of oilseed rape production by farmers (Lane 1984; Inglis *et al.* 1989, 1997).

Shooting of woodpigeons both for sport and for crop protection continues. However, the increasing acreages of oilseed rape have changed both its method and timing. Before the advent of oilseed rape, woodpigeons were scattered in winter in many small flocks over clover leys and pastures, and could be attracted within shooting range using decoys. When oilseed rape is present, however, woodpigeons tend to be concentrated in a few very large flocks that are inefficient to decoy since the time between disturbing one flock and attracting another may be prohibitively long. Shooting has therefore shifted to the late summer period (June–September) when the former pattern of distribution of many small flocks is now found on post-harvest stubbles and newly sown fields. This of course is the time when many woodpigeons are nesting, and summer shooting may therefore have contributed to the progressive decline in breeding success that has been observed (Inglis *et al.* 1994) as the population has increased. The density-dependent impacts of shooting and predation on breeding success are perhaps now more likely to constrain local population growth rates than is over-winter survival.

The woodpigeon's success and continued abundance on even the most intensively managed farmland is largely attributable to the fact that it consumes growing crops, and that modern, lightly wooded agricultural landscapes provide the perfect mix of nesting habitat and foraging ground. However, just as the Carlton studies have shown in the past, its future population trends are likely to be highly sensitive to the changes in cropping patterns that take place under the evolving CAP and world markets for arable crops.

Case study 10: Turtle dove

The recent fate of turtle doves in Britain is in stark contrast to that of woodpigeons, and provides an interesting case study of the differences in ecology that can cause two closely related species to differ so markedly in their responses to agricultural change.

Turtle doves are widespread throughout Europe, excepting only Scandinavia and the Ireland, Wales and Scotland. Throughout this range, they are associated with habitat mosaics in which wooded areas such as copses, groves and scrub used primarily for nesting are interspersed with open, agricultural habitats that provide a rich source of the weed seed and cereal grain upon which the birds rely for food. Pairs nest solitarily but non-territorially in scrub, large hedgerows or trees, laying clutches of just one or two eggs, but rearing up to two or three broods per season. Unlike woodpigeons, turtle doves are migratory, leaving their European breeding areas from late July to September to winter in the African Sahel, and arriving back in

Britain during early May. The current European population is estimated at 3.5–7.2 million pairs of which 1–2.5 million are in Russia, with a British population of just 44 000 pairs (BirdLife International 2004) (Table 5.2).

In Britain, turtle doves have always been restricted mainly to the lowlands of the south and east. Parslow (1973) recorded a spread during the nineteenth century into northern England, and Wales, extending into the south-east of Scotland during twentieth century, with East Lothian, Lancashire, the Welsh border counties, Carmarthenshire and Devon delimiting the northern and western limits of the species' range at its maximum, around the time of the 1968–72 *Breeding Atlas* (Sharrock 1976). Since then, the species' decline has been spectacular, with the CBC and BBS recording a decline of 82% since 1967, and a 66% decline on BBS survey squares between 1994 and 2007 (Table 5.2). Given the current population estimate, this implies that Britain may have lost roughly 150 000 pairs of turtle doves since the late 1960s, and the decline shows no sign of slowing. The impact on the species' breeding range has been considerable, with turtle doves becoming rare or disappearing altogether in south-west England, most of Wales, and almost everywhere north of a line from the Mersey to the Tees. By the time of the *1988–91 Breeding Atlas*, the bird was recorded in only three-quarters of the 10-km squares in which it was present in 1968–72, with high densities restricted to that part of England east of a line from York to Brighton (Gibbons *et al.* 1993).

From the perspective of understanding the cause of this decline, we are fortunate that the same Cambridgeshire study that contributed so much to our understanding of woodpigeon responses to changing agriculture also collected data on turtle doves (Murton *et al.* 1964; Murton 1968). This provided the foundation for the GWCT to conduct an intensive study of the breeding ecology and migration timing of turtle doves between 1998 and 2000 and, by comparison with Murton's data from the 1960s, to seek to understand whether the impacts of agricultural intensification on the species' breeding ecology could have driven the decline. In this context, it is important to bear in mind that the turtle dove is a species subject to exceptionally heavy recreational hunting pressure during migration through Mediterranean regions, with 2–4 million birds killed in autumn and a further 50 000 in spring (Boutin 2001). Drought and cutting (for charcoal) of acacia woodland in its African wintering grounds are also becoming progressively more severe (Jarry & Baillon 1991; Jarry 1994). These factors have thus also been considered potential causes of the species' decline. The GWCT's study was unable to return to Murton's former study area at Carlton as turtle doves are no longer found there, but was able to establish study areas in the fenlands on the Lincolnshire/Cambridgeshire border, and in north-west Suffolk, within a few tens of kilometres of the original site. After three years of research, the study had reached the following conclusions.

First, although there was little difference in the success of individual nests, turtle doves in the 1990s were found to have a shorter breeding season and consequently to produce little more than half the number of clutches and young per pair as formerly (Table 8.8). Specifically, where in the 1960s Murton (1968) found two distinct

Table 8.8 *Comparison of breeding success of turtle doves in eastern England in the 1960s and 1990s (means are presented ±1 standard error)*

	1960s	1990s	Significance of difference[a]
Sample size	72	116	
Clutches hatching (%)	46 ± 6	63 ± 5	**
Broods fledging young (%)	82 ± 7	69 ± 6	ns
Nests fledging young (%)	38 ± 6	43 ± 5	ns
Clutches per pair	2.9 ± 0.1	1.6 ± 0.1	**
Young fledged per pair	2.1 ± 0.3	1.3 ± 0.2	*

[a]*, $p < 0.05$; **, $p < 0.01$; ns, not significant.
Source: Browne & Aebischer (2004).

peaks of laying in June and August (reflecting first and second broods), the GWCT study found that this second peak of laying had disappeared; where a quarter of nesting attempts had begun in August in the 1960s, now only 5% did so (Browne & Aebischer 2004). This earlier end to the breeding season is also associated with a trend towards earlier departure, as indexed by records of migrant turtle doves from coastal bird observatories. Thus, although median spring arrival date had not altered between 1963 and 2000, median autumnal departure date had become earlier by eight days (Browne & Aebischer 2003a). A simple simulation model showed that, holding other parameters constant, the reduction in productivity caused by this curtailment of the breeding season could lead to a rapid population decline of 17% per annum (Browne & Aebischer 2004). A follow-up analysis of the BTO's nest record data confirmed the findings of the field study by showing that the success of individual nesting attempts did not change between the 1960s and 1990s (Browne *et al.* 2005). If breeding productivity has played a role in the decline, then it is through the reduction in the average number of nesting attempts per pair detected by the field study.

Second, Browne & Aebischer found that turtle doves fed mainly at 'man-made' sites (i.e. spilt grain, animal feed and grain stores), and were only infrequently recorded feeding at 'natural' sites on farmland. In fact, some radio-tagged birds were recorded flying distances of up to 10 km from their nesting areas to exploit artificial food supplies early in the breeding season (Browne & Aebischer 2003b). A large proportion of the local population was often concentrated in a very few feeding sites such as a patch of weedy, rabbit-grazed rape crop or weedy bulb fields, both of which were rare at the landscape scale. Diet analysis showed that wheat and rape averaged 61% of the seed eaten, in sharp contrast with Murton's 1960s study when weed seeds constituted over 90% of those eaten, with wheat and rape seeds making up only 5% (Browne & Aebischer 2003b).

The compelling conclusion from these studies is that the decline of turtle doves in Britain since the 1960s has been driven by the impacts of agricultural intensification on the availability of weed seeds. The main demographic mechanism is increasing limitation of the number of nesting attempts that individual females are able to make, though reductions in survival caused by hunting during migration and the degradation of wintering habitat may also have contributed (Aebischer 2002). The story is strikingly similar to that of another lowland farmland bird that historically has relied on the seeds of farmland weeds as its key food sources – the linnet (see below). Perhaps the one significant advantage held by linnets in recent years is that they are small enough to exploit ripening rape seed in pods on the standing crop, rather than depending on the short-term glut of ripe seed that becomes available on the ground after harvest. The fact that turtle doves seem increasingly dependent on unpredictable supplies of cultivated grain and rape seed associated with harvest and storage is of great concern and suggests that the well-established decline of the wild flora on lowland farmland (Robinson & Sutherland 2002) has reduced food availability for turtle doves to critically low levels over large areas. The trend towards increasingly secure storage of foodstuffs on farms, the abandonment of stock rearing (and the feedstuffs associated with these enterprises) over much arable farmland, and the almost complete disappearance of hay meadows and clover leys (much used for feeding by turtle doves in the 1960s) will only have exacerbated this problem.

The future for turtle doves on British farmland appears bleak, unless agri-environment schemes are able to provide areas of weed-rich habitat throughout the breeding season. England's new Environmental Stewardship scheme does offer such opportunities in the form of summer fallows following over-winter stubbles, uncropped wildlife strips at the margins of cereal fields, conservation headlands and wild bird cover crops. It remains to be seen, however, whether these particular management options will be adopted to a sufficient extent within the remaining range of the turtle dove in southern and eastern England to reverse declines.

Case study 11: Skylark

The skylark is arguably Britain's most familiar farmland bird, being both conspicuous and very widely distributed especially during the breeding season. Indeed, the bird was recorded in all but a few of Britain's 10-km squares during both *Breeding Bird Atlases* (Sharrock 1976; Gibbons *et al.* 1993) and the unmistakable song-flight of the male is a feature of most open habitats during the spring and summer, from grassy uplands to coastal marshes. Skylarks breed in all manner of habitats that provide short, grassy or herbaceous vegetation (typically 20–80 cm tall and of sufficiently open structure to allow access at ground level) in which the birds both nest and forage, although grass that is grazed uniformly very short and small fields enclosed by hedgerows or woodland are usually avoided (J. D. Wilson *et al.* 1997).

Breeding densities are highly variable and tend be highest in semi-natural habitats such as saltmarshes, dune and steppe grasslands and heathlands, rather than on farmland (Fig. 8.10), although simply by virtue of their area, farmland habitats

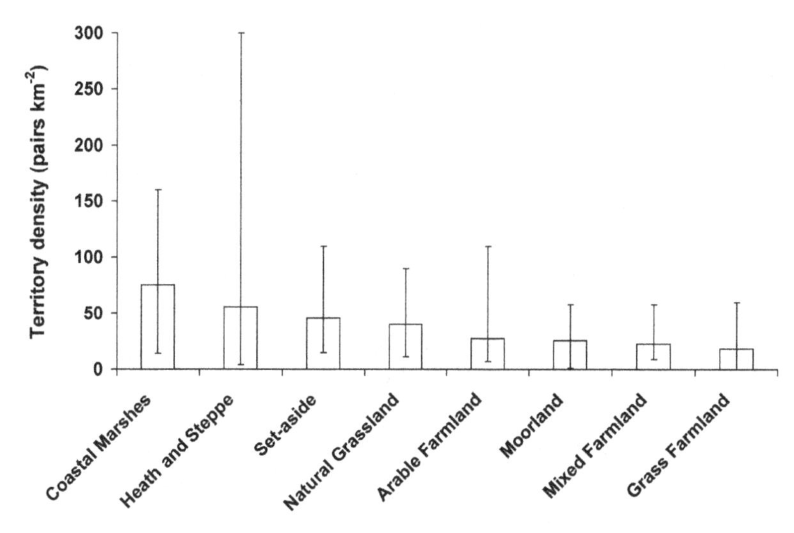

Fig. 8.10 Mean and range of skylark territory densities in different habitats, summarised from European studies. Data from Donald (2004).

support the majority of the British skylark population. A national survey carried out by the BTO in 1997 estimated that over 70% of an estimated British population of 1 million pairs of skylarks inhabited lowland farmland (with over 50% on arable ground) with most of the remainder in the uplands (Browne *et al.* 2000). On farmland, densities tend to be highest where cereal cropping predominates and lowest in areas dominated by intensive grassland, especially where high densities of grazing animals create swards too short for nesting and foraging, although even small areas of arable land within grass-dominated landscapes will result in much higher local skylark densities (Robinson *et al.* 2001). There are also arrivals of large numbers of skylarks along the east coast of Britain in autumn, although the extent which these remain to winter or pass on to wintering areas further south and west such as Ireland or Iberia remains largely unknown. Certainly, the number of exchanges of ringed birds between Britain and continental Europe is very small and not suggestive of a large wintering population of continental skylarks in Britain (Spaepen 1995; Gillings & Dougall 2002).

As recently as the 1970s, skylarks were so abundant on farmland as to be considered a pest due their grazing of young cereals, oilseed rape and sugar beet (Dunning 1974; Green 1980). However, following a long period of apparent stability in the breeding population, the number of skylarks recorded on CBC plots fell dramatically between 1978 and 1988. Numbers continued to fall subsequently at a slower rate and overall, by 59% between 1967 and 2005 (Table 5.2). In England, BBS survey results show that the decline continues, with a 21% fall between 1994 and 2007, although numbers in Scotland and Wales, where a smaller proportion of the population occurs on lowland farmland, have been stable over the same period (Risely *et al.* 2008).

Because both farmers and the general public were familiar with the species, the plight of the skylark came to symbolise the severe problems faced by birds and other wildlife in the farmed environment. The composite index of wild bird populations used by the UK Government since 1999 as one of its measures of the environmental sustainability of its policies and of the national quality of life was even dubbed the 'Skylark index' in media coverage at the time of its launch. In response, farmland populations of skylarks were studied intensively during the mid to late 1990s with work in Britain conducted by the RSPB, BTO, GWCT and Oxford University amongst others. These studies are summarised in Table 8.9, and their results confirm that changes in agriculture are responsible for the skylark decline.

Table 8.9 shows that skylarks face many of the same problems that have confronted lapwing populations on lowland farmland. Like lapwings, skylarks experience the high nest predation rates characteristic of ground-nesting species, and show morphological, behavioural and life-history adaptations to minimise predation risk. Skylarks will typically avoid nesting near tall structures (everything from hedgerows and woodland edges to farm buildings and electricity pylons: J. D. Wilson *et al.* 1997; Donald *et al.* 2001c; Milsom *et al.* 2001) that may harbour or provide vantage for predators. Nest sites are selected to offer good all-round visibility of approaching danger. Lark nestlings are unusually precocial for passerines and grow fast, with especially rapid leg development (Shkedy & Safriel 1992) so that the chicks are able to leave the nest at just eight or so days old, well before they can fly or feed independently. Also, like lapwings, skylarks nest on the ground in open field vegetation and, although this species will tolerate taller crops than lapwings, this sensitivity to vegetation structure is almost certainly in part a response to predation risk. Territory densities are usually highest in vegetation between 20 and 80 cm tall (the optimum being shorter where the vegetation cover is denser) (e.g. J. D. Wilson *et al.* 1997; Donald 2004), but then fall off rapidly with very few skylarks nesting in fields with vegetation over 1 m tall. Skylarks prey on a wide range of invertebrates when feeding chicks, and forage in sites combining rich invertebrate availability with relatively sparse, patchy vegetation cover that allows easy access at ground level, suggesting that accessibility of food is also a key factor underlying this species' sensitivity to vegetation structure. For example, Wilson (2001) found that skylarks collecting food for nestlings favoured set-aside, grass tracks and margins and organically managed grassland relative to their area whilst avoiding arable crops and intensively managed grassland, and Jenny (1990b) and Weibel (1998) found that skylarks selected wild-flower strips, stubbles, tracks and short grass whilst avoiding cereal crops. Odderskær *et al.* (1997a) followed radio-tagged skylarks and found that as crops of spring barley grew taller, foraging birds spent a progressively greater proportion of their time foraging in the unsown tractor wheelings ('tramlines') rather than the growing crop (Fig. 8.11), even though the latter held significantly greater densities of invertebrates.

Lastly, and most importantly in the context of the impacts of agricultural intensification, skylarks are capable of making several nesting attempts in one season to

Table 8.9 *Some recent studies investigating the ecology of the skylark in Britain and Europe*

Feature under investigation	Key findings	Studies of large-scale habitat associations	Intensive studies
Broad habitat preferences during the breeding season	• Preference for nesting in set-aside, cereals and grassy moorland • Higher densities on organic cereals and grassland than on conventionally managed fields • Importance of crop diversity in providing adequate spatial and temporal variation in vegetation structure to allow multiple nesting attempts within one territory • Avoidance of enclosed fields	Chamberlain *et al.* (1999), Chamberlain & Gregory (1999), Chamberlain *et al.* (2000), Browne *et al.* (2000), Henderson *et al.* (2000a), Chamberlain (2001), Kragten *et al.* (2008)	Schläpfer (1988, 2001), Jenny (1990a), Watson & Rae (1997), J. D. Wilson *et al.* (1997), Poulsen *et al.* (1998), Henderson *et al.* (2001), Donald *et al.* (2001c), Vickery & Buckingham (2001), Eraud & Boutin (2002), Kragten *et al.* (2008)
Habitat structure required during the breeding season	• Preference for vegetation height of 20–80 cm in crops and 15–25 cm on grassland • Need for bare/sparsely vegetated areas to provide access for foraging	Chamberlain *et al.* (1999)	Schläpfer (1988), Jenny (1990a, b), Odderskær *et al.* (1997a), J. D. Wilson *et al.* (1997), Wakeham-Dawson *et al.* (1998), Schön (1999), Weibel (1998), Donald *et al.* (2001c), Henderson *et al.* (2001), Milsom *et al.* (2001)
Diet of chicks	• Identified key invertebrate groups fed to chicks		Green (1978), Jenny (1990b), Poulsen *et al.* (1998), Elmegaard *et al.* (1999), Donald *et al.* (2001a)
Foraging habitat when chick rearing	• Identified importance of set-aside and grass tracks and margins		Jenny (1990b), Poulsen *et al.* (1998), Weibel (1998), Wilson (2001), Murray *et al.* (2002)
Nestling growth rates and body condition	• No differences between nests in sprayed and unsprayed crops, though nestlings fed on insect larvae tend to be in better condition as do those fed more diverse diets		Donald *et al.* (2001e)

(*cont.*)

Table 8.9 *(cont.)*

Feature under investigation	Key findings	Studies of large-scale habitat associations	Intensive studies
Nest success	• Nest success has increased during the period of population decline • Nest failure rates often high (>75%); predation and mowing of agricultural grassland are the main causes • Nest failure rates due to predation tend to be positively density dependent • Nest failures more likely close to field edges and crop tramlines where predators likely to be most active • Little evidence of impacts of agrochemical use on nest survival, though some evidence of lower productivity in nests in fields sprayed with insecticide when weather conditions poor	Chamberlain & Crick (1999)	Jenny (1990b), Odderskær *et al.* (1997b), J. D. Wilson *et al.* (1997), Poulsen *et al.* (1998), Weibel (1998), Donald *et al.* (2002a)
Seasonal productivity	• The number of breeding attempts per season has been reduced		Schläpfer (1988), Jenny (1990a, c), J. D. Wilson *et al.* (1997), Donald *et al.* (2002a)
Survival rates	• A local ringing study found low survival rates in the period 1980–85		Wolfenden & Peach (2001)
Habitat preferences in winter	• Preference for weedy stubbles (including naturally regenerating set-aside), and certain broadleaved crops • Prefer to feed in the centres of large fields	Gillings (2001) Gillings & Fuller (2001)	Wilson *et al.* (1996b), Buckingham *et al.* (1999), Robinson & Sutherland (1999), Buckingham (2001), Donald *et al.* (2001a), Vickery & Buckingham (2001), Hancock & Wilson (2003)
Diet in winter	• Mainly consume cereal grains and leaves of crops and other plants		Green (1978), Donald *et al.* (2001a)
Practical management solutions	• Unsown plots in cereal fields increase Skylark densities and breeding success		Morris *et al.* (2004)

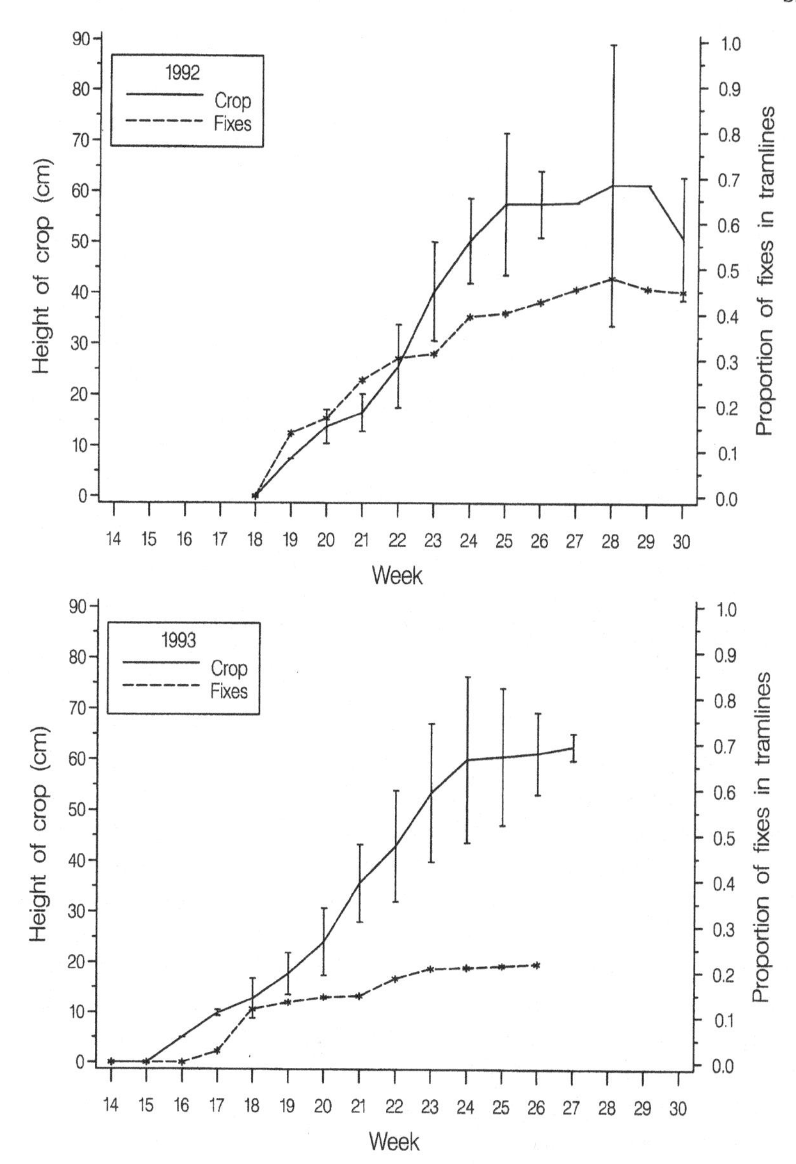

Fig. 8.11 Seasonal change in the proportion of fixes of radio-tagged skylarks in tractor wheelings (dotted lines) in relation to crop growth (solid lines) through the breeding season in each of two years (1992 and 1993). Week 14 = 1st week in April. Reproduced with kind permission from Odderskær *et al.* (1997a).

counter the loss of individual nests to predators. To do this within a single breeding territory requires that suitable field vegetation for nesting and feeding is available throughout a prolonged breeding season which extends from April until mid August. Landscapes providing this will usually support high densities of successfully breeding skylarks.

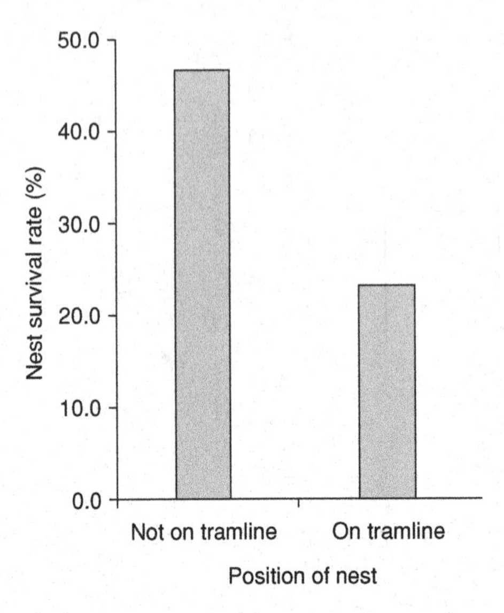

Fig. 8.12 Skylark nest survival rates in cereal crops classified according to whether the nest was built adjacent to a tractor wheeling ('tramline') or further into the crop. Reproduced from Donald (2004), with kind permission.

As with lapwings then, Marren's (2002) observation that 'everything that can go wrong has gone wrong' applies with almost equal force to skylarks. Mature autumn-sown cereal, oilseed rape and silage commonly grow to over a metre in height and combine this with a very dense, luxuriant vegetation cover that would not be found in semi-natural grasslands. Arable farmland is increasingly dominated by these crops, which usually grow too dense, too quickly to allow nesting later in the breeding season. Even where later nesting is attempted in cereals, the birds are often obliged to nest close to the unsown tractor wheelings, where they suffer almost double the predation rate of nests built further into the crop (Fig. 8.12) (Donald *et al.* 2002a; Donald 2004). Prior to the decline, spring-sown cereals were much more abundant in the landscape. As they are established and develop later than autumn-sown fields, they provided suitable habitat for later nesting attempts.

Modern agricultural grasslands are equally inhospitable; secure nesting cover is hard to find in heavily grazed and trampled pastures, and nest losses through destruction by machinery may be exceptionally high in fields that are repeatedly cut for silage. In Switzerland, Jenny (1990a) found that mowing of agricultural grassland accounted for 80% of all nest losses, including 95 of 98 nests in one study area. In contrast, on open saltmarsh or downland landscapes with low livestock densities, the availability of agriculturally unimproved and structurally complex swards throughout the breeding season support densities that are much higher than is typical of farmland (e.g. Milsom *et al.* 2001) (Fig. 8.10). One experiment illustrates perfectly the sensitivity of skylarks to grassland management (Wakeham-Dawson *et al.* 1998; Wakeham-Dawson & Aebischer 2001). This study took place on the

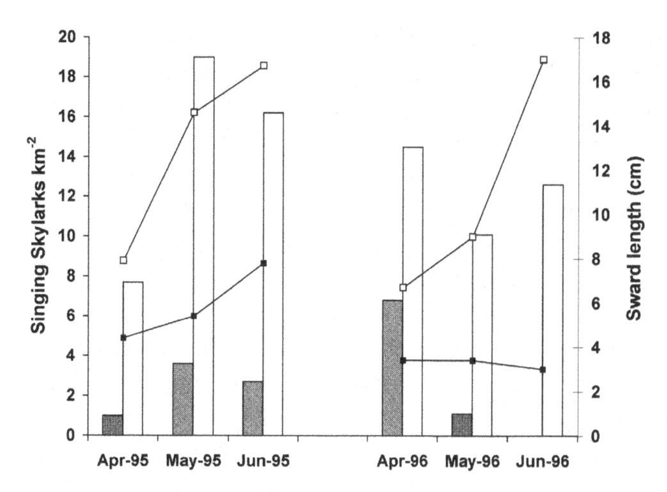

Fig. 8.13 Effect of varying sward length through controlled grazing on the density of territorial skylarks in arable reversion grassland in southern England. Solid bars (skylark density) and squares (sward length) show short-grazed fields; open bars and squares show long-grazed fields. Data from Wakeham-Dawson *et al.* (1998).

South Downs and South Wessex Downs ESAs in which farmers were paid to revert arable land to permanent grassland, either by sowing agricultural grasses such as perennial ryegrass, cocksfoot and clovers for grazing, or by restoring chalk grassland swards for conservation. Initial surveys suggested that territorial skylark densities were highest on chalk grassland reversion, in swards above 15 cm in height, and in fields without boundary scrub. Densities on grazed, agricultural grassland reversion were not higher than on non-reverted arable land. Twelve arable reversion fields (each of roughly 5 ha) were then selected and six grazed heavily to maintain a short sward (<10 cm) and six more lightly to maintain a long sward (15–25 cm) during the skylark breeding season. The results were striking, with singing territorial skylarks six times as abundant and skylark chick-food invertebrates (spiders, beetles, flies, and Hymenoptera) roughly twice as abundant on the longer-grazed than the shorter-grazed fields (Fig. 8.13).

At both landscape and field scales, conditions for breeding skylarks have deteriorated. Simpler crop rotations and reduced crop diversity on many farms as enterprises have become more specialised, have meant that breeding pairs are less able to take advantage of crops at different stages of growth for successive nesting attempts, simply by moving between fields within their territories. This means that pairs must either defend larger territories or abandon a territory and move on, thereby reducing the density of breeding skylarks that an area of farmland is able to support (e.g. Schläpfer 1988, 2001).

One pressure on skylarks that has probably not contributed to the population decline is predation. Although populations of some generalist predators such as foxes, badgers and corvids have increased, Chamberlain & Crick's (1999) analysis of skylark nest record cards shows that individual nest success rates have increased nationally since the 1950s. Density-dependent declines in predation rates as the population has fallen because of agricultural change probably explain this. The recent RSPB study of the breeding ecology of skylarks provides a striking demonstration of this mechanism in operation on a local scale (Donald 2004). On one farm in north Norfolk, this study found exceptionally high nesting densities (>1 pair per hectare) on set-aside fields supporting sparse vegetation regenerating on light, flinty soils, and these densities persisted through three consecutive years of study. In the first year, the farm was unkeepered, but in the second a gamekeeper began to control weasels, stoats, grey squirrels and hedgehogs over half of the farm, expanding to cover the whole farm in the third year. The effect on nest survival rates on the set-aside fields was spectacular, increasing from just over 10% to over 40% (Donald *et al.* 2002a), and the breeding season finished earlier as birds were not obliged to replace predated nesting attempts late into the summer. Critically, however, this response was seen only amongst the high densities of birds nesting on set-aside; on cereal fields on the same farm, where skylark densities were lower, no increase in nest success was seen. In other words, if skylark populations were to recover in response to future beneficial agricultural change, then rising predation rates might limit the extent of that population increase at high density, but it seems extremely unlikely that predation has played any direct role in driving the decline. Nevertheless, this illustrates that the design and deployment of habitat created under agri-environment schemes ought to take into account the unwelcome possibility of unwittingly creating predation traps.

Most British skylarks are relatively sedentary (Spaepen 1995; Dougall 1996; Gillings & Dougall 2002), although upland areas are largely vacated in winter as birds move to lower ground. Outside the breeding season, birds gather on stubbles, weedy fodder crops and newly sown fields (Wilson *et al.* 1996b; Donald *et al.* 2001b; Gillings & Fuller 2001; Hancock & Wilson 2003) where they feed on grain, weed seeds and fresh green plant material. It was their habit of grazing crop seedlings (especially those of rape, cereals and sugar beet) that led to skylarks being considered as crop pests in parts of eastern England in the 1970s (Green 1980). Ironically, this research showed that skylark damage to sugar beet seedlings was greatest where the more preferred seedling weeds had been removed by herbicide treatments. As in the breeding season, skylarks are usually found towards field centres, and select areas where short vegetation is interspersed with areas of bare ground (Buckingham 2001). Their varied winter diet and consumption of green plant material may mean that the reductions in over-winter seed supply that have contributed to the declines of many of the buntings, finches and sparrows have had less of an impact on skylark populations, though they can hardly have helped. As Robinson (2001) points out, even if declines in seed availability have not directly led to increases in mortality rate through starvation over winter, if birds are obliged to forage for longer or in more risky situations when seed supplies are scarce, then there may still be an indirect effect on survival rates. Intriguing in this respect is the fact that reductions of 70% in breeding numbers of skylarks were reported in unafforested peatland habitats in northern Scotland over just seven years (Hancock & Avery 1998). These birds leave these habitats outside the breeding season, and although their wintering areas are unknown, it seems likely that they winter on lowland farmland elsewhere in Britain. Given that there are no apparent changes to their breeding habitats that could explain such rapid declines, deteriorating winter habitat conditions on farmland are a possible cause. Even if deterioration in non-breeding season conditions have not been the main driver of the long-term population decline on lowland farmland, severe winters can certainly have an impact. The *Winter Atlas* (Lack 1986) revealed marked distributional changes both within and between winters, with a mass exodus in January 1982 following very harsh weather and probable influxes from the continent prior to this in December 1981, and in January 1984 (Gillings 2001). This coincided with a marked fall in skylark abundance on CBC plots between 1981 and 1985 (Siriwardena *et al.* 2001a). At the same time, a long-term study of a colour-ringed population of skylarks nesting in dune grassland on the Alt estuary in Lancashire, and wintering on nearby farmland, found that adult survival rates were low during the relatively hard winters of the early 1980s (averaging 39%) but increased to an average of 66% by the late 1990s (Wolfenden & Peach 2001).

Given the skylark's habitat preferences during the breeding and non-breeding seasons, it was widely anticipated that the species would benefit from the introduction of compulsory set-aside in 1993 (e.g. Wilson & Fuller 1992). A slowing of the skylark's decline during the 1990s may indeed have been partly attributable to set-aside. Several studies have revealed high skylark breeding densities on rotational set-aside fields (e.g. J. D. Wilson *et al.* 1997; Poulsen *et al.* 1998), and Gillings *et al.* (2005b) showed that skylark population declines during the BBS period (1994–2003) were reduced in magnitude in areas with greater quantities of over-winter stubble. Nonetheless, the amount of suitable set-aside habitat has been limited by the extent of the most beneficial options and the various rules governing its management, which have changed almost annually (Vickery & Buckingham 2001). Even at the peak acreages of set-aside that were present during the 1990s, it was perhaps too much to expect that adjusting the management of a measure deigned primarily for production control could alone reverse the national population decline of skylarks.

Management options likely to benefit skylarks are now incorp-orated in agri-environment schemes throughout Britain. Those providing spring-sown arable crops and summer fallows used for nesting, and over-winter stubbles and seed-bearing cover offering winter food are likely to be the most beneficial. In addition, the continued growth in the number of organic farms, which are known to support higher Skylark nesting densities in their later-sown and slower-growing cereals, and less frequently cut grass crops, is likely to benefit the species (J. D. Wilson *et al.* 1997). However, despite its decline, the skylark remains a very widespread species that still occurs on most farms so that conservation measures must be equally widespread if the national population is to recover (Evans & Green 2007). Given that expensive agri-environment agreements involving spring cropping or fallows, or organic conversion, are only likely to be possible on a proportion of farms, those interested in skylark conservation have been obliged to think more creatively in order to integrate the needs of this species with modern, intensive agricultural systems. Are there ways in which the management of some of the dominant crops on lowland farmland can be modified to accommodate the needs of skylarks? Exciting in this respect are trials of measures to modify the field-scale structure of autumn-sown wheat to provide nesting and feeding sites throughout the breeding season. A critical recent experiment led by RSPB compared the abundance and breeding phenology of skylarks on three experimental winter-wheat treatments replicated across 15 arable farms widely dispersed across lowland England, from Yorkshire to Wiltshire (Morris *et al.* 2004). Each treatment plot was at least 5 hectares in size and was either (i) provided with unsown patches (UP) of 4 m × 4 m at a density of two per hectare throughout the field (this is achieved simply by turning off the seed drill briefly during sowing), (ii) sown with wide-spaced crop rows (WSR) separated by 25 cm – double the usual distance, or (iii) was a control, conventional winter-wheat field. Both of the experimental manipulations were designed to create a more open cereal sward, and predicted to provide suitable nesting and foraging habitat for a greater part of the breeding season than conventional winter wheat. Morris *et al.* found that the UP fields supported a greater density of skylarks for longer than either the control or WSR fields. In particular, territory densities on the UP fields declined less than those on the control fields in the later part of the breeding season, and nests fledged approximately 0.5 chicks per attempt more on the UP fields than control fields. Nestling condition was also higher on both UP and WSR fields than control fields late in the breeding season. These encouraging results, coupled with the fact that unsown patches are cheap (£3–8 ha^{-1}) and easy to create (Morris *et al.* 2004), led Defra to include the provision of undrilled patches in winter cereal fields as an option within the new Entry Level Stewardship scheme in England in 2005. Set alongside the other 'skylark-friendly' management options available, there is for the first time the potential to see management to encourage skylark population recovery on many of England's arable and mixed farms. Management trials are also under way to trial practical ways of reducing skylark nest losses to cutting in crops of forage grasses (e.g. by raising cutting heights, altering cutting dates or leaving 'long-cut' patches). For the first time in a long time we can look forward with some optimism to the future of this most inspirational of songbirds.

Case study 12: Song thrush

The song thrush is one of Britain's most familiar birds, and its distinctive and melodious song, characterised by a rich diversity of repeated phrases, is well known to many who would not regard themselves as birdwatchers. Song thrushes breed in woodland, forestry plantations, farmland and gardens throughout Britain, excepting only Shetland. A moderately differentiated subspecies *hebridensis*, darker and with bolder spotting, occurs on the Outer Hebrides and the Isle of Skye (Forrester *et al.* 2007). A small proportion of Britain's song thrushes move south and west to winter in Ireland, France, Spain and Portugal, though most are sedentary. There is also a small influx to Britain in winter, largely involving birds which breed in the Low Countries but, overall, it seems likely that the great majority of wintering song thrushes are British breeding birds (Thomson 2002). Song thrushes are classical open-cup nesters, building a neat, mud-lined structure in trees, shrubs or climbing vegetation, or occasionally on the ground amongst thick vegetation. Three to five eggs are laid, and a pair may make several nesting attempts during a long breeding season from March to July, raising two, three or occasionally even four broods. A wide variety of invertebrates are taken as food, but song thrushes specialise in earthworms, especially when feeding their young. Fruit is also taken, and snails become very important in the diet in dry weather when earthworms are difficult to obtain. As we shall see, these dietary specialisations may have played a large part in explaining the recent declines of the species in Britain.

During the early part of the twentieth century, the song thrush was regarded as an abundant breeding bird, perhaps more common than the blackbird in some areas. However, declines since the 1940s, probably occasioned by the run of cold winters in the 1940s and early 1960s (Parslow 1973), were followed by a long period of almost continuous decline, in lowland England at least, lasting from the late 1960s to the late 1990s. A decline of 50% estimated by the CBC and BBS between 1967 and 2005 (Table 5.2), coupled with a current population estimate of 1 030 000 pairs, suggests that Britain may have lost over a million pairs of song thrushes over the last three decades of the twentieth century. Intriguingly, the last 10 years has seen evidence of a small recovery in some areas, with the BBS showing significant increases in Wales and some regions of England, but continuing declines in eastern England and around London (Chapter 5).

As we saw in Chapter 7, the song thrush has been the subject of some of the most intensive and sophisticated analyses of trends in demographic rates yet undertaken using the BTO's nest record and ring-recovery data sets. These show that whilst measures of individual nest success have, if anything, increased, trends in first-year survival rates, especially during the two months after fledging, are consistent with them having played an important role in driving the long-term population decline (Robinson *et al.* 2004). Until recently, however, there has been a lack of field research designed to understand what environmental factors may drive these demographic trends.

Between 1995 and 1998, RSPB undertook a detailed investigation of the breeding ecology of song thrushes in two contrasting study areas. The first was centred on mixed farmland and rural gardens in West Sussex in an area where CBC data showed that the population had remained stable since the 1970s. The second was on arable farmland in Essex where the population was declining rapidly. In these habitats, nesting song thrushes were strongly associated with hedgerows, woodland edges and gardens, the latter being particularly important in Essex where field sizes are larger, and hedgerow densities and woodland area lower than in Sussex (Peach *et al.* 2004a; see also Mason 1998, 2000). The first important result was that although measures of individual nest success did not differ between the two populations (both had an average clutch size of four eggs, nest survival rates of 25–30%, and 3.3 chicks fledged per successful nest), females laid significantly fewer clutches per season in the Essex population (2.5) than in the Sussex population (4.0). Consequently, total annual productivity in the Essex population (2.7 young fledged per pair per season) was only two-thirds of that in Sussex (4 young per pair per season) (Thomson & Cotton 2000). This difference accrued because, although song thrushes began nesting at the same time in both populations, the interval between nesting attempts was longer in the declining Essex population, with very few nesting attempts during June and July. The second important result was discovered when the same study radio-tracked juvenile song thrushes after fledging and found that post-fledging survival was lower in the first two weeks after fledging in Essex (<50%) than in Sussex (65%). This last result accords well with the results of Robinson *et al.*'s analyses of national ring-recovery data, and raises the possibility that the same environmental driver may underlie both the curtailment of the breeding season and the reduced post-fledging survival rate of song thrushes in intensively managed, arable landscapes.

This question was pursued further by the RSPB study, which went on to investigate in detail the foraging habitat selection by radio-tagged song thrushes in both study areas (Peach *et al.* 2002, 2004a) and the diet of adults and chicks (Gruar *et al.* 2003). In winter, song thrushes were found most often at the base of hedges and in wet ditches whilst in summer scrub, woodland edges, wet ditches and bare soil in gardens were all selected strongly relative to their area, and cereal fields were avoided. Grass fields and woodland hosted most song thrush foraging (53%) in the mixed Sussex landscape, but these habitats were scarce in Essex where 58% of all foraging observations were in gardens and arable crops. In all these habitats, earthworms were the preferred prey, especially in March and April, but snails and arthropods became the dominant foods by June and July, as soils dried and earthworms retreated deeper into the soil so becoming less available to foraging birds. This drying of the soils was most apparent in the arable study area (Fig 8.14) which already held relatively low earthworm densities, probably due to the damaging impacts of frequent cultivation and the lack of permanent pasture that is known to support high earthworm numbers (Tucker 1992). Dry summer soil conditions were associated with lower body weights of both adults and chicks (Gruar *et al.* 2003). All in all, the arable study area in Essex held a lower availability of high-quality nesting habitat (woodland and hedgerows) with a higher proportion of the population centred on gardens, had a lower availability of earthworm-rich foraging habitat (woodland edge and permanent pasture) and experienced a more rapid drying of soils. These factors combined to limit the availability of preferred foods for Song Thrushes, with measurable effects on adult and chick condition.

This lack of food in the Essex study area relative to the Sussex study area is a probable explanation for both the curtailed breeding season and reduced post-fledging survival of song thrushes. What agricultural changes have there been that could have caused lowland farmland in Britain to have become more like Essex than

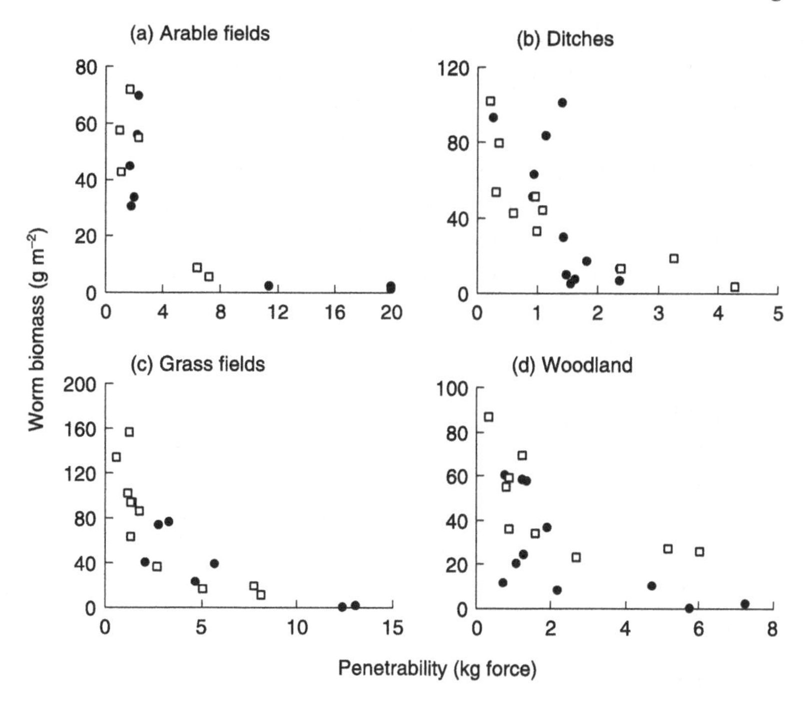

Fig. 8.14 Relationships between mean soil surface earthworm biomass and soil penetrability in four habitats sampled in the mixed farming (Sussex: open squares) and arable farming (Essex: filled circles) study areas during April and June 2000. Each point represents average biomass in the top 5 cm of soil in a single field, wood or ditch sampled during either April or June (some were sampled in both months). Reproduced from Peach *et al.* (2004a) with kind permission.

Sussex and driven the national decline of song thrush populations? Specialisation in arable agriculture and the associated loss of permanent pastures is undoubtedly part of the explanation, and may account for the fact that the decline of song thrushes that has been so severe in lowland England has been less evident to observers in the north and west of Britain where numbers may have changed little. However, the increasing efficiency of drainage of farmland may also have played its part by creating drier soils, earlier in the season. Rates of land drainage in England and Wales increased from the 1940s to the late 1970s, with approximately half of all twentieth-century land drainage occurring during the 1970s, especially on clay soils in arable areas of eastern England (Robinson & Armstrong 1988). Peach *et al.* (2004b) correlated drainage statistics from 30 areas in England and Wales covering the period of most intense, grant-aided under-field drainage activity (1971–80) with CBC trends from 347 census plots across these divisions. They found that the extent of drainage was a more important predictor of population change than rates of hedgerow loss over the same period; song thrushes declined more where under-field drainage was most intensive. Perhaps this finding should come as no surprise. As long ago as 1958 it was noted by Eric Simms that 'low sunshine levels and long damp conditions led to an extension of blackbird and song thrush nesting in north-west London', and it has long been suggested that summer rainfall can play a critical role in determining the length of the breeding season for thrushes (Simms 1978).

In summary, song thrushes on farmland require dense, woody vegetation for nesting (woodland shrub layer, hedgerows) close to damp soils rich in invertebrates, especially earthworms. Their recent decline – a phenomenon that has probably been concentrated in the arable-dominated areas of lowland England – has almost certainly been brought about by the effects on seasonal productivity and post-fledging survival of a reduction in the abundance and seasonal persistence of high soil invertebrate densities, in turn caused by loss of grassland combined with intensive under-field drainage. As Peach *et al.* (2004b) point out, current agri-environment schemes in Britain offer little to support the maintenance or creation of small-scale but seasonally persistent damp soils, except where management is directed at the field-scale raising of water levels for breeding waders. Policy initiatives that cater for the retention of damp soil conditions, at least at small scales, on intensively managed farmland (Bradbury & Kirby 2006) may do much to restore the fortunes of song thrush populations across lowland England.

Case study 13: Chough

Though nowadays usually pronounced 'chuff', the English name of this species derives from its former pronunciation 'chow', an onomatopoeic rendering of one of the bird's commonest calls. The loud, yelping, drawn-out 'chwee-ow' that carries for distances of many kilometres over the mountain valleys or coastal cliffs that are the principal breeding haunts of this species in Britain and elsewhere in Europe is instantly recognisable. The scientific name *Pyrrhocorax* translates literally as 'flame-coloured raven', a reference to the bird's striking, decurved red beak and legs. The call reflects the complex social organisation that binds choughs together as breeding pairs, family parties and larger flocks; the bill is a specialised tool used to obtain invertebrates from the surface or soil of sparsely vegetated ground. Both are important clues to an understanding of the biology of the species and its recent fortunes.

Choughs have attracted and fascinated naturalists wherever they occur in Britain and, as a consequence, many aspects of their ecology and behaviour have been well studied in several parts of their range. Although marked differences exist – notably in feeding behaviour and diet – that we will return to later, the generalities of the species' annual cycle are well established and are summarised here with particular reference to the long-term study of the population on the Hebridean island of Islay (Bignal *et al.* 1997).

Most choughs do not breed until their third year, and spend their first two years flocking with other sub-adult birds, foraging principally for soil-dwelling, surface-active or dung invertebrates, but also taking grain and fruits at times of invertebrate scarcity. Adult and larval beetles and flies are important foods year round, with a wide range of other invertebrates, including caterpillars, spiders, ants, kelp-flies, sandhoppers and earwigs also taken seasonally or locally (e.g. Roberts 1982). At night the birds attend socially structured communal roosts (Still *et al.* 1987), typically in caves, rock crevices or buildings. Breeding birds remain close to their nest sites and home range foraging area all year, though they may join flocks during severe weather, and often share their home range with adjacent pairs during mid-winter. By March, territorial defence has been re-established, with nests built in caves, rock crevices or buildings in late March or early April. Clutches of five are by far the commonest, and are incubated for 21 days from the laying of the last egg. Initially, young are fed by the female who regurgitates food provided by the male, but both parents later forage for the nestlings, which remain in the nest for six weeks. Moth caterpillars and cranefly larvae (leatherjackets) assume particular importance at this time as prey both for adults and as food for nestlings. After fledging the family group gradually ranges more widely, joining other family parties, and with the young birds eventually becoming independent of the parents to join local non-breeding flocks. Adult breeding performance tends to increase through life up to the age of 10 years, with the oldest birds surviving to 13 or 14 years old. The chough is evidently a highly social species and, although the implications of this for chough population dynamics have yet to be studied in detail, it seems likely that flocking, as in many other social species, may play an important role in helping birds to learn foraging techniques, identify high-quality foraging habitat and evade predators.

Although occasional long-distance movements do occur, choughs are largely sedentary throughout their European range. This combined with a complex social organisation, and the fragmented nature of their habitat and populations, has resulted in marked local specialisation in behaviour and diet, perhaps aided by cultural transmission of the kind famously exemplified by the spread of milk-bottle opening behaviour by tits. For example, two vagrant choughs studied in Cornwall in 1986 fed on earthworms (Meyer 1990), a habit not seen amongst Islay choughs (McCracken *et al.* 1992). Similarly, feeding on dung-pat invertebrates on Islay, on tide-wrack sandhoppers and kelp-flies on the Isle of Man and Bardsey Island and on mining-bee larvae on Colonsay (Roberts 1982, Bignal *et al.* 1997) are examples of local feeding specialisms. Perhaps most spectacularly of all for a species considered to be almost exclusively a ground-feeder, Piersma & Bloksma (1987) observed a flock of 350–400 choughs on the Canarian island of La Palma feeding on an infestation of moth caterpillars in Canarian pines. The birds used feeding techniques as varied as bulldozing heaps of pine needles on the ground, clinging to tree trunks and branches whilst probing under bark, and hovering amongst the branches to peck amongst bunches of needles.

The size of the chough population of Britain is well known from surveys carried out in 1982, 1992 and 2002, with additional surveys in Scotland at other times (Table 8.10). These surveys reveal a steady increase of at least a third in overall numbers over the 20 years from 1982 to 2002, although this conceals marked regional fluctuations (e.g. on Islay where the population declined from 95 pairs in

Table 8.10 *Recent population estimates for chough in Britain*

	Scotland		Wales		England		Total	
	Pairs	Flocks	Pairs	Flocks	Pairs	Flocks	Pairs	Flocks
1963	11		98		1	0	110	
1982	72	64	142	106	0	0	214	170
1986	105	115–130						
1992	88	62	150	151	0	0	238	213
1998	66	52						
2002	83		262		1	0	346	

Source: Data from Bullock *et al.* (1983), Monaghan *et al.* (1989), Bignal *et al.* (1997), Cook *et al.* (2001) and Johnstone *et al.* (2007).

1986 to 49 pairs in 1998 before recovering to 56–64 pairs by 2002), and some local extinctions and recolonisations. Overall, the breeding range has remained little changed: the Hebridean islands of Islay and Colonsay, and the Welsh coast from Anglesey to the Gower, with small numbers of birds breeding inland in the Welsh mountains. A single pair currently breeds on the Scottish mainland, and one pair recolonised Cornwall in 2002. The recent slight gains belie a long-term picture of decline. In the eighteenth century choughs bred around much of the coastline of Scotland and in many inland localities, as well as in several coastal English counties from Cumberland to Kent (Parslow 1973). The cause of the species' long-term decline and retraction of breeding range is unclear. In a review by Rolfe (1966), the impacts of human persecution of this confiding bird and a long run of cold winters during the middle part of the nineteenth century are suggested to be the main causes. Certainly the species is vulnerable to hard winters; the questionnaire survey carried out in 1963 by Rolfe (1966) found that the Scottish population declined from at least 30 pairs to 11 over the exceptionally cold winter of 1962/63, and that the total British breeding population was perhaps only 110 pairs in 1963.

Today, choughs in Britain and elsewhere in their breeding range are strongly associated with areas where open, short grass swards and bare ground are created by the grazing of livestock (Bignal *et al.* 1997; Blanco *et al.* 1998; McCanch 2000; Johnstone *et al.* 2002; Whitehead *et al.* 2005), and where agricultural grassland improvement through reseeding, drainage and fertiliser addition has been absent or limited (Bignal & McCracken 1996). These conditions are critical in combining both abundance and availability of the invertebrates upon which this species depends for food. For example, Johnstone *et al.* (2002) and Whitehead *et al.* (2005) were able to quantify breeding season foraging habitat selection for 14 breeding pairs in north Wales over three years. They found that unimproved, grazed coastal grassland and heathland and minor habitats with bare earth exposure such as paths and clod-diau (traditional stone-and-earth stock boundaries) were heavily selected relative to

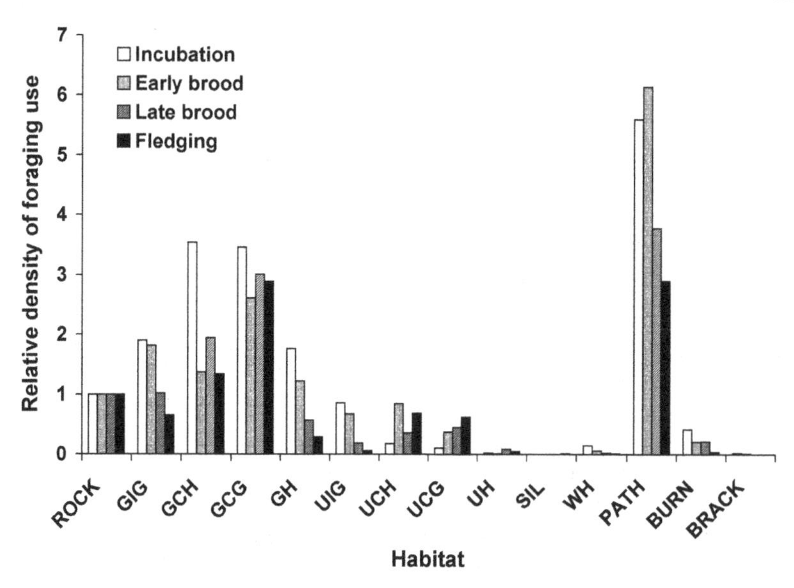

Fig. 8.15 Relative densities of use of different habitats by foraging choughs during the breeding season. ROCK, rock, cliff, scree, buildings and walls; GIG/UIG, grazed/ungrazed improved grassland; GCH/UCH, grazed/ ungrazed coastal heath; GCG/UCG, grazed/ungrazed coastal grassland; GH/UH, grazed/ungrazed dry heathland; SIL, silage; WH, wet heath; PATH, paths (very short swards or bare ground); BURN, burned areas of heath; BRACK, bracken patches. Data from Johnstone *et al.* (2002) and Whitehead *et al.* (2005).

their area, whereas grazed but improved agricultural grasslands were less used, and ungrazed heathland and grassland very little used at all (Fig. 8.15).

Management changes that either reduce invertebrate numbers or render them inaccessible through vegetation growth are likely to be inimical to chough. The above study demonstrates that feeding chough avoided improved agricultural grasslands, even when swards were short, probably because some key invertebrate foods such as ants and moth caterpillars are rarer or absent in fertilised, cultivated and species-poor swards. Equally, agricultural abandonment, silage production for in-wintered cattle, grazing restrictions to preserve cloddiau and conserve the flora of maritime heathlands and fencing to prevent loss of grazing stock over cliff edges are all recent management trends that have caused increases in sward height and encroachment of scrub and bracken. These changes will have removed foraging habitats for chough in Wales (Johnstone *et al.*, 2002; Whitehead *et al.* 2005).

As pointed out by Bignal & McCracken (1996), the chough is a classical example of a species dependent upon landscapes managed as 'low-intensity farming systems'; shifts in management either in the direction of grazing abandonment or agricultural intensification are likely to be detrimental to this species. In an agriculture policy context where recent increases in stock densities are widely viewed as both directly and indirectly damaging to conservation interests (e.g. Beintema & Müskens 1987; Fuller & Gough 1999), it is therefore particularly important

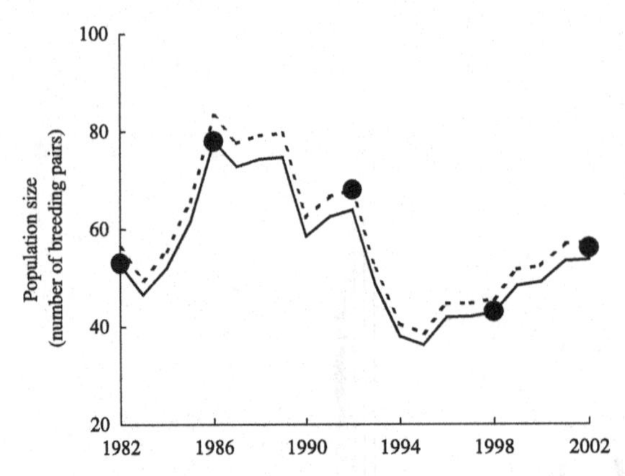

Fig. 8.16 Comparison between the 1982, 1986, 1992, 1998 and 2002 censuses of the Islay chough population (filled symbols) and the population trajectory modelled using estimates of year-specific demographic rates derived from empirical data on breeding success and survival. The modelled trajectory closely matches the census data across all years, suggesting that variation in the demographic rates that most affect population growth rate was accurately quantified. Overall, variation in population growth rate was caused primarily by variation in survival rather than breeding success. Variation in pre-breeding survival accounted for 56% of the total variation in population growth rate. Reproduced from Reid *et al.* (2004) with kind permission.

that grazing management options appropriate to breeding choughs are available within agri-environment schemes. These must be targeted both to the relatively restricted geographical areas in which choughs occur, and to the habitat and foraging specialisations of chough populations in different areas. However, management of breeding season foraging habitat is unlikely to be all that is necessary to secure long-term population and range recovery of choughs in Britain. Recent analyses of the demographic determinants of changes in the population growth rate of Islay choughs (Reid *et al.* 2004) showed that recent changes could be explained largely by variations in pre-breeding survival (Fig. 8.16).

Survival rates during the first year of life varied hugely, from 20% to 74%, in different years, but tended to be higher in years of lower rainfall and when the young were fledged in a good breeding season (Reid *et al.* 2003). Intriguingly these tended to follow warm preceding summers and dry preceding winters, suggesting interactions between food availability and management practice that are not yet understood. The mechanisms linking weather to variation in first-year survival and the potential importance of management of foraging habitats of the non-breeding flocks within which first-year birds establish themselves and acquire their life skills remain to be investigated, although it does seem that management directed towards improving pre-breeding survival of choughs may be key to their conservation (Reid *et al.* 2004). It is clear, however, that much remains to be learned if we are to understand how to manage agricultural landscapes in Britain to provide for the complex ecological and social needs of this absorbing species.

Case study 14: Linnet

Unusually for a songbird, but typically for the finch genus *Carduelis*, linnets rely largely on a diet of seeds, even when feeding their young. They nest as monogamous pairs in loose groups, wherever shrubs provide nesting cover on heathland, in scrub and in farmland hedgerows. Males defend their mates to prevent them from being copulated with by other males, but do not defend any larger territory (Drachmann *et al.* 2000). In southern England, singing males first occupy breeding habitat in late March and the first clutches (typically four or five eggs) are laid in early April, with the last being laid as late as mid August. Up to three broods may be fledged in one year, with adults flying distances in excess of a kilometre to collect seed for nestlings (Frey 1989), although many pairs may be limited to fewer nesting attempts by availability of seed sources. After the breeding season, linnets retain a need for shrubby vegetation to provide safe communal roost sites, and for seeds collected from low-growing herbaceous vegetation or from the ground. Birds congregate in flocks, sometimes thousands strong, on unharvested oilseed rape or linseed fields and later on crop stubbles (e.g. Wilson *et al.* 1996b), weedy winter fodder crops (Hancock & Wilson 2003) and saltmarshes (Brown & Atkinson 1996).

Linnets are classical partial migrants with part of the population remaining in Britain over winter, while others move south or south-west to winter in a narrow longitudinal band from western France, through central Spain to Morocco. Ringing studies show that some individuals may winter in different areas in different years, but that most British-breeding linnets are moderately site faithful, usually returning to areas within 20 km of those where they were fledged or had previously bred (Wilson 2002). However, in a colour-ringing study in Oxfordshire carried out between 1995 and 1999, Moorcroft (2000) found that not one of several hundred linnet nestlings was resighted on the farm from which it fledged, thus suggesting that dispersal from the immediate vicinity of the natal site may be considerable.

Linnets breed throughout Britain, excepting only Shetland and mountainous areas of northern and west Scotland. Inland breeding densities are highest in southern and eastern England, but coastal areas usually have higher densities than their neighbouring hinterland. The British population is currently estimated at 535 000 pairs (Table 5.2). Declines of 71% were recorded by the CBC between 1967 and 2005 (Table 5.2), and 89% by the BTO's Constant Effort Sites ringing scheme between 1984 and 2005 (Baillie *et al.* 2007). Although declines may have stabilised in Scotland and Wales, losses in England, which were very rapid until the late 1980s, have continued, with a 41% decline between 1994 and 2007 in BBS squares (Risely *et al.* 2008).

The causes of the recent decline of the linnet population in Britain have been studied using both demographic and ecological approaches. Siriwardena *et al.* (1999, 2000b) examined national trends in breeding success and survival rates using the BTO's nest record card and ring-recovery data sets. Moorcroft (2000) undertook detailed field investigations of the species' breeding and wintering ecology on Oxfordshire farmland. Changes in annual survival rates of linnets estimated from ring-recovery data were not correlated to changes in population trend (Siriwardena *et al.* 1999), so it seems unlikely that any major increase in mortality rate has driven the decline. Analyses of nest record cards, however, did indicate that nest survival rates at the egg stage were lower (64–67%) during periods of population decline than before or since (70–78%) (Siriwardena *et al.* 2000b). The Oxfordshire study found a marked seasonal trend in egg-stage nest survival rates which rose from just 47% in April, when many clutches were taken by carrion crows and magpies, to 86% at the end of the season in August (Moorcroft & Wilson 2000). Whether this seasonal decline in predation reflects better concealment of nests by vegetation or a change in behaviour on the part of the corvids remains unknown. Conversely, nestling survival rates remained high and almost unchanged through the season, with 82% of nests that hatched eggs going on to fledge young, and only 10% of nests experiencing deaths of individual chicks that may have been attributable to starvation. Overall, mean egg-stage nest survival rates were lower than those from nest record scheme analyses (63%), but with, on average, 4.2 nestlings fledged per successful nesting attempt, and with females making two or three nesting attempts in one year, nesting productivity was almost certainly sufficient to maintain a stable or increasing population.

All in all, it seems that Oxfordshire farmland provides good conditions for breeding linnets, but that the ability to raise second and third broods later in the breeding season, when predation impacts are lowest, may be essential in sustaining population levels. In fact, study of the diet of nestling linnets was to reveal more important clues about the ability of parents to rear these later broods, and provided the key to understanding the probable cause of population declines.

In both full-grown and nestling cardueline finches, including linnets, seeds are stored in gullet pouches covered by transparent skin. The proportion of the total seed volume made up by different seed types can thus be measured to provide a simple assessment of overall diet composition. This technique was adopted by Ian Newton, also in Oxfordshire, in studies in 1962–64 (Newton 1967), and provided a rare opportunity for historical comparison. The results revealed that nestling diet was now much less varied than it was in the 1960s, and that some seeds characteristic of less intensive farming environments (e.g. charlock, common sorrel and meadow buttercup) had disappeared from the diet to be replaced by seeds of plants known to thrive on intensively managed farmland (e.g. dandelion and sow-thistles) (Moorcroft *et al.* 2006) (Fig. 8.17). Most striking of all, however, was the dominance of the seeds of oilseed rape in nestling diet. These seeds are fed unripe to the nestlings and are easily identifiable in their gullet pouches. They first become available when the seed of autumn-sown rape crops begins to form in June, and continue to be available until the end of the breeding season in August, by which time large flocks of adults and fledged young may be seen feeding in rape fields.

Collection of similar data in Norfolk in 1999 tested the importance of oilseed rape in fuelling the later nesting attempts of farmland linnets. The agriculture of

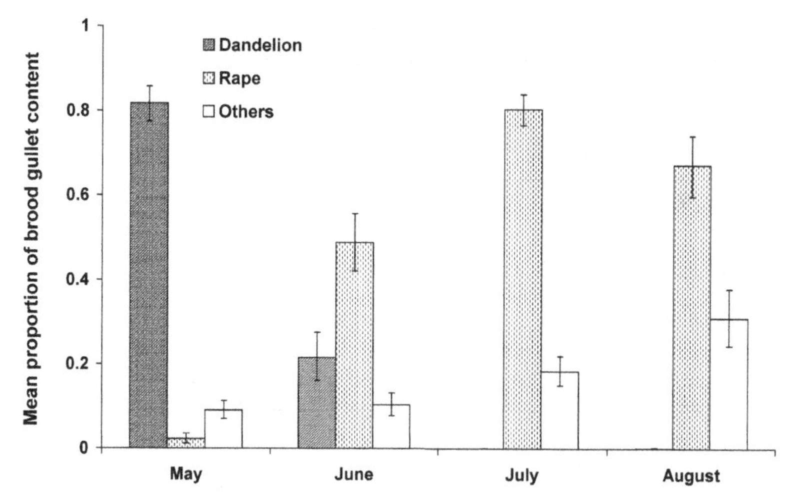

Fig. 8.17 Seasonal change in diet composition of nestling linnets (pooled across study sites) throughout the breeding season in central England (1996–99). Each food type is expressed as the mean proportion of brood gullet content over all broods in that month (May, $n = 87$; June, $n = 81$; July, $n = 76$; August, $n = 36$). Means and standard errors are calculated from arcsine transformations of the raw proportions.

Norfolk is intensive and arable, with fewer pastures (and hence fewer dandelions) than Oxfordshire, and oilseed rape is largely replaced as a break crop in arable rotations by sugar beet which does not provide a nestling seed source for linnets. If the availability of breeding-season seed sources is critical to the persistence of high densities of successfully breeding linnets, then the prediction was that the breeding success of linnets on farmland in Norfolk should be lower, with the birds perhaps present at lower densities than in Oxfordshire. The data collected in Norfolk in 1999 provided a striking confirmation of these predictions, with lower breeding densities, higher nest failure rates, a much higher incidence of nestling starvation and a shorter breeding season than in Oxfordshire (Table 8.11).

The relative stability of the linnet population since the late 1980s compared with the previous 20 years may well reflect the fact that the increasing availability of oilseed rape as a food source has compensated for declines in arable and grassland weeds upon which the birds once depended. Certainly, in Oxfordshire, the availability of oilseed rape on non-organic farms ensured that there were no detectable differences in breeding density or reproductive success between organic and non-organic farms in Moorcroft's (2000) study. Until recently, most oilseed rape would not have been grown were it not for the substantial CAP subsidy that it attracted. As part of ongoing CAP reforms, those subsidies have recently been withdrawn, but oilseed rape has continued to be widely grown, not just because it has emerged as an invaluable break crop within arable rotations, but also because of its growing importance as a biomass fuel crop. Agri-environment management to restore at least some of the populations of arable and pasture weeds upon which linnets fed in the 1960s may not be essential

Table 8.11 *Differences in breeding density and key reproductive parameters of linnet populations in study areas in Norfolk and Oxfordshire in 1998–99*

Measure	Norfolk	Oxfordshire	Difference significant?
Pairs per 100 m of suitable field boundary	0.05	0.10–0.15	$p < 0.01$
Daily nest failure rate	0.062	0.021–0.035	$p < 0.05$
Percent broods starved	0.10	0.014	$p < 0.001$
Percent broods with partial losses	0.57	0.06	$p < 0.001$
First third of clutches begun	Early May	Late April	$p < 0.001$
Last third of clutches begun	Early July	Late July/early August	$p < 0.05$

Source: Data estimated from graphed data in Moorcroft (2000). Significance values from original statistical tests in Moorcroft (2000).

to the recovery of linnet populations in areas where oilseed rape continues to be abundantly grown, but will certainly provide some security against any future policy changes that reduce oilseed rape acreages. Moreover, linnet populations are not recovering in numbers from the declines of the 1970s and 1980s across much of Britain, so more needs to be done if this species is to be removed from the Red List of birds of greatest conservation concern. It is possible that the trend towards intensive, mechanical trimming of some hedgerows and neglect of others has made many hedgerows less proof against predators than they once were, especially early in the breeding season. Improvements in hedgerow management may therefore help by reducing high early-season losses of eggs to corvids. Encouraging in this respect is that most of the entrants to England's Entry Level Stewardship scheme take up the one or more of the hedgerow management options which ensure that hedgerows are cut no more frequently than once in every two years and that not all of the hedgerows on the farm are cut in the same year. Equally, management to restore over-winter seed sources through retention of over-winter stubbles or provision of wild bird cover crops (e.g. Stoate *et al.* 2003) may improve over-winter survival rates and help to compensate for reduced breeding success. In this context, the study in Oxfordshire suggested that densities of at least 250 seeds m^{-2} of the preferred weed seed food sources are most likely to attract winter flocks of linnets, especially where sparsely vegetated ground makes the seeds easily accessible to the foraging birds (Moorcroft *et al.* 2002) (Fig. 8.18).

Case study 15: Cirl bunting

The cirl bunting is, by nature, one of our least obtrusive farmland bird species. Although superficially similar to the yellowhammer, the males' plumage and song are quite distinct. In Britain, the cirl bunting is predominantly a bird of mixed livestock and arable farmland, often associated with villages or farmsteads. Nest

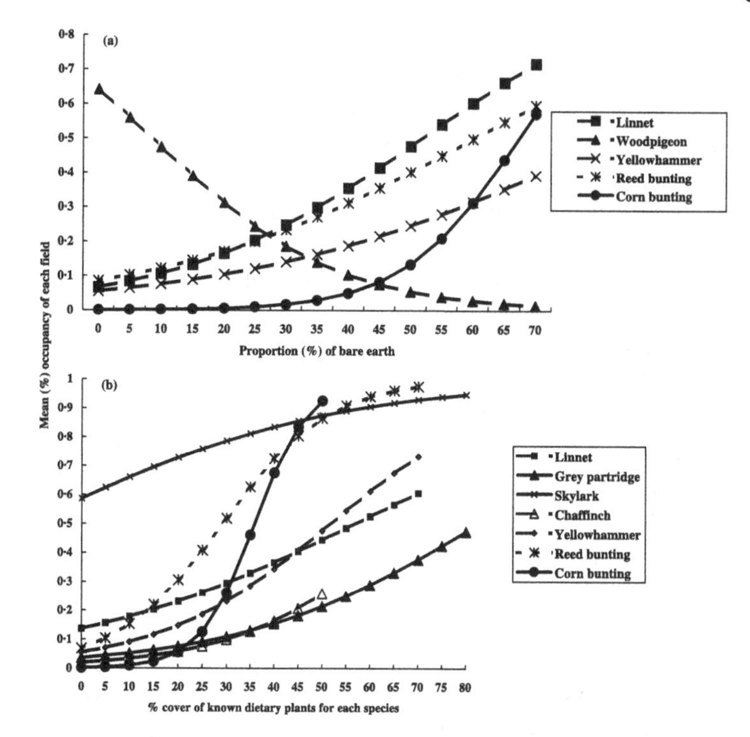

Fig. 8.18 The modelled effect of the percentage of (a) bare earth, and (b) cover of known food plants on probability of encountering seed-eating bird species on stubble fields in winter. Other predictor variables were held at mean values over the sample of 122 fields. See Moorcroft *et al.* (2002) for statistical details. Reproduced from Moorcroft *et al.* (2002) with kind permission.

sites are usually in dense hedgerows or scrub. The species is multi-brooded and, in common with its close relatives, tends to breed late in the season, with first eggs normally laid around mid May and breeding common until August, and even into September. Most clutches number three or four eggs. Nestling diet comprises a range of invertebrates, including caterpillars, spiders and ground beetles. As the season progresses grasshoppers and crickets become increasingly prominent in the diet, and during periods of rain, when invertebrates become less active and accessible, adults provide chicks with unripe 'milky' barley grain (Evans *et al.* 1997b). In October, family parties merge to form small wintering flocks. At this time of the year they are found almost exclusively on weedy stubble fields where they feed upon grain and the seeds of broadleaved weeds. They have also been recorded at cattle feeding stations and threshing yards and hay ricks. In recent years birds have been increasingly been taking advantage of garden bird feeding stations, and the seed foods provided there.

Although not documented as a British breeding species until 1800, by 1930 the cirl bunting was probably locally common in southern England. In the latter half of the last century the population collapsed, however, and by 1989 numbered no more than 120 pairs, more or less confined to south Devon (Evans 1992). It had thus become one of our most threatened farmland birds.

In 1988, RSPB began a research programme to diagnose the causes of the decline. There was no shortage of theories to test. The three most often cited were climate change, interspecific competition with yellowhammers and the loss of hedgerow elm trees. However, an initial study of habitat selection in winter quickly found that at this time of the year the birds were virtually entirely restricted to stubble fields as a foraging habitat (Evans & Smith 1992). Moreover, broadleaved weeds were important too: the greater the density of dicotyledonous plants, the more cirl buntings used the stubble (Evans 1997). We have seen in Chapter 2 how the area of stubble fields dropped dramatically throughout the 1970s as autumn-sown barley replaced spring varieties. There is also evidence that the quality of the remaining stubbles as a seed source has also fallen as the use of herbicides in arable weed control has increased (Robinson & Sutherland 1999).

The RSPB study also examined breeding ecology. It was shown that late nests have a higher chance of producing fledglings than those initiated before 1 July (Evans *et al.* 1997b). Most losses of chicks were due to starvation, particularly in wet weather, and predation. Nests that suffered predation tended to have chicks in poorer condition than those that went on to fledge. Indeed, there was no difference in chick condition between those that starved and those that were predated. This might suggest a causal link in that chicks in poor condition are easier to locate, perhaps because they beg more loudly. This link between poor nestling condition, begging intensity and predation risk has been demonstrated experimentally across a range of species (Haskell 1994; Cotton *et al.* 1996; Leech & Leonard 1997; Dearborn 1999).

The high success rate of late nests was associated with a high incidence of grasshoppers in the diet of the chicks, suggesting that the availability of orthopteran prey might be especially important in ensuring high breeding success. Although population trends of grasshoppers in Britain are not systematically documented, their habitat requirements are well known. In general terms, they require grassland with a tussocky structure comprising areas of long and short vegetation and bare earth for oviposition. Such a sward is produced by low-intensity grazing, particularly by cattle on permanent grassland. We have seen how virtually all agricultural grassland is now agriculturally improved through applications of inorganic fertiliser, drainage and frequent cultivation and reseeding practices that have enabled stocking rates to soar. The resulting swards are generally unsuitable for orthopterans and their populations on farmland have undoubtedly declined (e.g. van Wingerden *et al.* 1992).

Colour-ringing has demonstrated that cirl buntings are highly sedentary in Britain, and rarely move more than 2 km between wintering and breeding sites, at least at the low densities present during the early 1990s (Evans 1997). In the breeding season, territory selection typically involves areas comprising both low-intensity grassland and arable cropping within a 250-m radius (Stevens *et al.* 2002). These facts suggest that, in Britain, the cirl bunting is dependent on a low-intensity mixed livestock and arable farming system.

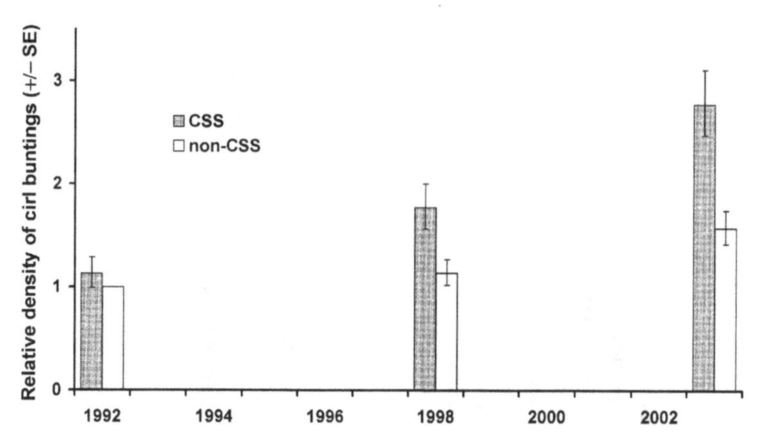

Fig. 8.19 Changes in numbers of cirl bunting territories recorded in 1992, 1998 and 2004 on land entering Countryside Stewardship Scheme (CSS) management agreements after 1992, and adjacent countryside which remained outside such agreements. Data from Wotton & Peach (2008).

As soon as the importance of stubble fields was realised, RSPB negotiated agreements with six farmers in key locations to leave stubble fields over winter, and four of these were used by up to 75 individual birds. In 1991, an agri-environment scheme for land outside ESAs, the Countryside Stewardship Scheme, was introduced in England. The South Hams of Devon was the target area, with the aim of reversion of the coastal arable areas to grassland. Had this objective been rigorously pursued, it would probably have led to the extinction of cirl buntings in Britain. Fortunately the staff of the then Countryside Commission who were piloting the scheme proved flexible and recognised the problem. Together with regional staff from the RSPB they developed a special project within the Countryside Stewardship Scheme which was launched in 1994. Farmers were therefore able to enter 10-year agreements to deploy management sympathetic to cirl buntings, including over-winter stubble fields following a reduced-input barley crop (Bradbury *et al.* 2008), grass margins around arable fields (to provide habitat for Orthoptera), and hedgerow and scrub management. In 1993 English Nature (now Natural England) began a Species Recovery Project for cirl buntings and joined RSPB to fund a project officer to oversee the targeting of special project agreements. The first stubble fields created under this scheme were in place by October 1992, and by 1999 120 10-year management schemes had been agreed creating 300 ha of over-winter stubble.

This strategy of deploying favourable management through an agri-environment scheme with an independent project officer to provide information to farmers, encourage them to apply for the scheme and help them with their applications has been highly successful. Between 1992 and 1998 the population on land under Countryside Stewardship Scheme agreements increased by 83%, but by only 2% on adjacent land outside agreements (Peach *et al.* 2001) (Fig. 8.19). In 1998 the

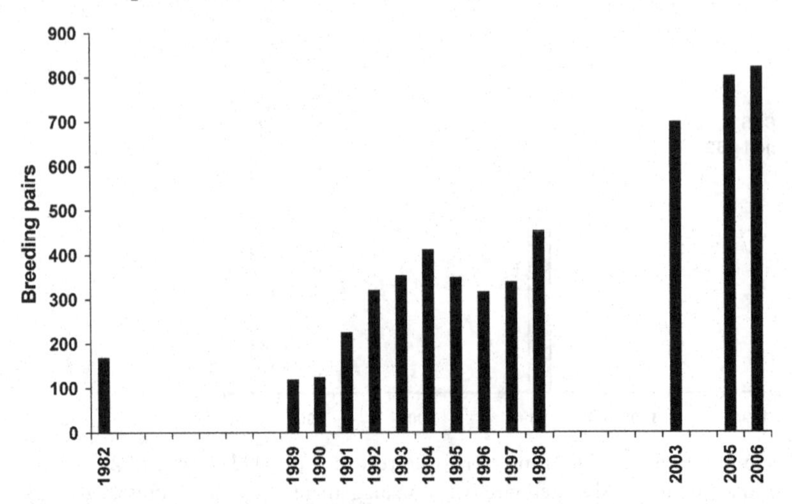

Fig. 8.20 Recent population trend of cirl buntings in the UK. Data from RSPB, Brown & Grice (2005) and Wotton *et al.* (2004).

population had recovered to 450 pairs (Wotton *et al.* 2000) and by 2003 had grown to an estimated 697 pairs (Wotton *et al.* 2004). The recent history of the cirl bunting population in Britain (Fig. 8.20) is one of the best examples of an agri-environment scheme targeted to the biological needs of species of critical conservation concern delivering its objectives (Evans *et al.* 2002). Countryside Stewardship has now been succeeded by Environmental Stewardship, the Higher Level component of which contains prescriptions for creation of suitable cirl bunting habitat. At the time of writing, not all farmers reaching the end of their 10-year Countryside Stewardship agreements have entered the new scheme. This is a cause for concern as 2006 and 2007 saw a net loss of management agreements for cirl buntings. In the medium term, the cirl bunting Biodiversity Action Plan contains aspirations to extend the species' geographical range beyond South Devon. Given the birds' poor dispersal capacity and large distances between areas with potentially suitable habitat, translocation is the best way of achieving these aspirations. In 2006 an ambitious project involving RSPB, Natural England and Paignton Zoo began to introduce cirl bunting fledglings which had been taken as chicks from South Devon and hand-reared, to a site in Cornwall. By May 2007, 26 birds of the 72 released were surviving and had formed at least nine pairs; encouraging progress given that this is the first time translocation of a passerine has been attempted in Europe.

Case study 16: Corn bunting

At the start of the twentieth century, the persistent jangling of singing male corn buntings would have been a familiar summer sound wherever cereals were grown. In some areas, they were considered pests for their habit of stripping thatch from corn ricks over winter, and there are several accounts of them being shot or trapped for the table (Shrubb 1997). The species bred from Shetland to the Isles of Scilly, and was absent only from high, inland, pasture-dominated ground. In Shetland, the species then reached the northern limit of its world range, yet was still sufficiently common to merit local names – 'docken sparrow' and 'cornbill' – and was found on crofting land even on the smallest, outlying Scottish islands, including St Kilda, Foula and the Out Skerries (Baxter & Rintoul 1953). Just 80 years later, the species was extinct on Shetland; a story of rapid and catastrophic decline that has been repeated throughout Britain during the twentieth century (Donald *et al.* 1994). By the turn of the millennium, corn buntings were effectively extinct in Wales, confined (barring a handful of birds) to two areas of Scotland (the coastal lowlands from Fife to Inverness, and the Western Isles), and were much reduced in both range and numbers across England. Nor are the declines confined to Britain. Similar declines and extinctions have been recorded in Ireland, Sweden (Jonsson 1992), Germany (von Busche 1989; Eislöffel 1997) and the Netherlands (Hustings 1997).

In Britain, Donald *et al.* (1994) identified two main phases of decline; the first during the 1920s and 1930s probably reflected the effects of the Great Depression (Chapter 1). Progressive abandonment of arable agriculture and conversion to rough pasture occurred throughout the period between 1870 and the Second World War, as the repeal of the Corn Laws in 1846 exposed British farmers to increasing volumes of imported grain and eventually caused a collapse in the wheat price (Shrubb 1997, 2003). Populations then recovered in some areas during the following three decades, as the Second World War resulted in the cultivation of almost 2 million hectares of grassland, and the restoration of protective farming encouraged the revival of ley-based mixed farming systems. The second phase of decline began in the 1970s and has continued ever since. The BTO's CBC revealed a decline of 86% between 1967 and 2005. A further 36% decline has been recorded by the BBS between 1994 and 2007 (Table 5.2). In 1993, a national survey coordinated by the BTO estimated the British population at 20 000 territories (Donald & Evans 1995). Taking this figure in combination with population trend information recorded before and after the date, the implication is that Britain lost over 50 000 corn bunting territories in the 20 years up to 1993 (Donald 1997), and has lost a further 10 000 or so since then. The population now stands at around 10 000 territories (Table 5.2), with 800–1000 of these in Scotland (Forrester *et al.* 2007).

Corn buntings nest on or close to the ground in dense vegetation. In Britain, fields of well-grown cereal or forage grasses are preferred (e.g. Brickle *et al.* 2000), although there are records of populations with other nesting habits. For example, in Cornwall, Ryves & Ryves (1934) found 48 of 54 nests in gorse or bramble bushes. Very recently, new research on the remnant Cornish population has found birds continuing to nest in low, tightly wind-clipped bushes of this kind (R. Setchfield pers. comm.). In the Outer Hebrides, 80% of nests are found in dune grassland, sheltered under clumps of hogweed (Hartley *et al.* 1995). In both of these situations, cereals were usually late-sown and most fields may simply not have provided sufficient cover at the time

Table 8.12 *Proportion of female corn buntings laying second clutches after fledging their first brood*

Study	Nests	Percent
Walpole-Bond (1938)	?	'common'
Ryves & Ryves (1934)	14	35.7
Thompson & Gribbin (1986)	27	7.4
Hartley & Shepherd (1994)	211	6.5
Brickle (1998)	120	8

nesting began. These considerations may also explain the often-observed but poorly understood preference for barley crops, since barley crops typically provide a denser 'field layer' of tillered shoots and leaves close to ground level than do wheat or oats. In comparison to most resident British breeding birds, the breeding season of corn buntings begins exceptionally late. In contrast to yellowhammers and reed buntings, few corn buntings lay eggs in May, particularly in northern areas, and the great majority of first clutches are laid in June (Yom-Tov 1992). Three or four egg clutches are the norm in Britain. Males are frequently polygynous, usually attracting two or three females, exceptionally more (Hartley *et al.* 1995). Most studies of corn buntings suggest that re-laying after nest failure is common, but although historical studies suggest that genuine second or even third broods were relatively common, all recent studies suggest that only a small minority of females lay second clutches after fledging a first brood (Table 8.12).

Outside the breeding season corn buntings are almost entirely granivorous and, like other granivores, flock on seed-rich habitats such over-winter stubbles. They have a rather more specialised diet than many other buntings, finches and sparrows, with grain predominant (Perkins *et al.* 2007), although seeds of other grasses and broadleaved agricultural weeds are also taken (Wilson *et al.* 1996a; Brickle 1998; Brickle & Harper 2000). Improved grasslands are usually avoided as winter foraging habitats due to their lack of seed sources although, intriguingly Mason & Macdonald (2000) noted winter use of coastal conservation grassland by corn buntings in Essex. This grassland had simply been mown in late July and it may be that subsequent growth and seeding of grasses and other wild plants had provided a useful seed source.

As with so many birds characteristic of lowland farmland, the decline has been coincident with a period of rapid, multivariate agricultural intensification, and it seems certain that at least some aspects of these changes have been responsible for driving recent declines. Until recently, a clear explanation of the fate of Britain's corn bunting population has been elusive. However, in March 1995, a two-day conference focused on the ecology and conservation of corn buntings brought together ornithologists from across Europe to review the level of understanding of causes of decline and make recommendations for future research, monitoring and conservation action (Donald & Aebischer 1997). Building on this impetus, Brickle (1998) studied the ecology of corn buntings breeding on the Sussex Downs, specifically with a view to understanding the ecological needs of the species, and the impacts of agricultural change on its decline. More recently, studies have begun to investigate causes of decline in the remaining populations in Scotland (J. D. Wilson *et al.* 2007a, b) and in Cornwall. How far has this research taken us towards understanding the reasons behind a century of population decline, and whether that trend can be reversed before this species faces the threat of extinction in Britain?

The specialisation of agricultural systems and the loss of cereal production from many northern and western farming and crofting areas is undoubtedly the main reason for the catastrophic range contraction of corn buntings through the twentieth century. On the Hebridean machair, a remnant population of just over 100 breeding corn bunting territories persists on those few islands where spring cereals are still grown as feed for out-wintered cattle. Even here, continuing population decline has been caused by an increasing trend for early harvesting of cereals – before full grain ripening – as arable silage and its storage in black plastic bales rather than traditional corn stacks. This effectively removes the key winter food source for the birds (J. D. Wilson *et al.* 2007b).

More widely across lowland Britain, the trend towards autumn sowing of cereals coupled with more powerful machinery has allowed post-harvest cultivations to be completed much more efficiently than in the past, and dramatically reduced the area of post-harvest stubble that persists through the winter. Much of the stubble that remains is sprayed after harvest with a broad-spectrum herbicide such as glyphosate and hence regenerates little weed cover that can provide a seed source over winter. At the same time, newly drilled spring cereals which provide an abundant food source at a time when other grain sources may have become exhausted, have also become less frequent. It may be no coincidence that light, easily worked and free-draining soils (e.g. those based on chalk, peat or sand) where spring sowing tends to persist are usually the areas where corn bunting populations have also persisted most successfully (Gibbons *et al.* 1993; Shrubb 1997). Unprotected grain stores in farmyards are virtually a thing of the past, stooks and rickyards disappeared with the advent of the combine harvester, and feed stands associated with out-wintered cattle have become scarcer. As Shrubb (1997) states, 'the declining availability of grain and seeds has been a progressive process since the 1830s'. This process has almost certainly contributed to the declines that have been seen even in areas retaining or specialising in arable farming because, outside the breeding season, dependence on grain and the stubbles, stackyards, cattle feed stands and freshly drilled cereal fields from which it could be obtained at different times of the winter (Donald & Evans 1994; Wilson *et al.* 1996b; Shrubb 1997; Brickle & Harper 2000) has made corn buntings highly susceptible to changes in agricultural practice that have deprived the birds of these feeding opportunities. Only the advent of rotational set-aside stubbles in the early 1990s has provided any counterbalance to the loss of seed sources from arable farmland. These stubbles are certainly exploited by corn buntings (e.g. Wilson *et al.* 1996b) and may have made some contribution to the slower rate of decline that was recorded during the 1990s.

In the breeding season, the rarity of second broods in recent years may at least partly explain declines in some populations. Given that the majority of corn buntings in Britain nest in cereal crops, and given that earlier sowing has advanced harvesting dates by up to a month (Donald 1997), the opportunity for successful late nesting in cereal crops has certainly diminished. So too in crops of forage grass; Shrubb (1997) reports that hay fields were the most frequently reported corn bunting nest site in the nineteenth century, perhaps because many were found by farmhands when these fields were cut from June onwards. Corn bunting nest losses to the hay harvest must always have been high, but with modern silage being cut as early as mid May, and with further cuts thereafter, these habitats are now as inhospitable for corn buntings as they have proved to be for corncrakes. In the south of England, the first silage cut is now so early that most corn buntings will not have started nesting, perhaps explaining why populations here are so concentrated in cereal crops. In north-east Scotland, however, where the first silage cut is later, a study found in 2004–05 that almost 30% of first nesting attempts were in silage, even though silage occupied only 15% of the total field area. Over 90% of these nests failed to fledge young, almost entirely as a consequence of nest destruction by silage cutting (J. D. Wilson *et al.* 2007a). An analysis of BTO nest record cards showed that nest losses to agricultural operations were three times as frequent after 1970 than before (Crick *et al.* 1994), and Brickle's study of corn buntings on the Sussex Downs showed that cereal harvesting caused high nest losses later in the season (Brickle & Harper 2002).

Changes in harvesting practice are not the only challenge that modern agriculture poses for nesting corn buntings. Brickle's Sussex Downs study found that both the probability of nestling survival and the condition of surviving nestlings fell with reduced abundance of key chick-food invertebrates in the vicinity of the nest, and that insecticide usage was a significant cause of reduced food abundance around nests in cereal fields. The facts that parent birds foraged further from the nest and that foraging trips were longer in duration when insecticides had been used, shed light on the probable cause of poorer nestling condition and survival (Brickle *et al.* 2000). The diet of nestling corn buntings consists primarily of invertebrates such as spiders, harvestmen, grasshoppers, caterpillars, sawfly larvae and adult beetles (Brickle & Harper 1999). These invertebrate groups are all themselves known to be sensitive to a variety of agricultural practices including insecticide application (directly toxic), herbicide and fertiliser application (removing food plants) and cultivation (removing food plants and egg-laying and larval habitats) (Wilson *et al.* 1999). The preference of corn buntings for foraging in uncultivated grassy field margins, unimproved grassland, set-aside and spring-sown cereals, as opposed to autumn-sown cereals and improved grassland (Brickle & Harper 2000) accords with the likely impact of agricultural intensification on their food supply. However, invertebrates do not constitute the only food source for nestlings. Many studies of nesting corn buntings (e.g. Watson 1992; Brickle & Harper 1999), though not all (Hartley & Quicke

1994), have observed that milky ripe cereal grain is also fed to nestlings, especially during wet or cold weather when invertebrate food may be harder to find. This habit has also been observed in cirl buntings and it is possible, as Brickle & Harper (2002) suggest, that the late onset of the breeding season of these two species may be connected with the need for ripening grain as a back-up food source. Barley crops may also be selected for nesting for this reason, as the first of the two main cereal crops to begin to ripen.

In summary, it seems likely that the severity of the corn bunting's decline in Britain stems from two facets of its ecology in particular. First, its specialisation on grain and other large seeds over winter has made it uniquely susceptible to multiple agricultural changes that have removed those food sources. This may be a particular problem for a relatively large seed-eating bird (40–60 g in weight) that is generally unable to feed from standing plants and relies on seed or grain lying on the ground, unobstructed by vegetation or snow cover. Second, its late breeding season and habit of nesting in crops of cereals or forage grasses means that in many areas harvesting dates preclude the rearing of second broods and, indeed, replacements for failed first nesting attempts. It is difficult to know which of these factors currently acts as the greatest constraint on corn bunting populations, but the same agricultural trend towards earlier sowing and earlier harvesting of cereal and grass crops underlies both.

If corn buntings are to have a future in mainland Britain, then it is likely to be on farms on light or peaty soils, supporting some spring-sown cereals (harvested as late as possible to provide for later nesting attempts) and late-cut hay or silage, grown with minimal pesticide inputs. Wide, uncultivated grass margins or central 'beetle banks' should be present to provide a rich source of invertebrate food for nestlings. Outside the breeding season unsprayed winter stubbles must be retained to provide grain and weed seed sources, preferably supplemented by strips of a cereal-rich wild bird cover to ensure a grain supply throughout the winter. In the remnant, dune-nesting breeding populations of corn buntings on the machair of the Western Isles it seems unlikely that breeding performance is being constrained by agricultural practice; here it is simply necessary to ensure that at least some cereals continue to be harvested ripe so that their grain can continue to provide a food source for corn buntings when fed to cattle over the course of the winter.

All of these management options are available in agri-environment and other management schemes throughout Britain. It remains to be seen whether they can be provided in combination, and at a large enough scale within agricultural landscapes to secure the future of this species within its current range, and to begin the long process of restoring it to the many areas from which it has been lost. Recent increases in Danish corn bunting populations in mixed farming areas with extensive cultivation of spring barley show what is possible (Fox & Heldbjerg 2008). As we write, a monitoring study being undertaken by RSPB is testing whether corn bunting populations in their remaining breeding range in north-east Scotland are responding positively to the provision of the management options outlined above as part of Scotland's Rural Stewardship Scheme. Preliminary data from monitoring

Table 8.13 *Summary of main agricultural causes of population change in the 16 case studies*

Species	Main agricultural causes of population change
Wintering geese	Increases permitted by intensification of grassland management and exploitation of arable crops as food
Sparrowhawk	Decline caused by direct toxic impact of bioaccumulated pesticide residues, with population recovery following withdrawal of these compounds from use
Black grouse	Decline caused by erosion of fine-scale habitat mosaics by grassland improvement, drainage and increasing grazing pressure, all leading to reduced invertebrate availability and increased predation risk. Afforestation of open moorland with exotic conifers created habitat in the short term but black grouse are excluded from plantations as the canopy closes
Grey partridge	Decline caused by indirect effects of pesticides on chick survival through reduction in insect populations, and effects of changing hedgerow management on availability of nesting habitat
Corncrake	Long-term decline caused mainly by destruction of nests and young when forage grasses are cut early and repeatedly. Conservation measures to delay cutting of meadows and provide dense cover throughout the breeding season have resulted in a marked population increase in the remaining core range in Scotland
Stone curlew	Long-term decline caused by loss of semi-natural grasslands and heathlands to intensive agriculture and forestry, compounded by reduction in area of spring-sown crops as alternative nesting habitat, and loss of nests on spring-sown arable fields to agricultural operations. Recent recovery permitted by nest protection and creation of suitable nesting habitat through agri-environment schemes and set-aside
Lapwing	Decline caused by the combined effects on availability of nest sites and nesting success of grassland intensification, loss of spring cropping and loss of mixed farming
Golden plover	No large-scale population trends known at the national scale, but local declines caused by effects of intensive sheep grazing and drainage on moorland vegetation mosaics, loss of habitat by conifer afforestation and effects of increasing predator populations adjacent to plantations
Woodpigeon	Earlier decline driven by loss of clover-based leys as a winter food source, but with subsequent increase permitted by exploitation of oilseed rape as an alternative winter food
Turtle dove	Decline driven by the combined effects of herbicide use, seed cleaning and efficient harvesting on the availability of seed supplies causing reduced breeding output
Skylark	Decline caused by loss of nesting and foraging opportunities in dense autumn-sown crops coupled with high rates of nest loss in intensive grass silage systems

(*cont.*)

Table 8.13 (*cont.*)

Species	Main agricultural causes of population change
Song thrush	Decline caused by loss of grassland in intensive arable areas, coupled with grassland intensification – notably field drainage – leading to reduction in soil invertebrate food supply and limitation of breeding productivity and post-fledging survival
Chough	Causes of long-term population fluctuations not fully understood, but both intensification and abandonment of grazing management of coastal grasslands lead to declines, through effects on the abundance and availability of soil invertebrates
Linnet	Long-term decline caused by effects of herbicides on weed seed availability, especially where tillage dominates; maintenance of numbers locally, probably aided by exploitation of oilseed rape
Cirl bunting	Decline caused by effects of herbicide use and loss of spring-cropping on availability of weedy, over-winter stubbles, combined with impacts of grassland intensification on availability of chick foods. Recent increase caused by reversal of these trends within agri-environment schemes
Corn bunting	Decline caused by same factors as for cirl bunting, but compounded by effects of earlier cereal harvesting and repeated silage cutting on nest success

of a smaller-scale scheme with similar management options, and focused on farms with breeding corn buntings – 'Farmland Bird Lifeline' – suggests that declines may at least have been stemmed (Perkins *et al.* 2008b).

Summary

The above case studies show that many aspects of agricultural change have played their part in bringing about declines (and in some cases increases) of populations of farmland birds (Table 8.13).

It is difficult to disentangle the effects of individual changes in agricultural practice on birds simply because farming practices are interdependent. For example, if a farmer chooses to increase the proportion of his cereal crop sown in the autumn, this may have knock-on effects on other parts of his crop rotation, on the types, amounts and application times of pesticides that he uses, and the probability that he will leave fields as stubble over the winter. Formal experiments are the ideal solution to this problem; holding other influential conditions constant whilst varying an experimental treatment of interest (e.g. pesticide application) to known degrees. The problem of course is that in order to measure the responses of wild birds to experimental treatments it is necessary to establish experiments at the scale of multiple fields or farms, and far more often than not, the costs and logistical complications of experimentation on this scale are prohibitive. In reality, most studies of the impact of farming practices on birds have sought to measure the strength of association between bird 'responses' (e.g. in territory location, nest success or foraging behaviour) and 'real-world' variation in farming practices. Chapter 9 moves on to review these studies from the perspective of some of the most influential changes in agricultural practice.

Plate 8.1 Male black grouse at the lek. This bird of 'edge', 'fringes' and 'early successional stages' continues to pose challenges to those seeking to understand how best to manage landscape-scale habitat mosaics of moorland, grazing and forestry for this species at the upland edge. Recent reversals of long-term population declines in the North Pennines and Wales give some cause for optimism. © RSPB.

Plate 8.2 Stone curlew attending its clutch, illustrating the very sparsely vegetated or bare ground on light chalk or sandy soils on which this species prefers to nest. Grazing abandonment on much downland and lowland heathland coupled with the switch from spring to autumn sowing of arable crops on nearby farmland robbed this species of much of its nesting habitat and caused a long-term decline. Protection of nests from agricultural operations in spring crops, restoration of grazing management and creation of fallow plots for nesting stone curlews in arable fields have combined to reverse the decline, although the population remains highly localised and fragmented. © RSPB.

Plate 8.3 Research has shown that the decline of the skylark on lowland farmland has been driven mainly by limitations to breeding output caused by dense, fast-growing winter cereal crops and loss of nests to silage cutting. This is one of the species that has benefitted most from set-aside, with high densities on many suitably vegetated fields. © RSPB.

Plate 8.4 Perhaps the classic example of a bird associated with low intensity pastoral agriculture throughout its European range, in Britain the chough is most closely associated with close-grazed, unimproved coastal grasslands from Cornwall to the Argyll Islands. Local feeding specialisations characterise different populations. Here a flock, including one colour-ringed bird, probes for sandhoppers on a beach on Oronsay. © RSPB.

9 · *Studies of changing agricultural practice*

In this chapter we combine evidence from both experimental and correlational studies, including the above species case studies, to synthesise what is known of bird responses to some of the most important elements of agricultural intensification. In a recent review of the main drivers of decline of lowland farmland bird populations in Britain, Newton (2004) identified herbicidal weed control, changes in timing of cereal sowing and harvesting, ploughing of winter stubbles, land drainage and associated intensification of grassland management, and increasing livestock densities as the main causes of loss of nesting and feeding resources. The structure of our discussion follows Newton's closely, although we consider the impacts of grazing on enclosed and upland grasslands separately, and also consider the impacts of changing management of field boundaries, especially hedgerows. Lastly, we also consider how agricultural change may have affected the susceptibility of some species to increased predation impacts and whether any such increases have contributed to population declines. This chapter is thus structured under the following headings:

(i) changing cropping cycles,
(ii) pesticide use,
(iii) agricultural improvement of enclosed grasslands,
(iv) upland grazing,
(v) field boundary management,
(vi) interactions between predation and agricultural management.

Changing cropping cycles

This heading subsumes several major changes in the annual cycle of management of tillage crops between sowing and harvest, and changes in tillage crop rotations between years which, either singly or in combination, have had dramatic effects on the value of tilled fields as nesting and foraging habitats for farmland birds.

The single most important change has been the development of hardy, high-yielding, autumn-sown cereal varieties, coupled with the development of herbicides specific to grass weeds (graminicides). Together, these have allowed autumn sowing of cereal crops to supplant spring sowing over large areas. Autumn-sown wheat and barley crops now dominate the arable landscape of Britain, especially on heavy soils where spring cultivation is difficult. The exception is where demand for low-nitrogen malting barley on light soils (e.g. in parts of eastern Scotland) maintains high levels of spring barley cultivation. This early cultivation and sowing after harvest of the preceding crop has brought with it a number of other changes. First, fewer crop stubbles are left into the autumn and winter. Second, the crop is in the ground for a longer period and is often subject to heavier overall pesticide inputs than would be the case for a spring-sown crop. Third, the crop has a head start at the beginning of the next growing season, so that there is rapid crop growth early in spring, leading to an earlier harvest. Recent climate warming, with a trend towards much milder winters, is almost certainly exacerbating this effect. In the south of England, autumn-sown barley is harvested in July, and wheat in early August, whereas harvesting of spring cereals may only start in August and continue into September. This relatively simple change in sowing time of our main cereal crops has had important impacts.

First, crop stubbles rich in seeding weeds and waste grain are known to support high densities of seed-eating birds (e.g. Wilson *et al.* 1996a; Buckingham *et al.* 1999; Mason & Macdonald 1999a; Hancock & Wilson 2003; Gillings *et al.* (2008) and see Chapter 6). The combination of better field drainage, more powerful cultivation machinery and the need to sow modern crops in autumn has greatly reduced the area of stubble persisting through the winter. The practice of post-harvest spraying of stubbles with broad-spectrum herbicides such as glyphosate, coupled with longer-term effects of herbicide use on arable seed banks, and more efficient harvesting has reduced the availability of seed food on much of what remains (Robinson 2001). It seems very likely that this widespread reduction in seed availability on arable farmland has been a main cause of decline of several seed-eating species. Peach *et al.* (2001) found that deployment of unsprayed stubbles under Countryside Stewardship had an independent positive effect on the cirl bunting population. An experimental study showed that over-winter survival rates of house sparrows could be increased from 39% to 65% by supplementary feeding of cereal grain at a farm where the population had fallen by 80% in the previous 30 years, whilst survival rates were unaffected by supplementary feeding on three other farms whose populations had remained stable (Hole *et al.* 2002).

At a much larger scale, the BTO examined the breeding population changes of a range of common farmland birds between 1994 and 2003 across more than 600 1-km squares in lowland Britain, and related these to the area of over-winter stubble. For both skylark and yellowhammer they found that populations were relatively stable in squares that held a large area of stubble (>20 hectares) compared to squares with smaller areas of stubble where large declines were evident (Gillings *et al.* 2005b) (Fig. 9.1). These results should be treated with some caution, however, since a range of other declining, seed-eating species (e.g. grey partridge, tree sparrow, linnet, reed bunting and corn bunting) did not show similar effects. Indeed, unexpectedly, tree sparrow populations showed more positive trends in squares that did not hold winter stubble than those that did.

Second, a number of ground-nesting species that require open, sparsely vegetated ground on which to nest (notably lapwing, stone curlew and skylark) are able to nest successfully in spring-sown cereals. In autumn-sown fields, crops grow so rapidly that nesting is either limited mainly to the early part of the breeding season (skylark) or frequently prevented altogether (lapwing and stone curlew) (J. D. Wilson *et al.* 1997, and see case studies above). The best-studied example is certainly that of the skylark where research has shown both that vegetation height and density limit skylark territory settlement, breeding success and foraging in cereal crops (e.g. Odderskær *et al.* 1997a; J. D. Wilson *et al.* 1997; Donald 2004). Experimental intervention has shown that creating unsown patches within modern autumn-sown wheat fields can

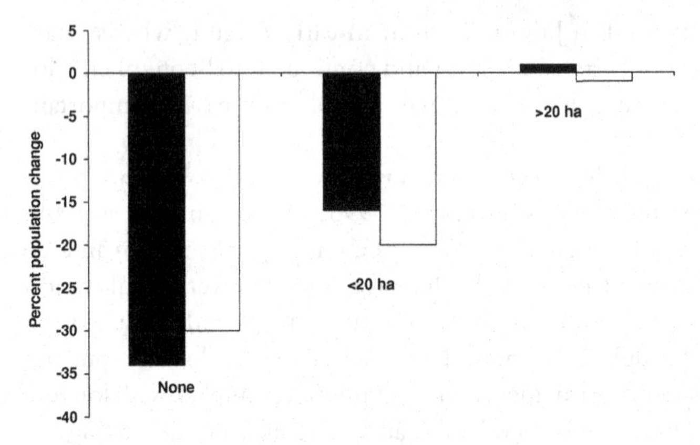

Fig. 9.1 Relationship between breeding population trend of skylarks (solid bars) and yellowhammers (open bars) between 1994 and 2003, and winter stubble availability per 1-km square. Data from Gillings *et al.* (2005b).

increase both skylark territory density and breeding success (Morris *et al.* 2004; and see skylark case study). Other studies suggest that similar processes may affect other species. For example, Morris *et al.* (2002a) found that yellowhammers foraging in cereal fields for invertebrate food for nestlings exploited crop patches that had shorter, sparser vegetation than those patches that were not visited. Even in crop stubbles, birds foraging for seed in winter show similar preference for areas where the stubble is open and sparse. Moorcroft *et al.* (2002) found that linnets, yellowhammers, reed buntings and corn buntings were all more likely to occur on stubble fields if a high density of preferred weed seed or grain was combined with a high proportion of bare earth (Fig. 8.18), and a follow-up experimental study by Butler *et al.* (2005a) showed that when stubble height was reduced from 14 cm to 6 cm, the probability of encountering both seed-eating and invertebrate-feeding passerine birds feeding on the stubbles increased. Additional experiments using captive birds in aviary conditions suggest that this preferential selection of more open sites with relatively short, sparse vegetation is likely to reflect both lower predation risk (i.e. a higher probability that the foraging bird will see an approaching predator and be able to take evasive action) as well as easier access to food (Butler & Gillings 2004; Whittingham *et al.* 2004; Butler *et al.* 2005b).

Third, early harvesting of autumn-sown crops may also cause problems for nesting birds at the other end of the breeding season. For example, both Montagu's harriers and corn buntings have relatively late breeding seasons, and nest on the ground in mature cereals. In Spain, Arroyo *et al.* (2002) found that cereal harvesting could destroy up to 60% of Montagu's harrier nestlings and lead to rapid population decline in the absence of conservation measures to protect individual nests or delay harvesting. Although their traditional nest sites are on heaths and wetlands, the arable fields around the Wash have supported a high proportion of Britain's breeding Montagu's harriers since the species resumed regular breeding here in the mid 1970s. The successful breeding of these birds has only been possible by locating nests and liaising with local farmers to ensure protection of nests from agricultural operations. This usually involves delaying the harvest, raising the combine blades over the nest or temporarily removing the young. There is good evidence of increasing rates of loss of corn bunting nests to agricultural operations in recent years, and the early harvesting of modern cereals, at least in southern England, also means that very little habitat is available for late nesting attempts once the cereal harvest is under way from early July onwards (see Crick *et al.* 1994; Brickle & Harper 2002; and corn bunting case study).

Not all of the major changes in cropping that have taken place during recent agricultural intensification have been wholly detrimental to birds. A good example is the arrival of oilseed rape as a major component of lowland arable systems in Britain. Almost unknown as a crop in Britain before the 1970s, its area peaked at over half a million hectares in the late 1990s as, supported by EU subsidy, it replaced ley grassland as a break crop in cereal rotations across large areas of arable farmland. The population effects of oilseed rape in providing a new winter food (its leaves) for woodpigeons, and in providing a crucial breeding season food source (its seeds) for breeding linnets and turtle doves in environments where traditional weed seed foods have become scarcer have been documented by the case study accounts. However, a wide variety of other bird species also exploit oilseed rape fields both for feeding and nesting (e.g. Stoate *et al.* 1998; Watson & Rae 1998; Burton *et al.* 1999; Proffitt 2002). This may be because the availability of rape seed is complemented by a relatively rich invertebrate fauna (e.g. Holland *et al.* 2002; Moreby & Southway 2002). Notably, Gruar *et al.* (2006) found that reed bunting territory densities in a Nottinghamshire study area were four times greater in fields of oilseed rape than cereals or set-aside and a study by Burton *et al.* (1999) found that adult reed buntings also carried out over 75% of their foraging to provision nestlings within the rape crop even though the crop itself represented under 40% of available foraging habitat. Not only has oilseed rape provided important seed resources for linnets and turtle doves at a time when agricultural intensification has diminished the availability of other foods, it has also provided important new nesting habitat for reed buntings when agricultural drainage has reduced the availability and quality of traditional marginal wetland habitats for this species on farmland. Not all is rosy, however. Replacement nesting attempts and second broods of reed buntings are still in the nest in July when most oilseed rape crops are either swathed prior to harvesting or left standing but sprayed with an herbicidal desiccant such as glyphosate. Burton *et al.* (1999) found that although no reed bunting nests survived the mechanical swathing of the crop, spraying did not destroy nests and appeared to have no effect on the subsequent survival of broods to fledging.

Pesticide use

The catastrophic environmental impact of bioaccumulation of persistent organochlorine pesticides during the 1950s and 1960s is certainly the most widely known example of the impact of industrial agriculture on wildlife and was brought to global attention by Rachel Carson's *Silent Spring*. In Britain, severe declines in populations of peregrine falcons (Ratcliffe 1980) and sparrowhawks (Newton 1986; and see case study) became a cause célèbre for conservation. The eventual withdrawal of these chemicals and subsequent recovery of raptor populations was one of the notable conservation success stories of the twentieth century. Nonetheless, the development and industrial application of pesticides has continued apace. Direct toxicity to vertebrates is now largely a thing of the past in Britain, except where products are used illegally or inappropriately (Burn 2000). However, farmers continue to have an array of products available to them which are highly effective in removing both grass and broadleaved agricultural weeds (herbicides), and invertebrate pests (insecticides and molluscicides). These range from products that are targeted to specific pests to others that have broad-spectrum effects and so kill a wide range of non-target species. For example, an application of glyphosate to a set-aside or stubble field kills virtually all growing plants.

From the perspective of birds, the use of these products (mostly on tillage crops but to some degree on grassland too), removes potential food supplies either by killing invertebrates or the plants whose seeds are exploited as food or by more complex food-chain effects that occur when herbicide application removes invertebrate food plants. These indirect effects were poorly understood for many years. Even as late as 1997, a major review for the Joint Nature Conservation Committee concluded that although there was good evidence of correlations between farmland bird declines and increasing pesticide use, only for grey partridge (see case study) was there incontrovertible evidence that pesticide use reduced food supply and nestling survival rate and led to reduced population density (Campbell *et al.* 1997). However, the same review also found that there was substantial evidence that known declines in the abundance and diversity of a wide range of invertebrate and plant food groups of importance in the diet of farmland birds had been caused by aspects of agricultural intensification including pesticide use (Wilson *et al.* 1999). There was, therefore, strong evidence that increasing pesticide use could be implicated in the declines of several other species.

One of these species was the yellowhammer, and during 1999–2003 further studies by the Central Science Laboratory, the GWCT and RSPB set out to assess in more detail the indirect effects of pesticides on the ecology and demography of this species. It was chosen because nestling diet is dominated by arthropods collected from crops as well as field margins (Morris *et al.* 2001), because the population has declined rapidly since the late 1980s on arable farmland (Kyrkos *et al.* 1998), and because a detailed study of breeding ecology in Oxfordshire suggested that current breeding productivity was too low to sustain populations on arable and mixed farmland (Bradbury *et al.* 2000). One part of the study examined correlations between insecticide applications to cereal fields and yellowhammer foraging habitat selection and breeding success across eight arable farms in Oxfordshire and Lincolnshire. This found that invertebrate abundance was lower in cereal fields sprayed with insecticide in summer and that breeding yellowhammers foraged for invertebrates less in such

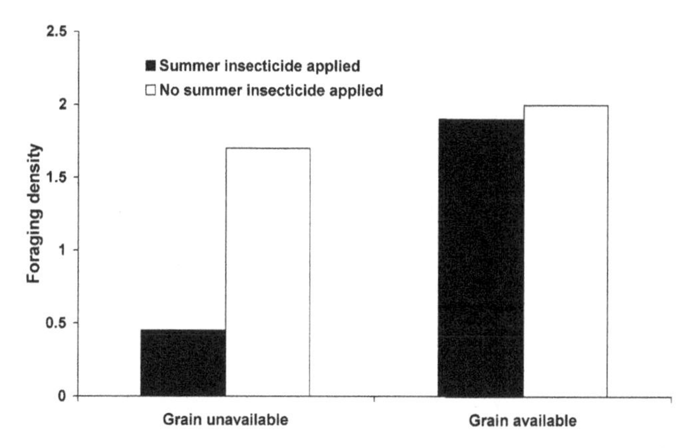

Fig. 9.2 Foraging density of yellowhammers in cereal fields with and without summer insecticide applications. The lack of effect of insecticides on foraging frequency once the cereal grain has begun to form reflects the fact that yellowhammers also visit the crop to collect grain from this point in the season, the availability of which is unaffected by insecticide application. Modified from Morris *et al.* (2002b, 2005) with kind permission.

fields than unsprayed fields or fields sprayed only in the previous autumn and winter (Fig. 9.2). Moreover condition of nestlings was reduced in nests adjacent to fields that received two or more insecticide applications compared to those that received none or just one. Although this did not lead to any detectable effect on starvation rates of nestlings (Morris *et al.* 2002b, 2005), fledglings in poorer condition may have reduced survival prospects. In the second part of the study, two of four 100-hectare blocks of arable farmland in North Yorkshire were subjected to an increase in insecticide applications above normal levels, from an average of 0.4 applications to each cereal crop in the two control blocks to 1.7 in the treated blocks. The probability of brood reduction was significantly increased at those nests where a higher proportion of the area enclosed by a 200-m foraging radius around the nest had been sprayed with insecticide within the previous 20 days (Boatman *et al.* 2004) (Fig. 9.3). This effect resulted from the reduction in abundance of arthropods important in the diet of nestling yellowhammers as a result of insecticide application, and consequent reduction in nestling condition (Hart *et al.* 2006).

The fact that a recent study of breeding corn buntings in Sussex found similar insecticide-induced reductions in abundance of chick food, nestling condition and nestling survival prospects (Brickle *et al.* 2000; and see case study) suggests that indirect effects of insecticide use on the breeding success of buntings on farmland may be general.

There have been no studies demonstrating the indirect effect of herbicide use on bird populations through reduction in seed availability. This may be because the many other causes of reduction in seed availability on farmland (e.g. autumn sowing, efficient harvesting, and ploughing of stubbles immediately afterwards, and bird- and rodent-proof grain storage) would make it extremely difficult to distinguish

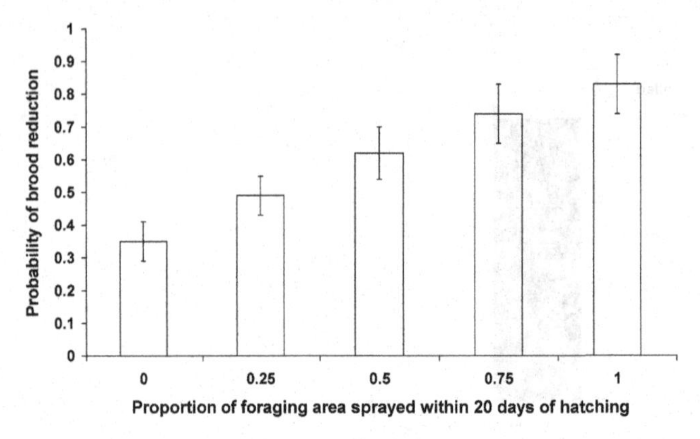

Fig. 9.3 Effect of summer insecticide spraying in a 200-m foraging radius around yellowhammer nests on probability of brood reduction. Data from Boatman *et al.* (2004) which see for further details.

the contribution of herbicides. Also, the fact that for most species the effects of herbicide-induced reduction in seed availability would be to reduce survival rates rather than nest success, makes these effects more difficult to detect. However, as Robinson & Sutherland (2002) showed (see Fig. 6.6), seed densities declined dramatically in arable soils in Britain over the course of the twentieth century, in no small part due to increasing use and efficacy of herbicides.

The development of genetically modified, herbicide-tolerant (GMHT) crops during the 1990s provided the potential to further increase the efficacy of herbicide treatments in reducing arable weed seed banks and, as a side effect, weed seed and invertebrate food sources for birds. By 1998, four such crops (sugar beet, maize and spring- and winter-sown varieties of oilseed rape) had been assessed as safe in terms of human health and direct environmental impact. In each case, the GMHT crop varieties incorporated genes from soil bacteria which conferred resistance to the herbicide glyphosate – a broad-spectrum, post-emergence product which controls a wide variety of broadleaved and grass weeds, but has no residual activity and has a low risk of toxicity to soil invertebrates and microbial communities (Firbank 2003). Glyphosate is widely used in conventional agriculture, for example to spray off vegetation cover at the end of a set-aside period, to prevent weed regeneration in crop stubbles prior to cultivation and sowing of the next crop, and to desiccate oilseed rape crops prior to harvest. Little was known of the effects of the herbicide management of these crops on farmland wildlife. However, modelling studies predicted that the enhanced ability to control broadleaved weeds using glyphosate in GMHT crops could accelerate rates of reduction in the populations and seed banks of a range of arable weeds and other organisms associated with them (e.g. Watkinson *et al.* 2000). Because of the intense public concern and debate surrounding the wider environmental impact of genetic modification in agriculture, the UK Government commissioned experimental Farm Scale Evaluations to test the effects of the introduction of GMHT crops on farmland wildlife. Farmland biodiversity (weed and invertebrate populations) was compared by splitting arable fields at over 60 sites and growing a conventionally managed crop in one half of the field and a GMHT crop in the other. This was repeated separately for crops of beet, maize and both spring- and winter-sown oilseed rape (Squire *et al.* 2003). The results for weed populations showed that overall seed densities in the seed bank were reduced by approximately 20% in GMHT beet and spring-sown rape, relative to conventionally managed crops, an effect that over time would cause further large decreases in arable weed population densities (Heard *et al.* 2003a, b). No similar effect was found for maize, and indeed there was weak evidence that seed banks might increase in these crops under GMHT management. Similarly, results for invertebrate populations found that GMHT management tended to reduce captures of soil-surface-active invertebrates, and some aerial groups such as butterflies and bees in spring-sown oilseed rape and beet, but increase counts of some groups – notably springtails (Collembola) – in maize and winter-sown oilseed rape (Brooks *et al.* 2003; Haughton *et al.* 2003; Bohan *et al.* 2005). The reason for the marked difference in outcomes between beet and oilseed rape on the one hand, and maize on the other, probably lies in the fact that conventional maize management relied on a battery of persistent and toxic pesticides such as atrazine (now banned in the EU) rather than suggesting that GMHT management of maize is unusually benign. A subsequent study took the Farm Scale Evaluation data on seed 'rain' (quantities of weed seed falling from growing plants on to the field surface), specifically focusing on weed species whose seeds are known to be important in the diet of farmland birds (Gibbons *et al.* 2006). For 16 of 17 seed-eating bird species, the rain of seeds important in the diet was lower in GMHT than in conventionally managed crops in spring-sown rape and beet, and the same was true for 10 of 17 species in winter-sown oilseed rape. Across all three crops, there was only one species for which rain of seeds important in the diet was greater under GMHT crop management. Again, results for maize were dissimilar with rain of seeds important in the diet higher under GMHT crop management for seven species and the reverse being the case for none. Overall, there seems little doubt that for oilseed rape and beet crops at least, the replacement of conventional crop varieties with GMHT management would have the potential to cause yet further substantial reductions in the availability of key seed and invertebrate foods for birds in arable systems, much as was predicted by Watkinson *et al.* (2000).

Agricultural improvement of enclosed grasslands

The impacts of the agricultural improvement of grassland on wildlife are just as profound as those wrought by the use of pesticides on tillage crops. Agricultural improvement entails the establishment and maintenance of a sward dominated by plants preferred as forage for grazing stock. Typically this involves sowing ryegrasses and clovers (and sometimes also cocksfoot, timothy, meadow fescue and tall fescue)

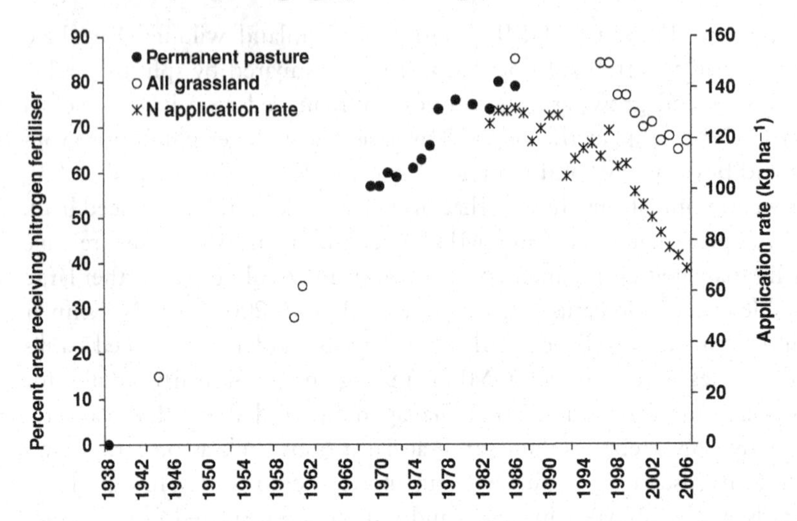

Fig. 9.4 Trends in nitrogen usage on agricultural grassland in England and Wales. Circles refer to the percentage of the total area dressed with nitrogen fertiliser. Stars show trend in application rate since the early 1980s. Falls in nitrogen usage since the mid-1990s reflect measures to reduce nitrate pollution to watercourses and other non-agricultural habitats. Data from Fuller (1987) and the British Survey of Fertiliser Practice (www.defra.gov.uk).

at the expense of bents, red fescue and a wide variety of broadleaved plants (Fuller 1987). Agricultural productivity of this grassland is then further increased through drainage, the use of herbicides to control perennial weeds at sward establishment and application of inorganic nitrogen, phosphate and potash fertilizers and organic slurries. As a consequence of these changes, farmers have been able to support higher stocking densities of cattle and sheep on pasture, and to replace a single cut of forage grass from hay meadows with multiple cuts of grass stored as silage (Vickery *et al.* 2001).

In the late 1930s, most grassland outside the arable rotation existed as permanent, semi-natural pasture, either genuinely old grassland, or land long reverted to grassland after the agricultural depression that had begun during the 1870s. It received little or no fertiliser (Green 1982; Fuller 1987). However, from the Second World War, agricultural improvement of grassland (as indexed in Fig. 9.4 by intensity of nitrogen fertiliser application) proceeded apace, with peak intensity of lowland grassland management being achieved during the late 1980s and early 1990s. Likewise, rates of grant-aided under-drainage of both arable and grassland fields also peaked in the 1970s and 1980s, leading to the agricultural landscapes of Britain becoming amongst the most intensively drained in Europe (Green 1979; Robinson & Armstrong 1988). By 1984, the then Nature Conservancy Council published a report concluding that this loss of semi-natural grassland on fertile, lowland soils was the biggest of all of Britain's losses of habitat of nature conservation interest in the previous 40–50 years (Nature Conservancy Council 1984). Of the

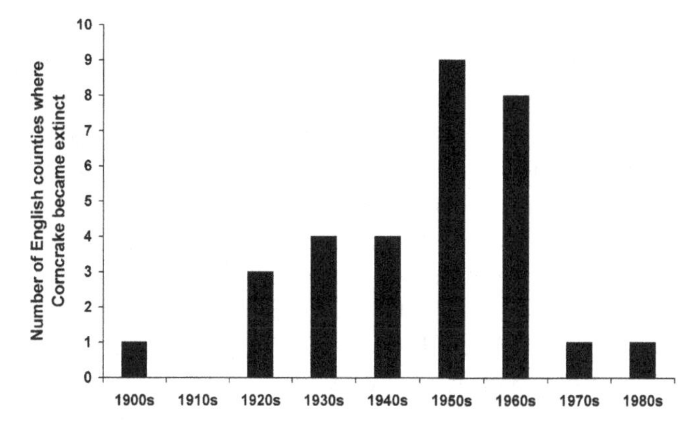

Fig. 9.5 Progressive extinction of breeding corncrake populations in English counties during the twentieth century. Data from Brown & Grice (2005).

area of unimproved lowland grassland that existed before the Second World War, just 3% remained unaffected by agricultural intensification by the mid 1980s.

Whether through ploughing and reseeding or the impacts of drainage and fertiliser application, the botanical losses caused by intensification of grassland management have been severe (e.g. Marrs 1993b; Kirkham *et al.* 1996), and in the *New Atlas of the British and Irish Flora*, Preston *et al.* (2002) regard the impoverishment of the grassland flora through intensive agricultural management or conversion to arable cultivation as being the most profound botanical change of the twentieth century. Even during the 1990s, the Countryside Survey continued to record reductions in grassland plant diversity in conditions of increasing nutrient availability (Haines-Young *et al.* 2000).

There are three major ways in which agricultural improvement of grassland in Britain has affected birds: first, the effect of change from single-cut hay to multiple-cut silage management on birds nesting in meadows (i.e. fields of forage grass); second, the effect of drainage and changes in grazing management on the value of grassland as a breeding habitat for waders; and third, the combined effects of the various components of grassland intensification on the abundance and availability of invertebrate and seed foods for birds foraging in agricultural grasslands.

From hay to silage: effects on meadow-nesting birds

The twentieth-century demise of the corncrake population in England and much of the rest of Britain (see case study in Chapter 8) is a spectacular example of the rapid and catastrophic effect of destruction of nests by earlier and more frequent mowing operations in grass fields. The progressive extinction of corncrakes in English counties matches very well the accelerating intensification of grassland management after the Second World War (Fig. 9.5). Similarly, although skylarks and corn buntings persist as a nesting species in silage meadows, nest loss rates to cutting can be very high (Jenny 1990a; Donald 2004; J. Wilson *et al.* 2007a).

Other, less well-studied species have probably been victims of the same changes. For example, widespread losses of breeding yellow wagtails from the upland hay meadows of the Yorkshire Dales and North Pennines have probably been caused by nest destruction as a result of earlier cutting associated with conversion to silage management (Smith & Jones 1991; Court *et al.* 2001). There is a marked association of the remaining breeding birds with meadows managed under the ESA scheme, with late cutting dates designed to conserve botanical diversity. The whinchat was widespread throughout lowland England during the first half of the twentieth century (Alexander & Lack 1944). By the early 1970s, however, the species had declined substantially over much of the Midlands (Parslow 1973) and by the time of the *New Atlas* (Gibbons *et al.* 1993) had, with the exception of a few lowland heathland and downland areas such as the New Forest, Breckland and Salisbury Plain, become almost exclusively a bird of the uplands. The cause of decline has been little studied in Britain, but detailed studies in Switzerland provide important clues. Here, whinchats were once widespread, but are now restricted to mountain and subalpine grasslands. A long-term study by Müller *et al.* (2005) found that mowing of meadows had advanced by up to 20 days between 1988 and 2002 and that the area was a mosaic of traditionally managed hay meadows and pastures with low fertiliser inputs, and heavily fertilised pastures and silage fields. Here, breeding success of whinchats now depended heavily on mowing date, with as few as 5% of nests fledging young in years and at sites where mowing took place early and as many as 78% in years and at sites where mowing was later. At a lower-altitude study area where early mowing was more frequent, breeding success was too low to sustain the population in nine of 15 years, and breeding population size was strongly correlated with breeding success in the previous year. A further study by Britschgi *et al.* (2006) found that not only had intensification of grassland management made whinchats more vulnerable to nest loss through early mowing, but that intensively managed pastures and silage fields held a lower abundance and diversity of arthropods, especially of the larger sizes favoured as nestling food (see also Beintema *et al.* 1991; Blake *et al.* 1994, 1996). Parents in these habitats fed a lower biomass of arthropod food to nestlings and needed to fly further to collect food, so that fledging success was lower, even for nests which survived mowing. Britschgi *et al.* regard the whinchat as a good indicator species for 'nature-friendly' management of agricultural grasslands and, in that context, the virtual disappearance of whinchats from lowland Britain probably tells us just how 'nature-unfriendly' modern, intensive grassland management can be. On the Salisbury Plain Training Area (SPTA), the long history of military use has helped to preserve the largest remaining semi-natural grassland landscape in lowland Britain, with agricultural management limited to low-intensity grazing. Much of the area is now designated under the EU Birds Directive as an SPA for stone curlews, as well as breeding populations of quail and hobby, and wintering hen harriers. However, the fact that this dry, chalk grassland landscape holds populations of approximately 14 600 pairs of skylarks, 8800 pairs of meadow pipits, 580 pairs of whinchats and 260 pairs of grasshopper warblers (Stanbury 2002) shows what intensively managed agricultural grasslands may have lost. Of these species, only skylark would be likely to be found breeding in such modern landscapes, and at much reduced densities.

Other species whose declines were so early and so rapid that they are barely remembered as 'farmland birds' may also have succumbed to grassland intensification. For example, the reasons behind the demise of the red-backed shrike in Britain remain mysterious. Nonetheless, the species was a widespread countryside bird during the nineteenth century, at least in the southern half of England and, like the whinchat, it is strongly associated with low-intensity agricultural grasslands in much of its remaining European breeding range (e.g. Vanhinsbergh & Evans 2002; Brambilla *et al.* 2007). Also like whinchats, red-backed shrikes take large arthropod prey, so perhaps the detrimental effect of intensification of grassland management on availability of large arthropod prey also affected this species. It is certainly plausible that its progressive decline in Britain through the twentieth century and final extinction in the 1990s are, in part, due to grassland intensification.

Grassland intensification and breeding waders

Wet agricultural grasslands in Britain support a distinctive assemblage of breeding waders including oystercatcher, lapwing, snipe, curlew and redshank as well as, very locally, black-tailed godwit and ruff. This association with wet soil conditions has been shown by several studies in a variety of grassland systems (e.g. Green 1988; Green & Robins 1993; Milsom *et al.* 2000, 2002; O'Brien 2001, 2002; Small 2002). Wet soil conditions through the spring and summer are essential in providing soft soil substrates suitable for foraging by both adults and chicks. Beyond this, the specific requirements of the species reflect different preferred ranges of vegetation height and structure, and the relationship between water management and the abundance and accessibility of invertebrate prey (Wilson *et al.* 2004; Durant *et al.* 2008a; Schekkerman *et al.* 2008). For example, both lapwing and redshank favour fields with shorter swards and retained standing water and take invertebrates from the soil surface and vegetation (Beintema *et al.* 1991; Johansson & Blomqvist 1996; Small 2002; Ausden *et al.* 2003; Devereux *et al.* 2004; Smart *et al.* 2006; Durant *et al.* 2008b), whilst snipe and curlew select taller, more tussocky vegetation in which to nest, and favour soft soils, though not necessarily standing water, to probe for invertebrate food (Green 1988; Green *et al.* 1990; O'Brien 2001). The combination of drainage, coupled with reseeding and stocking at high grazing densities has thus taken a severe toll on breeding wader populations of agricultural grasslands, at least in the lowlands. Surveys of breeding waders of lowland wet grassland (i.e. grassland below 200 m and potentially subject to waterlogging or flooding in the absence of agricultural intervention) in 1982, 1989 and 2002 across England and Wales revealed large declines of all species except oystercatcher (Fig. 9.6). In addition, there has been a marked contraction in the distribution of the declining species to nature reserves and other protected sites. Populations of lapwing and curlew have also declined

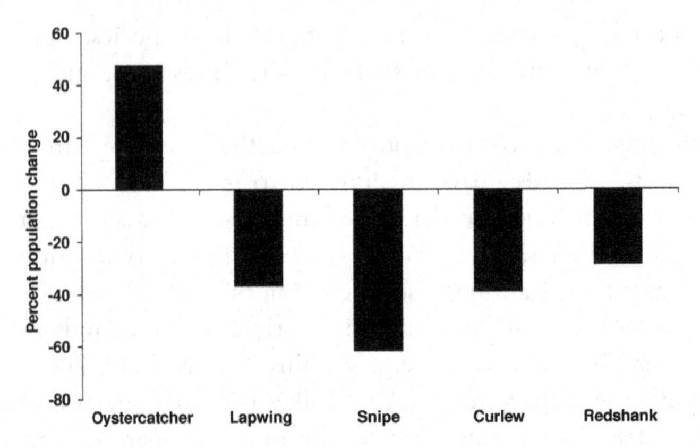

Fig. 9.6 Population changes of five species of breeding waders on lowland wet grassland in England and Wales between 1982 and 2002. Data from A. M. Wilson *et al.* (2005).

dramatically in recent years in Northern Ireland, with losses of approximately 60% of both species between 1987 and 1999 (Henderson *et al.* 2002). Survey work also suggests that increased grazing intensity on saltmarshes is the cause of a 23% decline in the abundance of breeding redshank in this habitat in Britain between 1985 and 1996 (Brindley *et al.* 1998; Norris *et al.* 1998). The only species to buck these declining trends, oystercatcher, appears to favour relatively intensively managed, short-grazed cattle pasture (Baines 1988) and has also spread to nest successfully in arable fields in northern England and Scotland.

At the time of the 1968–72 *Breeding Atlas*, lapwing and redshank and, more locally, snipe could all reasonably have been regarded as breeding birds of Britain's wider agricultural countryside. As in the case of whinchat in the previous section, this is no longer the case, at least in the lowlands. On marginal upland farmland in northern England and, especially, Scotland, all of these species remain more widespread, although even here there are worrying signs of recent decline (O'Brien *et al.* 2002; Taylor & Grant 2004). Recent BBS results from Scotland show declines of 38% in lapwing and 48% in curlew populations between 1994 and 2007, and even a 27% decline of oystercatchers (Risely *et al.* 2008).

Beyond the immediate impacts of drainage, additional effects of intensification of grassland management in causing wader declines probably vary between species. Certainly, the same trends towards replacement of late-growing, single-cut hay meadows by heavily fertilised, fast-growing, multiple-cut silage creates swards that are inimical to nesting by waders. Those that do nest more likely to suffer losses in exactly the same way as corncrakes, yellow wagtails or whinchats (e.g. Baines 1990; Kruk *et al.* 1996). Increased nest loss may also occur on grazed grassland as higher stocking densities lead to higher rates of trampling (e.g. Beintema & Müskens 1987; Green 1988; Hart *et al.* 2002), and reductions in arthropod size and abundance which accompany intensive grassland management may further reduce breeding success (Beintema *et al.* 1991; Blake *et al.* 1994). Equally, on marginal upland farmland where there is often a polarisation between intensification of grassland management in some areas and agricultural abandonment in others, breeding waders may suffer at both ends of the spectrum because most species are unable to persist at very low or zero grazing levels when wet pastures become infested with blanket cover of soft rush (O'Brien 2001).

Less well understood are the potential effects of intensive agricultural management on the susceptibility of wader nests to predation. Certainly, high rates of nest predation are common amongst breeding wader populations (e.g. Beintema & Müskens 1987; Baines 1990; Grant *et al.* 1999; Ottvall 2005), and high densities of grazing stock reduce heterogeneity in the sward and may make nests more detectable to predators (Baines 1990; Whittingham & Evans 2004). Moreover, there is some evidence that predation effects may be worsened where populations have already been reduced by habitat deterioration. For example, lapwings exhibit communal mobbing defence of their nests against predators (Elliot 1985), and there is some evidence that the effectiveness of this defence is reduced when nesting aggregations are reduced in size (Berg 1996; Seymour *et al.* 2003). Until recently, however, there had been no published studies of the effect of experimental manipulation of predator numbers on breeding success or population density of grassland waders. However, an eight-year cross-over experiment to test the effect of fox and crow control on the breeding success and population trends of lapwings across 11 lowland wet grassland sites has filled this gap (Bolton *et al.* 2007). Across a total sample of 3139 lapwing nests, the study found no overall effect of predator control on the failure rate of nests, despite the fact that across all sites, predator control reduced adult fox numbers by 40% and territorial carrion crow numbers by 56%. However, this overall result concealed marked variation between sites with good evidence that predator control was able to reduce nest loss rates where predator densities were high. Overall, it seems that the role of predation as a driver of long-term change in breeding wader populations may by highly site- and time-specific, and even this long-term experiment failed to find any evidence of an effect of predator control on lapwing population trends across the 11 sites. We return to the more general topic of the relationship between agricultural change and predation impacts later in the chapter.

Effects of agricultural improvement on value of grasslands as a foraging habitat for birds

The impact of intensive grassland management on the availability of invertebrate and seed foods for birds has been reviewed recently by Wilson *et al.* (1999), Vickery *et al.* (2001) and McCracken & Tallowin (2004), and most of the following account draws on this work.

The botanical impoverishment of grassland associated with agricultural improvement itself results in a net reduction in the diversity and abundance of invertebrates (Fenner & Palmer 1998) and a reduction in the availability of seed that is further compounded by early cutting of silage crops before seed set (Marshall & Hopkins 1990). For example, van Wingerden *et al.* (1992) considered grasshopper abundance and diversity to be a highly sensitive indicator of agricultural nitrogen load in grasslands; although the fertiliser had no direct impact, its effect was to increase vegetation density, limit soil surface insolation and temperature, and thus reduce the probability that grasshoppers were able to complete their life cycle successfully in comparison with less densely vegetated natural grasslands. High rates of fertiliser application may be more directly detrimental to other invertebrate groups, including earthworms and beetles (e.g. Edwards & Lofty 1982). Several studies have shown that cutting of grassland as occurs more frequently under intensive silage management depletes the invertebrate fauna. This can occur either directly (Curry & Tuohy 1978) or indirectly, as in the removal of nectar sources for butterflies (Feber *et al.* 1996), even though newly cut fields may provide a temporary flush of invertebrate food for birds both in the cut sward and the newly exposed soil surface (e.g. Devereux *et al.* 2006). As Morris (2000) puts it, 'tall grassland supports more species, individuals and a greater diversity of arthropods than short swards'. However, not all effects are negative. Some generalist herbivorous invertebrates such as sap-feeding bugs may benefit simply from the increased volume of nutrient-rich vegetation available to feed from (e.g. Andrzejewska 1976) and high densities of bibionid and tipulid flies and their larvae, important prey items for birds, may be associated with heavy applications of organic fertilisers (dung or slurry) in heavily stocked grassland systems (e.g. D'Arcy-Burt & Blackshaw 1991; McCracken *et al.* 1995).

The effect of improved drainage of agricultural grassland on its quality as a foraging habitat for birds has been less well studied. The negative impacts of soil drying on species that rely on probing soft soil for invertebrate food such as waders are clear enough (e.g. Green 1988). These also extend to other species that rely for food on soil-dwelling invertebrates, including song thrushes (Peach *et al.* 2004a; and see case study) and starlings (Olsson *et al.* 2002). A recent study has shown that yellow wagtails select grassland breeding territories associated with short swards and areas of bare earth where winter flooding has given way to shallow-edged ponds or wet ditches (Bradbury & Bradter 2004). However, an experimental study at Southampton University found that shielding the ground from spring (May) rainfall in agricultural fields reduced the abundance of a wide variety of arthropods for over three months after treatment, whilst experimental irrigation had positive effects (Frampton *et al.* 2000). The authors note the potential relevance of their findings for understanding the impacts of agricultural or climate change on farmland birds. Given the widespread and intensive drainage of British agricultural landscapes (Robinson & Armstrong 1988), it does seem plausible that this change may have had a very general effect in reducing both the abundance and availability of invertebrate food for farmland birds.

Where agricultural improvement of grassland leads to high stocking densities, this produces uniform, short and densely tillered swards, especially where sheep are the main grazers (e.g. Orr *et al.* 1988; Tallowin *et al.* 1989). Seed production from these grasslands tends to be limited to a few species that can tolerate frequent defoliation and trampling, and the abundance and diversity of invertebrates is limited by the structural simplicity of the sward. Where cattle are the main grazers then their trampling, patchy deposition of dung and more selective grazing can combine, if the density of animals is moderate, to yield more heterogeneous swards with a relatively diverse invertebrate fauna (e.g. Morris 2000). The presence of livestock dung itself supports a fauna of dung-dwelling invertebrates – notably flies and beetles – which are important prey items for a wide variety of farmland birds, including corvids, wagtails, starlings and hirundines. Concern has been expressed that the treatment of cattle with anthelminthic veterinary products, notably avermectins, may have detrimental effects on this invertebrate fauna through insecticidal effects of residues excreted in the dung of treated animals (McCracken 1993). However, although several studies have detected effects of avermectins on the abundance and growth rates of insect larvae inhabiting cow-pats (e.g. Madsen *et al.* 1990; Lumaret *et al.* 1993; McCracken & Foster 1993; Strong & Wall 1994; Webb *et al.* 2007) there is no strong evidence that these local effects have any lasting impact on the invertebrate populations or the birds that feed on them (McCracken & Tallowin 2004). Recent studies of the breeding colonies of barn swallows at livestock farms demonstrate the importance of the invertebrate fauna associated with livestock production as a food source. In Denmark, cessation of dairy farming in favour of arable production led to a decline in barn swallow abundance of almost 50% associated with reduced clutch sizes, reduced frequency of second clutches and a reduced recruitment rate of yearling birds to the breeding population (Møller 2001), whereas no change was observed on farms where dairy farming was retained. Similar results have been obtained in Northern Italy (Ambrosini *et al.* 2002). In Britain a study in Oxfordshire found that aerial invertebrate abundance over cattle-grazed pasture fields was two to four times greater than that over silage or cereal fields (Fig. 9.7). Pasture fields also hosted approximately twice as many foraging barn swallows as either silage or cereal fields (Evans *et al.* 2007). At a national scale, Henderson *et al.* (2007a) have recently obtained similar results by examining the habitat associations of barn swallows encountered on BBS transects. They found that the single most important correlate of the presence of swallows was cattle.

Although no study monitoring the population consequences of loss of dairy farming has been carried out in Britain, long-term population monitoring does suggest that barn swallow declines are more likely to have been seen in areas of eastern England where arable cropping has become dominant at the expense of livestock farming (Evans *et al.* 2003b; Robinson *et al.* 2003). Starlings too favour grazed pasture over arable fields as a foraging habitat (Whitehead *et al.* 1995; Bruun & Smith 2003) since the former usually hold much higher densities of leatherjackets and earthworms (Tucker 1992). Studies in Sweden have shown both that colony size and breeding success are positively related to the availability of pasture close to colonies, and that fledging success tends to be more consistent between years when more pasture is available (Smith & Bruun 2002). In Britain,

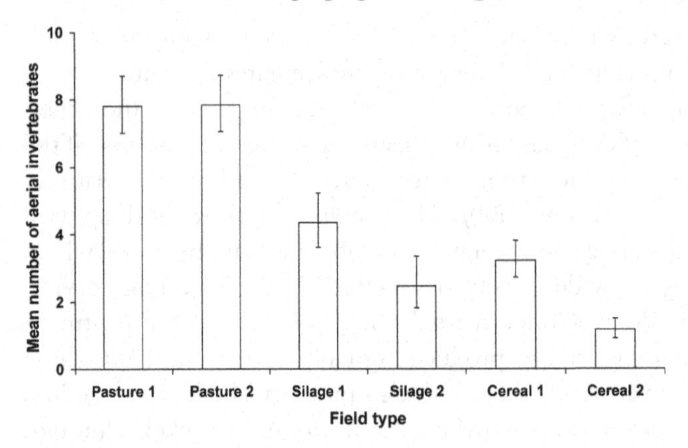

Fig. 9.7 Aerial invertebrate abundance over three field types early (1) and late (2) in the barn swallow breeding season. Data from Evans *et al.* (2007).

there is certainly local evidence of starling declines in areas where farmland has become dominated by arable production as, for example, in Lincolnshire (Feare 1994). However, a recent analysis of long-term starling population trends in Britain has found that the steepest rates of decline are now in areas of mixed and livestock farming (Robinson *et al.* 2005b). The reasons for this are unclear, although the trend towards indoor wintering of cattle may have rendered some traditional food sources (e.g. grain-based cattle feed) less accessible. Another strong candidate explanation must be that as a species dependent on probing the ground for soil invertebrates, the starling is another long-term victim of increasing 'drying out' of farmland as a consequence of more efficient field drainage coupled to more rapid evapotranspiration losses of soil water through highly fertilised vegetation (Garwood 1988). Further research to test these hypotheses for the decline of starlings in grassland areas would be valuable, especially given that recent analyses of ring-recovery data for starling suggest that reduced survival rates during the first year of life may be driving the population decline (Freeman *et al.* 2007).

During the winter, several studies have examined the use of agricultural grasslands as foraging habitats by birds, notably those by Tucker (1992), Wilson *et al.* (1996b), Perkins *et al.* (2000), Barnett *et al.* (2004), Atkinson *et al.* (2005), Buckingham *et al.* (2006) and Gillings *et al.* (2008). The first two studies encompassed several field types in addition to grassland, and demonstrated that a wide variety of soil-invertebrate-feeding birds (corvids, thrushes, lapwings, golden plovers, starlings and black-headed gulls) favoured pastures over silage leys or arable fields for foraging. Tucker's study also sampled soil invertebrates and showed that pastures supported considerably higher densities and biomass of soil-dwelling invertebrates such as earthworms and fly and beetle larvae than did the other field types. Radio-tagging studies of woodcock have also shown that earthworm-rich agricultural pastures are an important foraging habitat throughout the year (Hirons & Bickford-Smith 1983; Duriez *et al.* 2005; Hoodless & Hirons 2007). All these studies demonstrate the importance of even intensively managed grassland as a winter food source for some farmland and woodland birds, perhaps especially in areas where arable farming predominates. Although there are differences of detail, the other studies largely concur in finding that those birds feeding on soil invertebrates favour actively grazed fields with relatively short swards and areas of bare earth that provide immediate access to the soil surface. Short swards are also known to be selected by waders feeding on grassland in winter (Milsom *et al.* 1998). The two studies that also considered breeding season patterns of foraging use of grassland (Atkinson *et al.* 2005; Buckingham *et al.* 2006) found that these generalities held through the year. Interestingly, the studies by Barnett *et al.* (2004), Atkinson *et al.* (2005) and Gillings *et al.* (2008) found evidence of greater use of grassland by thrushes, corvids and starlings when these were more intensively managed, perhaps suggesting that populations of generalist phytophagous, soil-dwelling larvae may respond positively to high root biomass in nutrient-rich grassland. Barnett *et al.* suggest that grassland improvement may be one of the causes of recent increases in numbers of some generalist corvid species in Britain. Some herbivorous bird species may also benefit from nutrient-rich intensive grassland, as shown by studies of pasture selection in brent geese by Hassall & Lane (2001) and Hassall *et al.* (2001). In contrast, the only species which are consistently associated with ungrazed grassland in winter are snipe (Wilson *et al.* 1996b; Barnett *et al.* 2004), which may benefit from the cover afforded by the longer sward, and seed-eating passerines such as skylark, linnet, goldfinch, yellowhammer and corn bunting, which are able to exploit seeding grasses in ungrazed fields (Mason & Macdonald 2000; Perkins *et al.* 2000; Buckingham *et al.* 2006). These species too, however, still prefer an open sward with areas of bare earth. A detailed study of wintering skylarks on arable reversion grassland in the South Downs and South Wessex Downs ESA found that the birds preferred swards longer than 10 cm and were more likely to be found on fields with an open sward structure (Wakeham-Dawson & Aebischer 1998).

Upland grazing

The huge increase in sheep and (in Scotland) deer grazing in the uplands since the Second World War (see Chapter 4) has stimulated substantial conservation concern over the possible impact of changing upland vegetation mosaics and associated food resources on upland breeding bird populations. Because upland vegetation response to changing grazing pressure is gradual and difficult to subject to controlled manipulation over large areas, most studies (e.g. Haworth & Thompson 1990; Brown & Stillman 1993; Stillman & Brown 1994; Tharme *et al.* 2001; Pearce-Higgins & Grant 2006) have examined spatial associations between bird distributions and variation in vegetation composition and structure.

Taking these studies together, there is a clear group of species that select moorland with substantial heather cover to provide nesting cover. These include red grouse, merlin, hen harrier, short-eared owl, ring ouzel and twite. Loss of this cover through increased grazing pressure or abandonment of grouse moor management in favour of sheep or deer grazing has certainly locally reduced the availability of nesting habitat for these species. In addition, heavy grazing may remove supplies of moorland berry crops such as those of bilberry and crowberry which are important late-summer foods for species such as grouse and ring ouzels (Appleyard 1994; Picozzi & Hepburn 1986; Welch 1998), and may indirectly affect breeding success of top predators such as golden eagles through impacts on availability of grouse and mountain hare prey (Watson *et al.* 1992). However, with the exception of red grouse, all of these species also make extensive use of grassland swards for foraging. The three raptor species seek prey such as meadow pipits and voles which are more abundant in grass swards (e.g. Redpath *et al.* 2002), whilst ring ouzels take earthworms and twite seek out seeding plants in grassland (Brown *et al.* 1995; Burfield 2002; Raine 2006). Skylarks, meadow pipits and wheatears all avoid extensive areas of heather cover and tend to be found at highest densities on grassy moorland (Smith *et al.* 2001; Pearce-Higgins & Grant 2002, 2006).

In summary, at the national scale, loss of dwarf shrub heath in favour of grassland as a result of increased grazing pressure in the uplands is unlikely to have been the primary or sole cause of population decline for any other species than red grouse (Thirgood *et al.* 2000a). However, populations of black grouse (Calladine *et al.* 2002), ring ouzel (Buchanan *et al.* 2003; Sim *et al.* 2007a) and twite (Raine 2006) may have suffered from the loss of nesting habitat and specific food sources at a local scale. Similarly, at higher altitudes, the effects of higher red deer and sheep grazing pressure, combined with nitrogen deposition, have reduced the areas of montane, wind-clipped heather and *Racomitrium* (woolly fringe-moss) heath favoured by nesting dotterel in favour of grassland (e.g. Thompson & Brown 1992; van der Waal *et al.* 2003; Ratcliffe 2007).

Recent studies suggest that as well as simply the relative proportion of dwarf shrub heath and grassland, the structural heterogeneity of the moorland vegetation mosaic is also very important in determining breeding bird densities. In a study across 10 upland blocks in southern Scotland and northern England in 1999–2000, Pearce-Higgins & Grant (2006) found that golden plover were present at higher densities wherever grazing or burning created open swards, whether the vegetation was dominated by heather, grasses or sedges, and that densities of snipe, curlew, stonechat and whinchat were all associated with moorland that provided a mosaic of tall and short vegetation. In the case of whinchat, stands of bracken are a particularly important component of the moorland vegetation mosaic (e.g. Allen 1995), so that large-scale bracken control management should take into account possible detrimental effects on this species. Structural variety in vegetation mosaics may also offer benefits where birds use different habitats at different stages of the breeding cycle or can exploit different seasonal patterns of invertebrate availability in different vegetation types. For example, golden plovers (see case study) use different vegetation types at different stages of the breeding cycle, and feeding chicks select a variety of invertebrate-rich foraging habitat types including wet flushes, short grassland, bare peat and stands of bilberry and crowberry. Meadow pipit nests may be located at the edge of blanket bog and drier heath or grass vegetation to allow the birds to exploit different patterns of prey availability in each habitat (Coulson & Whittaker 1978; Coulson & Butterfield 1985).

In general, just as on lowland farmland (J. D. Wilson *et al.* 2005), structurally diverse swards provide nesting opportunities and cover from predation (e.g. bracken, mature heather, ungrazed grass swards) in association with shorter, more open swards to provide good access to soil- and surface-dwelling invertebrates. The case study of golden plover illustrates this well. Likewise, the following three case studies reveal the extent to which the modification of moorland vegetation mosaics through grazing can have important effects on the abundance and demography of moorland breeding birds of diverse ecology: black grouse, hen harrier and meadow pipit.

Black grouse. Experimental reductions in sheep grazing intensity (to roughly 33–50% of pre-treatment levels) implemented by the North Pennines Black Grouse Recovery Project (Calladine *et al.* 2002; and see black grouse case study) were associated with increases in densities of both displaying male black grouse and females during the brood-rearing period, relative to reference sites. However effects on females were most marked at sites where grazing reduction occurred on smaller patches of ground. Given that previous studies had indicated that grazing reductions could lead to big increases in availability of invertebrate food sources for the grouse (Baines 1996), there is an indication here that black grouse too may benefit from management to create a mosaic of longer- and shorter-grazed swards.

Hen harrier and other vole predators. Studies of early breeding season habitat use by a declining population of hen harriers on Orkney found that hunting of male harriers was strongly associated with rough grassland with a litter layer, a habitat maintained by low-intensity grazing. This provides good habitat for Orkney voles, a key harrier prey item (Amar & Redpath 2005). Females similarly avoided areas of high heather cover and of intensively managed and grazed pasture. Areas of rough grassland supporting high vole densities have declined on Orkney in association with agricultural intensification and increased sheep densities, and these changes may have contributed to the observed decline of the hen harrier breeding population by limiting the amount of food that male harriers are able to supply to females before egg-laying and during incubation. Similarly, in the North Pennines, Wheeler (2008) has found that exclusion of sheep grazing from moorland increased field vole densities by between 50% and 150%, and estimated that grazing reductions of this kind across the North Peak ESA could have marked positive effects on the densities of vole predators such as kestrels, barn owls, little owls and short-eared owls.

Meadow pipit. A recent study examined the effect of experimental manipulation of grazing intensity on breeding abundance of meadow pipits on sheep-grazed moorland in Scotland. Replacement of high intensity sheep grazing with lower intensity grazing by a mix of cattle and sheep proved capable of increasing pipit densities by roughly 30% in just two years, whilst plots that were ungrazed or grazed only by sheep yielded no such increase (Evans *et al.* 2006b). Again, the probable mechanism is that reduced-intensity, mixed grazing increased the structural heterogeneity of vegetation and the abundance and biomass of foliar arthropods, thus increasing the availability and accessibility of key food resources to foraging pipits (Dennis *et al.* 2005, 2008).

The practice of strip muir-burn on managed grouse moors is designed specifically to create structural variety in heather cover to offer optimum combinations of nesting cover and feeding habitat for grouse and their chicks. Thus stands of mature heather provide cover from predators and weather, and optimum nesting cover (Smith *et al.* 2000), while burned areas provide access to young, nutritious heather shoots. As Palmer & Bacon (2001) found, both adult female red grouse and their young broods showed strong associations with the heterogeneous mixes of grass and heather and with the edges between burned and unburned heather patches which provided both protection and foraging opportunity.

Other species with similar habitat needs also benefit from muir-burn, including curlew (Robson 1998) and golden plover (Whittingham *et al.* 2001b). Several other species that nest in heather also tend to be associated with grouse moors, including black grouse, merlin, hen harrier, short-eared owl and ring ouzel (Thompson *et al.* 1997). Accordingly, in a specific comparison of breeding densities of 11 upland birds on heather moorland managed for grouse shooting, and similar moorland with no grouse management, a joint RSPB and GWCT study found that densities of golden plover, lapwing and curlew (as well as red grouse) were substantially higher on moors managed for driven grouse shooting than on non-grouse moors. Non-grouse moors held higher densities of species that are either associated with other vegetation covers (e.g. meadow pipit, skylark and whinchat) or reduced in numbers by intensive predator control on grouse moors (e.g. carrion crow) (Tharme *et al.* 2001) (Fig. 9.8). On grouse moors, however, the potential habitat benefits of creating spatial heterogeneity in vegetation structure through muir-burn are difficult to disentangle from the additional effects of intensive predator control. Nests of ground-nesting bird species are vulnerable to predation by exactly those predators (e.g. foxes, corvids and mustelids) that are controlled by gamekeepers for the benefit of grouse stocks. Studies of golden plover and curlew both suggest that increases in nest predation by generalist predators can in some circumstances cause population declines or prevent population recovery (e.g. Parr 1992; Grant *et al.* 1999; Pearce-Higgins & Yalden 2003b). It is therefore likely that this aspect of grouse moor management can sometimes add to benefits generated by habitat management, although all too frequently illegal control of raptor populations on the same moorlands continues to limit the population size and breeding range of protected species such as hen harrier and golden eagle (Etheridge *et al.* 1997; Green & Etheridge 1999; Whitfield *et al.* 2003, 2004a, b).

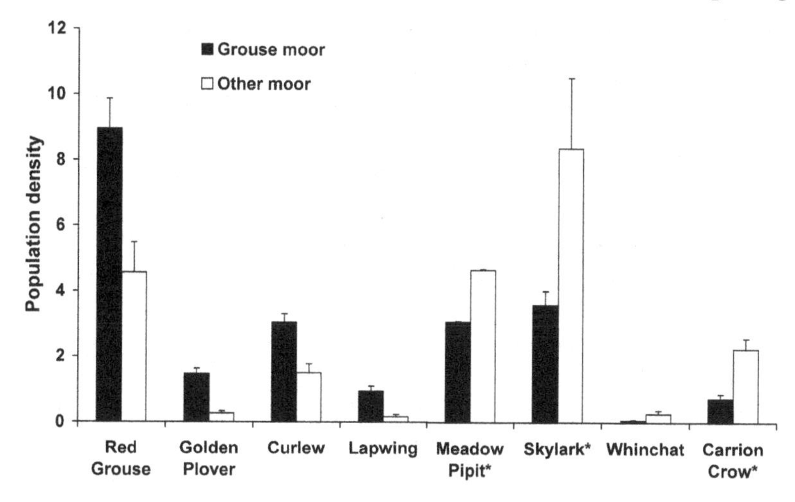

Fig. 9.8 Mean population densities (+SE) (km^{-2} except for meadow pipit where ha^{-1}) of breeding birds on grouse moors and non-grouse moors. * for these species densities are of individuals; for other species, breeding pairs. Data from Tharme *et al.* (2001).

The other major upland land-use whose effects may interact with those of agricultural grazing in affecting moorland breeding birds is conifer afforestation. Blanket afforestation of large areas of the uplands of course changes bird communities utterly (Avery & Leslie 1990). For example, afforestation during the 1980s was estimated to have caused declines of breeding golden plover, greenshank and dunlin in the Caithness and Sutherland peatlands of the order of 17–19% (Stroud *et al.* 1987; Avery & Haines-Young 1990). Despite this, as we saw in Chapter 4, plantations at the pre-thicket stage may attract high densities of a range of moorland breeding species to areas in which vegetation cover and associated invertebrate and small-mammal populations respond to the withdrawal of grazing animals. Less well understood are the ecological impacts of forestry on surrounding, unafforested moorland. As far as birds are concerned, forestry may reduce the availability of open foraging ground for raptors such as golden eagles (Whitfield *et al.* 2001). It may also act as a source of predators such as foxes and corvids which ground-nesting birds either avoid or to which they succumb (Parr 1992), or it may change grazing patterns on adjacent moorland thus affecting vegetation structure and composition for birds. Avery (1989), Stroud *et al.* (1990), Parr (1992), and Hancock *et al.* (2009) found evidence that plantation forest reduced breeding wader densities on adjacent moorland and in some cases that nest predation rate increased, but effects were generally small and varied between sites. Similarly, Buchanan *et al.* (2003) found some evidence that ring ouzel declines on moorland in Scotland between the late 1980s and late 1990s were greater on moorland surrounded by forest cover, although the underlying mechanism remains unknown. One possibility noted by Derek Ratcliffe (1990, 2007) in the heavily afforested Southern Uplands of Scotland is that managers of grazed moorland adjacent to forestry often reduce or cease burning operations close to plantations for fear of the fire running through the forest. Other areas may be so isolated by forestry that active grazing management ceases. The consequence of this may be rank growth of heather or grass cover which eventually renders the habitat unsuitable for nesting and foraging by many ground-nesting and feeding birds.

Overall, it is clear that agricultural grazing of sheep, in conjunction with forestry and management for grouse shooting, is a key driver of change in moorland vegetation mosaics, and hence the bird community likely to be encountered in a given upland area (e.g. Pearce-Higgins *et al.* 2008). There is little evidence, however, to suggest that grazing has been the primary driver of widespread population decline for any bird species, with the possible exception of red grouse. However, some species make use of agricultural habitats on both sides of the enclosure line. For example, upland breeding waders such as golden plover (see case study) and curlew make much use of enclosed moorland edge pastures to forage for soil invertebrates and indeed improved, close-cropped pastures may make high-quality foraging habitats (Fuller & Gough 1999) for these species. Other species may fare less well, and a prime candidate here is the twite which nests in mature heather or dense bracken at the moorland edge but relies heavily on seeding weeds of adjacent farmland as a food source for both adults and nestlings. Although over 90% of the British twite population nests in Scotland, the small and declining population of around 100 pairs in the English South Pennines has been more intensively studied. Here there is good evidence that loss of seed supplies has followed heavy grazing and improvement of moorland edge pastures, and early cutting of hay and silage meadows. Together with the impacts of increased grazing pressures, burning and bracken control on the availability of suitable nesting habitat (e.g. Brown *et al.* 1995; Reed 1996; Brown & Grice 2005) these changes may have contributed to causing a decline which now threatens the English twite population with extinction (Raine 2006). Recent studies of a high-density twite population on the Western Isles of Scotland (N. Wilkinson *et al.* unpublished) also show that the birds are restricted to moorland edge nesting habitats within easy reach of seed-rich foraging habitats on the adjacent grazed and cultivated machair. Here perhaps is another species whose fortunes may be determined by grazing practices and grassland management on both sides of the moorland wall.

Field boundary management

The wholesale removal of hedges and associated uncropped habitats from the agricultural landscape since the Second World War (see Chapter 1), especially in eastern areas where conversion to arable agriculture deprived hedges of their stock-proofing function, is one of the most visually dramatic aspects of agricultural intensification. As Pollard *et al.* (1974) put it, 'hedge removal from large areas of eastern England has changed the enclosed landscape beyond recognition'. As we saw in Chapter 7, an analysis of CBC data in the 1980s found that the abundance of the majority of breeding bird species on farmland CBC plots was explained by measures of the availability of non-crop habitat features such as hedges, woods and ponds. In this context, how can it be that loss and degradation of these marginal habitats has been of only secondary importance in causing farmland bird population declines, as concluded by Gillings & Fuller (1998)? Were Murton & Westwood (1974) correct when, in an early review of the effects of agricultural change on birds, they came to the bold conclusion that 'hedgerows appear to be suboptimal habitats which have become a red herring so far as the real issues affecting the welfare of birds in Britain are concerned'. The answer is 'no' for two reasons. First, since the 1970s the accelerating effects of agricultural intensification on the availability of bird food resources in the agricultural landscape and the availability of nesting opportunities for field-nesting species have probably simply outstripped degradation of marginal habitats as the primary cause of bird population decline, and masked its effects. If we consider Table 8.13 then the trend of all of the case study species (whether an increase or a decrease) except sparrowhawk is most convincingly explained by impacts of changing agricultural practice on the availability of food or the opportunity for and security of nesting attempts in open-field habitats. Similarly, 19 of the 20 Red List species of high conservation concern (Gregory *et al.* 2002) which could be regarded primarily as farmland birds rely wholly or partly on cropped or grazed agricultural land for either nesting or feeding, or both. In this context, it is no surprise that recent studies such as those of linnets (Moorcroft 2000) and yellowhammers (Kyrkos 1997; Bradbury *et al.* 2000) have concluded that the present-day availability of suitable hedgerow breeding habitat is unlikely to be limiting breeding populations of these species despite the historical losses of these habitats. However, there are certainly cases where degradation of field boundary habitats has been a key cause of decline. For example, loss of well-vegetated hedge bases and field margins as nesting habitat has contributed to the decline of grey partridges (Potts 1980; and see case study). Loss of hedgerow elms to Dutch elm disease in the 1970s and 1980s had a severe limiting effect of nest site availability for hole-nesting species such as barn owl (Osborne 1982), and loss of food resources and secure nest sites in large hedgerows as a consequence of increasingly severe mechanised hedge management is a probable contributing cause of bullfinch declines (Proffitt 2002). Second, Murton & Westwood (1974) based their conclusions partly on the fact that bird species richness on their own Cambridgeshire study site held up despite 'drastic changes in land use' during the 1960s, including substantial hedgerow removal. They also noted the fact that then recent studies of wrens and great tits (see Chapter 4) had suggested that hedgerows were suboptimal breeding habitats which individuals of these two species would readily vacate for higher-quality woodland territories if the opportunity arose. This overlooked the fact that the Calton study area did experience marked declines in breeding bird density despite maintaining its species richness (O'Connor & Shrubb 1986). Furthermore, whilst great tits and wrens are classically woodland species, there are a wide range of other species ('scrub'-nesters) for which hedgerows and associated herbaceous cover are the only suitable breeding habitat on most farmland – providing secure nest sites and ready access to foraging habitat in adjacent fields and field margins and for which woodland certainly does not provide a more enticing alternative habitat (Fuller *et al.* 2001). Grey partridge, turtle dove, whitethroat, linnet, yellowhammer, cirl bunting and reed bunting are all examples.

So, even though recent effects of agricultural intensification on open-field resources for birds have masked the effects of hedgerow loss and management in driving population change for most farmland birds, there is no doubt that changes in field boundary management have been influential. Understanding how best to reinstate and manage hedgerows and other marginal habitats is one of the most cost-effective ways in which management for birds and other wildlife under agri-environment schemes has recently begun to reverse the effects of agricultural intensification (Chapter 10). In that context, a series of studies of the associations between breeding bird populations and field boundary structure, management and composition carried out during the 1980s and 1990s (Arnold 1983; Osborne 1984; Green *et al.* 1994; Parish *et al.* 1994, 1995; Macdonald & Johnson 1995; Sparks *et al.* 1996) and recently reviewed by Hinsley & Bellamy (2000) have been of great value. In summary, the following characteristics of hedgerows and associated habitats are critical:

(1) Taller, wider, hedges support a greater diversity and abundance of birds, and may provide greater nest security against avian nest predators (e.g. Chamberlain *et al.* 1995). However, there are a small number of species (notably linnet, yellowhammer and whitethroat) which favour short tightly trimmed hedgerows (e.g. Stoate *et al.* 1998; Bradbury *et al.* 2000; Eaton *et al.* 2002). Rotational management of hedgerows, including leaving some to develop taller, scrubbier structures (especially along green lanes) will provide a variety of hedgerow structures on individual farms and offer the best variety of habitats for breeding birds, as well as increasing invertebrate diversity and abundance (Maudsley 2000). Minimising summer or autumn cutting is critical (Sparks & Martin 1999), not only for breeding birds but also to ensure a supply of hedgerow fruit for bullfinches, redwings, fieldfares and other winter thrushes. Not all species benefit from hedges. Corn buntings may value them as little more than song posts, whilst some open-field-nesting species such as skylark and lapwing avoid small fields enclosed by hedges (e.g. J. D. Wilson *et al.* 1997; Sheldon *et al.* 2007), probably because these field edge habitats are a source of predation risk.

(2) The presence of standard trees, including dead trees, can greatly increase the number and diversity of birds occupying hedgerows, particularly hole-nesters and species more characteristic of woodland such as owls, woodpeckers, tits and treecreepers. Where large, old, ditched or banked hedgerows are doubled along farm tracks or green lanes then a habitat more akin to a linear patch of scrub or young woodland is created and this may attract species such as nightingales, blackcaps, garden warblers, marsh and willow tits that are not usually found in farm hedgerows (e.g. Arnold 1983; Fuller *et al.* 2001). In a recent study in Cheshire, Walker *et al.* (2005) found that for 14 of the 15 most abundant breeding

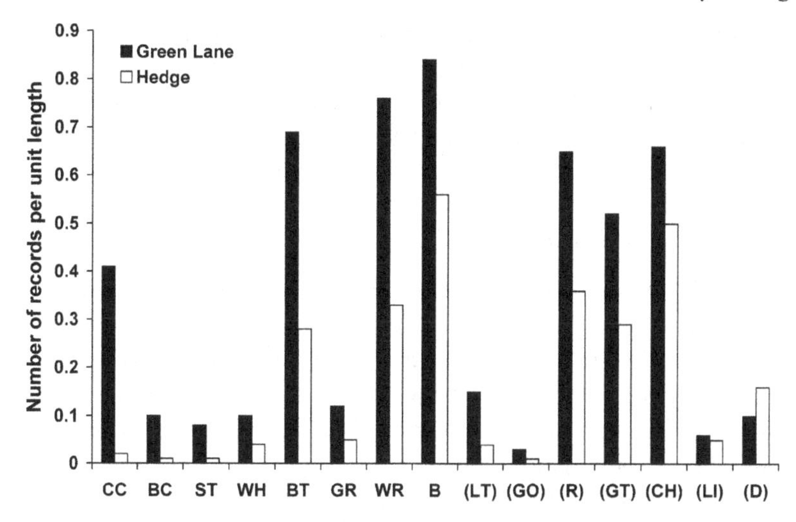

Fig. 9.9 Mean count of 15 songbird species per 50 m of green lane and 100 m of single hedgerow. Difference between two field boundary types significant at $p < 0.05$ except for species in brackets. CC, chiffchaff; BC, blackcap; ST, song thrush; WH, whitethroat; BT, blue tit; GR, greenfinch; WR, wren; B, blackbird; LT, long-tailed tit; GO, goldfinch; R, robin; GT, great tit; CH, chaffinch; LI, linnet; D, dunnock. From data in Walker *et al.* (2005).

small passerines, abundance estimates were higher in green lanes than in single hedgerows and that for eight of these species this difference was significant (Fig. 9.9).

(3) Dense herbaceous vegetation cover at and around the hedge base increases bird diversity and abundance by providing a combination of nest sites and favouring abundance and diversity of invertebrates as potential food sources (Maudsley 2000) for a wide range of species. For example, the availability of dense, concealing marginal vegetation adjacent to hedge bases is critical for successful nesting and chick-rearing by grey partridges (Rands 1986, 1987). Similarly, yellowhammers and whitethroats use both dense hedge-bottom vegetation and the hedge itself for nesting and make extensive use of hedge base and field margin vegetation for feeding (Bradbury *et al.* 2000; Morris *et al.* 2001; Stoate & Szczur 2001; Eaton *et al.* 2002; Perkins *et al.* 2002).

(4) Bird species richness and abundance tend to increase with the woody plant species diversity of the hedge which may itself be related to hedge age and the availability of seed, fruit and invertebrates.

(5) Ditches and ponds associated with hedges and field corners can be valuable in providing food sources for birds that rely on aquatic invertebrates or soft ground for accessing soil invertebrates, such as sedge warbler, reed bunting (Brickle & Peach 2004), woodcock (Duriez *et al.* 2005) and song thrush (see case study). There is some evidence (Parish *et al.* 1994, 1995) that ditches are associated with occupation of hedges by linnets and goldfinches, which may reflect the value of a water source for these species which feed almost wholly on seeds.

A recent study of the comparative wildlife value of ponds, ditches and streams in agricultural landscapes in southern England found that well-managed ponds and other small areas of aquatic habitat can make an important contribution to biodiversity at the regional level (Williams *et al.* 2003).

(6) Hedgerows may have additional value in providing connectivity between woodland habitats for woodland species such as tits, treecreepers, nuthatches and robins (Opdam *et al.* 1985; Hinsley *et al.* 1995; Clergeau & Burel 1997).

Until the recent advent of agri-environment schemes, however, management of hedges where they were not removed tended either towards neglect and dereliction – resulting in conversion to gappy lines of scrubby trees – or hard, annual, mechanical trimming reducing the value of the hedge as nesting cover and preventing regeneration of new standard trees from the hedge. Hedge base vegetation tended to be removed by a combination of pesticide spraying, grazing or cultivation into the hedge base and ditches were frequently neglected or filled (Watt & Buckley 1994; Haines-Young *et al.* 2000). The next chapter will show how our now detailed knowledge of bird associations with hedgerow structure and management has been used in conjunction with tighter hedgerow management regulations and opportunities offered by agri-environment schemes to improve field boundary management for birds and other wildlife.

Interactions between predation and agricultural management

A frequent focus of debate between land managers and conservationists in Britain centres on whether predation impacts from increasing populations of generalist predators in agricultural systems have been overlooked as a cause of many bird population declines in comparison with the effects of agricultural change. Certainly it is true that populations of many predatory birds that exploit agricultural habitats, including magpies, carrion crows, buzzards and sparrowhawks, have all increased in recent decades (e.g. Gregory & Marchant 1996; Gibbons *et al.* 2007a). Over the same time period, populations of many predatory mammals, including foxes, badgers and grey squirrels have also increased although changes in those of stoats and weasels are less certain (G. Wilson *et al.* 1997; Whitlock *et al.* 2003; Game Conservancy Trust 2004). The causes of these increases are varied. Relaxation of gamekeeping pressure, removal of organochlorine pesticides from the environment and increased legal protection of raptors have all contributed to long-term increases in populations of some predatory bird species, whilst populations of some species (notably foxes and corvids) may have benefitted from aspects of intensive agricultural management and game-rearing (Gibbons *et al.* 2007a) that may have increased food availability for these species.

Many studies have been carried out to assess the population-level impacts of predation on bird populations, and these studies have been well reviewed (e.g. Newton 1993, 1998; Gibbons *et al.* 2007a; Park *et al.* 2008). The outcomes form two distinct groupings.

Table 9.1 *Examples of recent studies illustrating limitation of populations of ground-nesting birds by predation*

Species	Reference	Evidence
Curlew	Grant *et al.* (1999)	82–95% of nests failed, with predation accounting for 90% of failures. Net rates of reproductive output were sufficient to account for observed declines
Golden plover	Parr (1992, 1993); Harding *et al.* (1994)	A local decline to extinction in north-east Scotland coincided with cessation of predator control, nearby afforestation, increased nest predation rates, reduced reproductive output and local increases in populations of generalist predators
Red grouse	Redpath & Thirgood (1997); Thirgood *et al.* (2000a, b)	Strict protection of raptors (especially Hen Harrier and Peregrine) on a Dumfriesshire grouse moor allowed their populations to increase. Spring raptor predation removed 30% of breeding Red Grouse and, by the end of the six-year study, Hen Harriers removed 45% of grouse chicks annually. This predation was probably additive to other forms of mortality
Skylark, meadow pipit	Amar *et al.* (2008)	Populations declined more at sites where hen harrier, peregrine and merlin numbers increased than at other sites. Predation was sufficient to remove up to 40% of the June meadow pipit population and 34% of the skylark population. However, breeding waders were not similarly affected

(1) Ground-nesting birds

For a range of studies of ground-nesting prey species – notably gamebirds and waders – there is growing evidence that predation can limit populations and, in some cases, drive population declines (e.g. Macdonald & Bolton 2008a). For example, we have already discussed the experimental studies carried out on Salisbury Plain which showed convincingly that populations of generalist predators were able to hold spring densities of grey partridges at much lower levels than were reached in areas where those predators were removed (grey partridge case study; Tapper *et al.* 1996). Conversely, an eight-year study of the effect of fox and carrion crow control on breeding success and population size of lapwings (Bolton *et al.* 2007) produced equivocal results, with no consistent effect on lapwing population trends and improvements in nest success only apparent at sites where initial predator densities had been very high. Although these are currently the only large-scale experimental studies that have been completed in Britain, other non-experimental studies provide evidence that predation can in some circumstances limit populations of ground-nesting birds (Table 9.1).

The catastrophic impact of introduced predator species on island avifaunas is well known, and the high-density breeding wader populations of the Hebridean machair (Chapter 4) provide an example of this impact in Britain. Hedgehogs were introduced to the Western Isles by humans during the 1970s. Studies of the predatory impact of hedgehogs on wader nests found that although hedgehogs encountered wader nests by chance whilst foraging for other foods (e.g. earthworms), their opportunistic predation of eggs was sufficient to account for up to 60% of nests, with dunlin and redshank being the most susceptible species (Jackson & Green 2000). Subsequent experimental studies revealed that if hedgehogs were removed and excluded from a large, fenced area of machair, nesting success of waders inside the fenced areas was increased by approximately 2.4 times that in adjacent, unfenced areas where hedgehog densities were high (Jackson 2001). By the year 2000, hedgehogs occupied all of South Uist and Benbecula but remained absent from much of North Uist and associated smaller islands. A full survey of breeding waders on the machair of all these islands found that numbers had declined overall by almost 40% in the area occupied by hedgehogs but had increased slightly in hedgehog-free areas (Jackson *et al.* 2004). Population declines of lapwings, redshank and dunlin were most marked, whilst those of oystercatcher were unaffected, perhaps because hedgehogs are unable to break into the larger eggs of this species. Habitat change was unable to explain these population trends and it seems probable that they had been driven mainly by the increasing impact of predation by hedgehogs.

(2) Lowland songbirds

Potential effects of predation on songbird populations divide roughly into two groups: first, impacts on breeding success caused by nest predators such as corvids and mammals, and second, impacts on survival of full-grown birds of predation by raptors, primarily sparrowhawks. One of the most powerful studies to investigate these effects used 30 years of data from almost 300 CBC plots to assess whether the rates of population change of 23 songbird species varied depending on whether or not either magpies or sparrowhawks were present (Thomson *et al.* 1998). The results were unequivocal in that out of the 46 comparisons (23 songbird species and two predator species) in only two cases did the songbird species show a more negative population trend in the presence of a predatory species than in its absence (Fig. 9.10). This number is fewer than expected by chance alone and hence provides no evidence that sparrowhawks or magpies were driving population declines.

The disappearance of sparrowhawks from much of the British countryside during the era of organochlorine pesticide use (see sparrowhawk case study) also provided an opportunity to test whether their absence and subsequent return had any impact on songbird populations. Newton *et al.* (1997) studied songbird populations at Bookham Common, Surrey, annually for over 30 years (1949–79), with sparrowhawks absent for 13 years during the 1960s and early 1970s. They found that nine of 13 species studied increased during the course of the study, but that only one (song thrush)

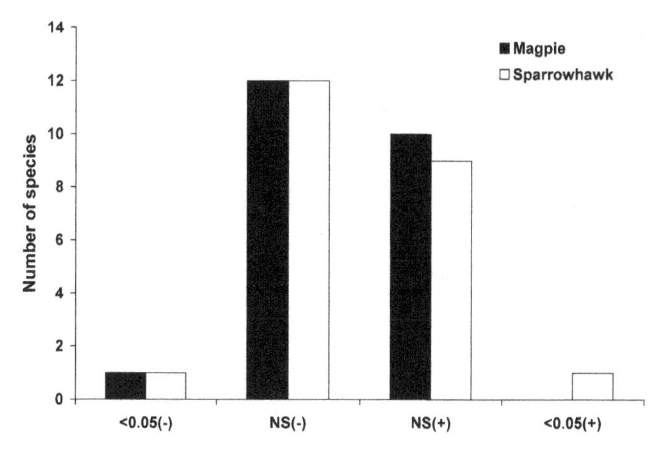

Fig. 9.10 Relationships between rates of population change and presence of magpies or sparrowhawks after controlling for nationwide variation in the rate of songbird population change. Significant relationships given by <0.05, non-significant by NS. Relationships are ether positive (+) or negative (–). Data from Thomson *et al.* (1998).

showed any evidence of a decline in numbers after recolonisation by sparrowhawks. Similarly, in the long-term study of great tit population dynamics at Wytham Wood in Oxfordshire, Perrins & Geer (1980) found that although sparrowhawks took up to a third of young great tits during the summer, fluctuations in great tit numbers remained unrelated to changes in sparrowhawk populations, and indeed great tit numbers tended to increase over the 40-year study period as the wood matured. Lastly, Chamberlain *et al.* (2008a) found no strong evidence that change in numbers of birds in gardens in winter was related to sparrowhawk presence.

Three other studies have used the BTO's nest record card database to test whether variation in nest failure rates of songbirds have been associated with increasing magpie populations. Gooch *et al.* (1991) found no such relationship over a 20-year period (1966–86) when magpie numbers were increasing rapidly, and also found no evidence that songbird nest failure rates were higher in areas where magpie densities were high. Groom (1993) did find that high densities of magpies in urban parkland in Manchester were associated with very low blackbird nest survival rates (only 5% produced fledged young), with magpie predation as the main cause of nest failure, so that the locally stable blackbird populations may only have been maintained by immigration. Similarly, Paradis *et al.* (2000) found that incubation-stage failure rates of blackbirds and song thrushes tended to be higher in areas where magpies and jays were more widespread. None of these studies, however, went on to assess whether there was any association between corvid numbers and predatory impact, and longer-term population change of the prey species.

Overall, there is some evidence to suggest that increasing populations of generalist predators may have played a direct role in contributing to the reduced breeding success that has driven population declines of some ground-nesting breeding birds.

Locally, high predator densities may also depress songbird nesting success and it is known that songbirds may avoid nesting in the immediate vicinity of nests of raptors such as kestrels or sparrowhawks (Suhonen *et al.* 1994; Norrdahl & Korpimäki 1998). However, the evidence suggests that the predatory impact of corvids on songbird breeding success or of raptors on mortality rates has played no significant role in driving large-scale population change of songbirds in Britain.

Agricultural change and the impact of predators on bird populations will often interact. For example, when breeding at high densities, lapwings exhibit communal nest defence, successfully mobbing predators to defend nests. Where agricultural change has reduced lapwing populations to much lower densities (see case study), then these effects may be compounded by much greater vulnerability to predation (e.g. Berg *et al.* 1992; Seymour *et al.* 2003; Stillman *et al.* 2006; Macdonald & Bolton 2008b). Here there is the potential for positive feedback which could drive local populations to extinction. In this lapwing example, the increased impact of predation is brought about when agricultural change reduces the prey population. In other cases, agricultural change may increase the activity or numbers of predators. For example, where forests are fragmented by agricultural activity, increased opportunities for generalist predators along the greater length of forest edge can lead to greater predation impacts on forest species (e.g. Andrén & Angelstam 1988; Andrén 1992; Paton 1994; Kurki & Lindén 1995; Kurki *et al.* 1997, 1998; Chalfoun *et al.* 2002; Thompson *et al.* 2002; Summers *et al.* 2004).

At a smaller scale, the effects of limitation of nesting or foraging opportunities that are driven by agricultural change may be compounded if in seeking to compensate for this through behavioural change, birds are exposed to greater predation risk. For example, we have already seen that when skylarks are obliged to nest close to tractor wheelings due to luxuriant crop growth, then nest predation rates increase (skylark case study; Donald 2004), and it is also known that where food supplies are reduced and nestlings are hungrier, then louder begging may result in higher predation rates (cirl bunting case study). A further example is provided by the bullfinch. Bullfinches rely on seed foods throughout the non-breeding season. Reductions in weed and hedgerow seed availability on farmland, coupled with reduction in availability of key woodland seed sources such as bramble due to increased deer browsing (Kirby 2001; Perrins & Overall 2001; Sage *et al.* 2004), are likely to have reduced winter food availability across a wide range of the habitats that this species exploits (Marquiss 2007). However, the species is also relatively vulnerable to sparrowhawk predation, especially when feeding near the ground (Selas 1993; Gotmark & Post 1996). For example, in eastern Scotland Marquiss (2007) found that bullfinches were more vulnerable to sparrowhawk predation in late winter when other seed foods are scarce, and the birds were obliged to feed far from cover on seeds of moorland heather. In this context, both Proffitt (2002) and Newton (2004) have suggested that the combined impact of reduction of food resources coupled with progressive restriction of safe foraging habitat to areas closer to cover as sparrowhawk populations have increased could have been important in reducing the carrying capacity of many farmland habitats for bullfinches, thus driving the observed population decline.

Perhaps one of the most subtle examples of this form of interaction also reveals how widespread it may be. It is well known that body mass in songbirds, especially in winter, varies as a mass-dependent consequence of a trade-off between starvation risk and predation risk. Thus birds that feed successfully and gain mass (as stored fat) have a higher chance of surviving overnight, but a higher risk of predation either due to reduced escape capabilities or increased time exposed to predation risk whilst foraging. Conversely, birds that feed less tend to experience lower predation risk but higher starvation risk (e.g. Lima 1986; Houston *et al.* 1993; Witter & Cuthill 1993). Gosler *et al.* (1995) found that this trade-off was reflected at the national scale in the masses of great tits captured for ringing. They discovered that, after controlling for other influential variables (e.g. body size, sex, age, time of day and temperature), the birds were heavier in years and at places where sparrowhawks were absent (again due to the effect of organochlorine pesticide use) and lighter in years and at places where hawks were present. Since then similar large-scale relationships between mass and measures of predation risk have been established for blackbirds, starlings and house sparrows (Macleod *et al.* 2005, 2006, 2008). These results imply that where such trade-offs are found, especially in declining species, this may indicate that predation risk is exacerbating the limitation on population size imposed by reductions in food supply (Macleod *et al.* 2007).

Evans (2004) notes that formal studies of interactions between agricultural habitat change and predation, and their effects on bird populations remain scarce. Such studies would help to design conservation measures that recognise the potential interactions (Morris & Gilroy 2008), increase the effectiveness of habitat-based interventions (e.g. Whittingham *et al.* 2006) and reduce the need for interventions that focus on predator removal. In view of the evidence reviewed above, we share Evans' view that further study of habitat management for bird conservation in agricultural habitats which recognises the role played by interactions with predation is a fruitful avenue for new work.

Summary

In arable farming systems, changes in the timing of sowing and harvesting of cereal crops have had pervasive and generally detrimental effects on farmland bird populations. The modern dominance of autumn-sown varieties limits both the availability of over-winter stubbles as seed-rich foraging habitats for granivorous species, and rapid crop growth in spring limits nesting opportunities for ground-nesting species such as lapwings and skylarks and, locally, Montagu's harriers and stone curlews. Earlier harvesting of autumn-sown crops also critically limits the length of the breeding season for corn buntings, especially in areas where spring-sown crops as an alternative nesting habitat are scarce or absent. Intensive use of insecticides on arable crops, especially in spring and summer, has contributed to the decline of grey partridges and there is increasing evidence that it can limit the breeding success and nestling condition of other species, notably buntings. Genetically modified herbicide-tolerant crops are not yet widely grown commercially in Britain, but experimental trials suggested that their introduction would be capable of yet further reducing the availability of key invertebrate and weed seed food sources for birds in arable systems.

Not all changes in arable rotations have been wholly damaging for farmland birds. Increasing oilseed rape acreages have provided a winter food source for woodpigeons which has reversed earlier declines brought about by the loss of clover-ley-based rotations. Similarly, rape seed has provided a crucial new food source for species such as linnets and turtle doves in areas where other seed foods have become scarce, and may have played a part in slowing population declines of linnets. In many areas, oilseed rape fields may now be important nesting habitat for reed buntings, though the success of the birds in this crop depends on the timing and method of harvesting.

Agricultural improvement of lowland grasslands by drainage, intensive fertilisation and reseeding with a limited flora of competitive grass species has had effects as profound as those in arable systems. Whereas unimproved hay meadows may historically have been occupied by a rich assemblage of breeding and foraging birds including quail, corncrake, curlew, skylark, meadow pipit, yellow wagtail, whinchat, grasshopper warbler, twite and corn bunting, only skylarks routinely seek to nest in intensively managed, multiply cut silage meadows where even this species suffers high rates of nest loss. Even where meadows are only cut once as in upland and northern areas, increasingly early cutting may have catastrophic effects, as exemplified by yellow wagtails in the hay meadows of the Pennine Dales and corn buntings in silage fields in north-east Scotland. Drained and heavily grazed pastures have now largely lost their breeding wader populations, especially in the lowlands, although effects can increasingly be seen in enclosed upland grasslands and in-bye ground too.

Intensification of grassland management also generally reduces foraging opportunities for birds. Species-poor, intensively managed meadows and pastures support a limited diversity and abundance of foliar invertebrate populations compared to unimproved grasslands, with populations of large species such as Orthoptera particularly badly affected. This affects species whose breeding success depends on access to large invertebrates, such as whinchats, cirl and corn buntings and, historically, red-backed shrikes. Similarly, heavy grazing or multiple cutting greatly reduces seed set from grasslands and can make some grassland landscapes almost devoid of seed sources for many granivorous species, other than in boundary habitats. Nonetheless some generalist herbivorous invertebrates do respond positively to agricultural improvement of grassland, and this may have benefitted species such as golden plover, woodcock, corvids and starlings feeding on earthworms, leatherjackets and other arthropod larvae, and barn swallows feeding on insects associated with cattle-grazed pastures. In otherwise arable landscapes, pockets of agricultural grassland remain very important as foraging habitats for a wide variety of species that feed on soil invertebrates whose densities are reduced by tillage. These include corvids, thrushes, plovers, gulls and starlings. Overall, the biological impoverishment of lowland grasslands through agricultural improvement is one of the most profound conservation losses of the twentieth century, and its effects can be seen widely across taxonomic groups, including flowering plants, insects, mammals and birds.

In the uplands, increases in stocking densities of sheep have had marked effects on the botanical and structural composition of moorland vegetation with losses of dwarf shrub cover, especially heather, in favour of grass-dominated swards. The only species whose national population decline is likely to have been driven by this change is red grouse, but species such as black grouse, ring ouzel and twite are likely to have experienced losses of nesting and foraging habitat at least locally. Generally, wherever intensive grazing and drainage in the uplands have led to loss of botanically and structurally diverse mixes of heath, grassland and blanket bog vegetation in favour of more uniform short-grazed grass-dominated swards, then the richness of the moorland breeding bird assemblage is likely to suffer. Conversely, carefully managed grazing regimes remain an essential management tool if diverse upland vegetation mosaics capable of sustaining Britain's unique and internationally important moorland breeding bird assemblages are to be maintained.

The effects of agricultural management of fields may have overtaken outright loss of boundary habitats such as hedgerows as a driver of change in bird populations on farmland in recent decades. However, loss of well-vegetated hedge bases and field margins has contributed to grey partridge declines, and loss of food resources and secure nest sites in large hedgerows as a consequence of mechanised hedge management has probably played a part in bullfinch declines. Management of hedgerows and associated marginal features such as wet ditches, uncultivated field margins, hedgerow trees and ponds remains central to agri-environment management designed to restore nesting and foraging opportunities for a wide range of farmland species, especially in landscapes where the fields themselves are now intensively managed.

The role of increasing populations of generalist predators in causing declines of bird populations in agricultural habitats has been a matter of long debate for conservation scientists and land managers. Evidence suggests that predation may in some circumstances limit populations of ground-nesting species such as waders and gamebirds, but it is extremely unlikely that increasing predation rates have played any direct role in songbird population declines. However, there is increasing evidence of interactions between predation impacts and habitat management such that increasing predation rates may be caused by, and exacerbate, the effects of agricultural change that have driven prey population declines. Equally, the effectiveness of habitat management solutions to bird population declines driven by agricultural change might be limited if those populations are now being held at low levels by abundant populations of generalist predators. Further study of habitat management for bird conservation in agricultural habitats which recognises the role played by interactions with predation is a fruitful avenue for new work.

From the most intensively under-drained arable or pasture fields to the 'gripping' of moorland to improve grazing quality, drainage has characterised the agricultural 'improvement' of the British countryside for many decades. Indeed, grant-aided under-drainage during the 1970s and 1980s has made British agricultural landscapes some of the most intensively drained in Europe. With the exception of the impacts of drainage on the ecology of breeding waders, the effects of this drainage on birds in agricultural systems in Britain have been rather poorly studied. However, there is increasing evidence of the importance of small-scale wet features on farmland for species as diverse as song thrushes, starlings, yellow wagtails and tree sparrows (Field & Anderson 2004), and populations of many key invertebrate food sources for birds are known to be limited by water availability. Given predictions of drier summers across much of lowland Britain under future climate scenarios, the management of water on farmland may need to be more central to development of agri-environment management for farmland birds in future than it has been in the past. We consider this issue further in the context of wider management of farmland for biodiversity and other ecosystem services in Chapter 10.

Plate 9.1 Over-winter stubbles can be a valuable source of weed seed and grain for many farmland birds. The earlier that these stubbles are ploughed in, the greater the reduction in food supply for these species. © RSPB.

Plate 9.2 The corn bunting is a good example of a species for which the main drivers of population decline have differed in different areas of Britain. © RSPB.

Plate 9.3 Intensively grazed landscapes of species-poor, drained and improved grassland support a limited biodiversity and limited nesting and feeding opportunities for birds. Gulls, corvids and starlings feeding on earthworms and soil-dwelling arthropod larvae may be the only species regularly seen in abundance. © RSPB.

Plate 9.4 Studies in Switzerland concluded that the whinchat was an excellent indicator species of 'nature-friendly' management of agricultural grasslands. The virtual disappearance of this species from the enclosed agricultural landscape of Britain is equally an indicator of just how 'nature-unfriendly' modern agriculture has become. The abundance of whinchats and many other species on Salisbury Plain, Britain's last large, contiguous area of unimproved lowland grassland, shows what we have lost. © RSPB.

Plate 9.5 The red grouse is perhaps the only species for which a national population decline has been caused by increased intensity of sheep grazing in the uplands. In part, this also reflects a change in land-use as the number of moors managed for driven grouse shooting has declined. © RSPB.

Plate 9.6 The bullfinch is a species for which there is some evidence that predation risk from Sparrowhawks may have exacerbated the effects of agricultural intensification in reducing the carrying capacity of modern farmland. © RSPB.

10 · *What future for birds and agriculture in Britain?*

In a recent review of technological and policy drivers of European farming, Buckwell & Armstrong-Brown (2004) identified three agricultural eras spanning the past 3000 years. The era of pre-industrial agriculture began in Britain with the first efforts to grow crops and rear domesticated stock, and lasted until the seventeenth century. The second era – industrial agriculture – spans the successive application of horse and then fossil fuel power, coupled with biological, chemical, mechanical and communications advances, and the financial muscle of state support to drive increases in crop and animal production. This book has sought to provide an account of the profound effects of this era of industrial agriculture on wildlife habitats and their associated bird species. Chapter 1 showed how agriculture in Britain responded to changing socio-economic conditions, new agricultural techniques and technologies, and evolving Government policies. Especially since the Second World War, these have driven British agriculture to levels of mechanisation, specialisation, management intensity and production that would have been unthinkable to earlier generations. The following chapters introduced three broad components of agriculturally managed landscapes – the field, the field boundary, and semi-natural heathlands and grasslands – and their bird communities, as they stand today. Chapter 5 reviewed the wealth of information available on the population trends of farmland birds in Britain, and Chapters 6–9 considered the recent growth of research evidence that has given us clear insights into the impacts of agricultural intensification on wildlife, in general, and bird populations, in particular.

Buckwell & Armstrong-Brown's proposal is that increasing concern over the environmental costs of industrial agriculture may be leading to a third agricultural era; one that seeks to maintain the necessary levels of agricultural production, but in ways that are consistent with the delivery of the other 'public goods' desired by society, and with public subsidy directed at their provision. These public goods include clean air and water, healthy food and attractive rural landscapes rich in wildlife. The rapid development of explicitly 'agri-environmental' policies by UK and European governments since the late 1980s (see below), and the evidence base to support their implementation, certainly provides some justification for this optimistic view. So, in this last chapter we seek to do three things.

First, we set the context by reviewing the role of industrial agriculture and its environmental impacts in stimulating a progressive marriage between agriculture and nature conservation policy in Britain – from the formation of the Nature Conservancy Council (NCC) in 1949 to the sustainable development policies and agri-environment schemes that influence the management of agricultural landscapes today. Second, with respect to bird populations, we ask how effective the dawn of this third agricultural era has so far been in restoring past losses. In particular, we review the effectiveness of agri-environment schemes in reversing population declines of birds in agricultural habitats, and further changes that may improve the effectiveness of such schemes in future. We conclude with a look forward at the future prospects of the agri-environment era in the context of three drivers of change with a global reach – changing climate, globalisation of agricultural trade and increasing human populations. These pressures could have far-reaching consequences for the management of Britain's agricultural systems and their bird populations and, in the context of these challenges we highlight gaps in our scientific understanding of relationships between birds and farming that remain to be filled.

The history of conservation policy in Britain and the role of agricultural change: 1945–1987

The catastrophic impacts on wildlife of the technology and subsidy-driven revolution in agricultural production following the Second World War (see Chapter 1) forced the pace of development of nature conservation as a political and institutional force in Britain. Government had been persuaded by the idea that some land might be 'set aside' for nature conservation since the 1940s, and in 1949 established the NCC. The NCC was to 'provide scientific advice on the conservation and control of the natural flora and fauna of Great Britain', 'establish, maintain and manage nature reserves in Great Britain', and 'organize and develop the scientific services related thereto' (Stamp 1969). The underlying philosophy was based on the allocation of land to primary uses – agriculture, forestry, housing, industry – with 'a little for nature' (Smout 2000). Immediately, areas of land of special nature conservation interest could be scheduled as Sites of Special Scientific Interest (SSSI). Such sites were notified to local planning authorities who nonetheless retained a free hand in deciding whether or not development on these sites should proceed. Moreover, neither agricultural improvement nor afforestation constituted developments requiring planning permission, and the NCC's only recourse in these cases was compensation. The resources at their disposal were, however, paltry by comparison with the subsidies available from both the then Ministry of Agriculture and the Forestry Commission to encourage increased production. SSSIs therefore suffered heavily at the hands of agriculture and forestry to the extent that by 1980 it was estimated that 8% of SSSIs had suffered damage during the preceding 12 months (Marren 2002). Mellanby (1981) estimated that across 3750 SSSIs covering 1.16 million hectares, the annual rate of loss to other land uses was 4%. In 1978, a corner was turned when nature conservation arguments led by the NCC, RSPB and the Council for the Protection of Rural England (CPRE) were upheld at a (rare) public inquiry. The Southern Water Authority was prevented from securing Ministry of Agriculture grant-in-aid to drain the Arun floodplain at Amberley Wild Brooks in Sussex for the purposes of grazing improvement. This case established the precedent that the drive for agricultural improvement did not have automatic and universal priority over nature conservation interests. In 1981 the Wildlife and Countryside Act finally gave some statutory protection to SSSIs, requiring landowners to give advance notice of 'potentially damaging operations' on SSSIs (Marren 2002), and NCC and National Park authorities to compensate landowners for profits foregone in maintaining SSSIs. In many ways, this was only a small step in the right direction, since local planning authorities could still permit developments that damaged or destroyed SSSIs, though it did engender some co-operation between conservation agencies and landowners in SSSI management. Nonetheless, damage to and loss of SSSIs continued throughout the 1980s and 1990s. Only when – in the middle of the 1997 general election campaign – Friends of the Earth and the Sussex Wildlife Trust made headline news of the partial ploughing of Offham Down by a farmer attracted to generous (if temporary) EU subsidies to grow linseed, was the political momentum created that led to the Countryside and Rights of Way (CRoW) Act of 2000. This and the Nature Conservation (Scotland) Act of 2004 gave the nature conservation agencies the powers to prevent neglect, to refuse consent for action likely to damage SSSIs without risk of large compensation claims, and to prosecute if necessary. Public bodies now have a duty to take reasonable steps to protect SSSIs in carrying out their primary functions, and most importantly, management agreement funding shifted in emphasis from payments to prevent damage to payments for positive management. Indeed English Nature's pioneering Wildlife Enhancement Scheme had pre-empted this in being introduced ahead of the CRoW Act. The planning system can still permit development on SSSIs, but the bar is now set higher for the would-be developer than it has been before. As Peter Marren (2002) put it, 'for the first time in their half-century existence, SSSIs have become protected sites'. Throughout that time, agricultural intensification, more than any other form of development, did the greater part of the ecological damage, and forced the pace of legislative change.

Because the focus of nature conservation in post-war Britain was so squarely on the designation, legal protection and management of land for wildlife on SSSIs and nature reserves, the conviction that wildlife conservation objectives should be accommodated on actively farmed land was slow to gain ground. However, spectacular effects of agricultural intensification on wildlife – such as the complete loss of sparrowhawks from large areas of the countryside, and the disappearance of hedges and flower-rich grassland – were well documented and helped to raise public awareness of the impacts of modern farming on wildlife, as did the clarion call of Rachel Carson's powerful *Silent Spring*. As early as 1969 the RSPB organised a conference which for the first time set out ways in which productive farming and wildlife conservation might be reconciled *on the same ground* (Barber 1970). The report notes the importance of hedgerow protection, planting of farm woodlands and the moderation of pesticide use to minimise effects on non-target species; all themes which remain central to environmental management on farmland today. This initiative led to the formation of the Farming and Wildlife Advisory Group (FWAG) to help and advise those farmers interested in finding room for wildlife in their farming operations.

Despite these early efforts, ultimately more important in slowing the 'engine of destruction' (Marren 2002) was the accumulation of food surpluses during the 1980s. These surpluses, first of milk, then of cereals and meat were all fuelled by the price guarantees of the European Economic Community (EEC), of which the UK became a member in 1973, and were paid irrespective of whether there was a market for the produce. However, as the huge intervention stores built up (the infamous European food 'mountains' and 'lakes' of the 1980s), so also did the challenges to a system in which as Rackham (1986) put it, 'we contrive at the same time to subsidise agriculture much more than any industry, *and* to have expensive food, *and* a ravaged countryside' (see also Mabey 1980 and Shoard 1980).

Cereal surpluses were tackled by the introduction of set-aside, first introduced in 1988. In 1992, receipt of cereal subsidies was made conditional on farmers leaving up to 18% of their arable ground out of production ('set-aside'). Although not designed as a conservation mechanism, the fallowing of several hundred thousand hectares of arable ground in some of the most intensively farmland landscapes in Britain was gradually harnessed to deliver some wildlife benefit through modifying management requirements (Wilson & Fuller 1992; see Chapter 6). Although some assessments of the conservation value of set-aside have been unduly dismissive (see Marren 2002), the perception that farmers were being paid 'for doing nothing' was understandable, and the effectiveness of the intervention in reducing cereal output was undermined when farmers ploughed compensation payments back into increasing output elsewhere on the farm. Set-aside schemes are, as their name implies, production control mechanisms and any wildlife benefit that the management of set-aside has yielded has been additional to this primary purpose. In 1984, however, an influential House of Lords report on *Agriculture and the Environment* and the 1986 Agriculture Act prompted the then Ministry of Agriculture, Fisheries and Food (MAFF) to introduce, in 1987, a scheme to encourage – and fund – farmers to adopt agricultural practices that would safeguard and enhance parts of the country of particularly high landscape, wildlife or historic value. This Environmentally Sensitive Area (ESA) scheme proved long-lived and, by the turn of the millennium, approximately 10% of England's and 20% of Scotland's agricultural land was subject to ESA management. The arrival of ESAs in 1987 marks the point at which Government first began to offer positive financial support to mitigate the environmental effects of agricultural intensification by managing agricultural land for goods other than food and fibre – in this case, the conservation and enhancement of wildlife habitats, landscape and historic features. The era of agri-environment policy-making had begun.

If membership of the EEC tied the UK to the production surpluses which hastened the birth of 'agri-environment' thinking, it also tied it to European legislation of great significance for bird conservation. The 1979 EEC Directive on the conservation of wild birds (often known as the Birds Directive) required Member States to classify Special Protection Areas (SPAs) for two groups of birds: certain species rare or vulnerable to habitat change that were listed on Annex I of the Directive, and all migratory species not so listed. By the early 1990s (Stroud *et al.* 2001) the UK had a well-developed network of terrestrial SPAs covering a wide range of semi-natural and coastal habitats including upland and lowland heaths, machair and saltmarsh. Broadly, sites were selected to encompass those areas with high proportions of species' populations (whether breeding, wintering or migrating), but the selection process was also informed by secondary criteria such as seeking to ensure wide coverage across a species' range, designating areas characterised by high breeding productivity, and favouring areas of natural or semi-natural habitats. Although SPA designation does not preclude development, it does provide a high level of protection with strict procedures and tests to be met by proposed developments located either on sites or affecting them (Stroud 2002). Management of SPAs is focused on the requirements of the species for which the site has been selected, and is supported by funding available to landowners from the statutory conservation agencies in each UK country.

A wide range of species characteristic of grazed, semi-natural habitats now benefit from SPA designation and associated conservation management across Britain, including wintering wildfowl, hen harrier, golden eagle, merlin, peregrine, corncrake, stone curlew, most wader species, short-eared owl, nightjar, woodlark, Dartford warbler and chough. Because, by virtue of their ecology, the breeding populations of many of these species tend to be widely dispersed with a relatively low proportion of their populations covered by SPAs (the raptors are notable examples in the above list), an integration of designated site management with management of the wider agricultural countryside through agri-environment schemes is critical. This is a particular concern for species that make extensive use of cropped habitats during their life cycle. These habitats tend to have been excluded from SPA site designations based strict adherence to the guideline favouring restriction of SPA boundaries to natural or semi-natural habitats, irrespective of a species' ecological needs. Although there are exceptions, such as the designation of areas of agricultural grassland and arable crops for breeding stone curlews, conifer plantations for nightjar and woodlark, and an area of largely semi-improved pasture for wintering bean geese, many current SPA boundaries exclude areas of agricultural land on which a qualifying species may be dependent. A good example might be exclusion of in-bye pasture – a key foraging habitat for moorland-breeding golden plover – from an upland heathland SPA for which this is a qualifying species.

The agri-environment era: 1987 to the present

The EU first exerted its influence on the development of agri-environment policy in Britain with the so-called 'MacSharry' reforms of the CAP in 1992, which made agri-environment schemes available to all Member States, and offered joint European funding. In Britain, the first generation of 'stewardship' schemes were launched – the Countryside Stewardship Scheme in England (actually first launched as a pilot scheme in 1991), Tir Cymen in Wales and the Countryside Premium Scheme in Scotland. A key difference of these schemes was that they were not geographically restricted in the manner of ESAs, but were available to all farmers. However, the focus remained on the management of semi-natural habitats of high wildlife, landscape or historic value, such as calcareous and upland grasslands, heathland, unimproved pastures and hedgerows.

By the mid 1990s, three strands of evidence combined to persuade Government of the need for an agri-environment scheme focused on intensively managed, arable landscapes. These were: first, the widespread and severe declines in farmland bird populations, second, the potential wildlife benefits of well-managed set-aside in intensive arable landscapes (e.g. Wilson *et al.* 1995), and third, a recognition that a production-control mechanism was not well suited to providing a sustainable basis for wildlife recovery (rates of compulsory set-aside had fluctuated between 5% and 18% of the arable area). So, in 1997 RSPB, English Nature and the Game Conservancy Trust submitted proposals to MAFF for an 'Arable Incentive Scheme'. After receiving Ministerial approval, further consultation resulted in the development of the pilot Arable Stewardship scheme which was trialled in two areas of lowland England from 1998 to 2000, after which the successful options were incorporated within the Countryside Stewardship Scheme. These options were tailored to arable land and included over-winter stubbles, summer fallows, low-input spring crops and crop headlands, uncultivated field margins, beetle banks and cover crops designed to provide over-winter seed sources. Some were also introduced into several ESAs known to hold important populations of declining farmland birds (e.g. South Downs and Breckland).

Funding agri-environment management on productive arable land is an expensive business, and the 1999 reforms of the CAP (the 'Agenda 2000' reforms) made a critical contribution in this regard. First, they made agri-environment schemes compulsory across EU Member States, and second, they allowed Member States to redirect a proportion of their agricultural subsidy budget into various rural development initiatives, including agri-environment schemes. For the first time subsidy support for Europe's farmers was substantially 'decoupled' from production and diverted to a wider range of rural development measures, including agri-environment schemes (Barnett 2007). Accordingly, in autumn 1999, the UK Government announced its intention to progressively switch up to 4.5% of the subsidy budget into rural development (Evans *et al.* 2002), with agri-environment schemes as the main beneficiary. Since then, financial support for stewardship schemes has generally increased throughout Britain, with substantial increases in the area of land managed under agri-environment schemes (Fig. 10.1).

Also in 1999, the UK Government adopted 'wild bird populations' as one of 15 'Quality of Life' indicators of progress in its sustainable development strategy. Soon afterwards, MAFF adopted an undertaking to reverse the decline in farmland birds[1] by 2020 as one of its Public Service Agreement targets – the undertakings given by UK Government departments to improve their performance. This effectively formalised the UK Government's commitment to bird conservation on agricultural land. Reversing the decline in farmland birds is now viewed by UK Governments (especially in England) as a measurable surrogate for assessing the success of its

[1] The Farmland Bird Indicator – a composite index of farmland bird abundance in England, based on the trends of 19 typical farmland birds recorded by the BBS.

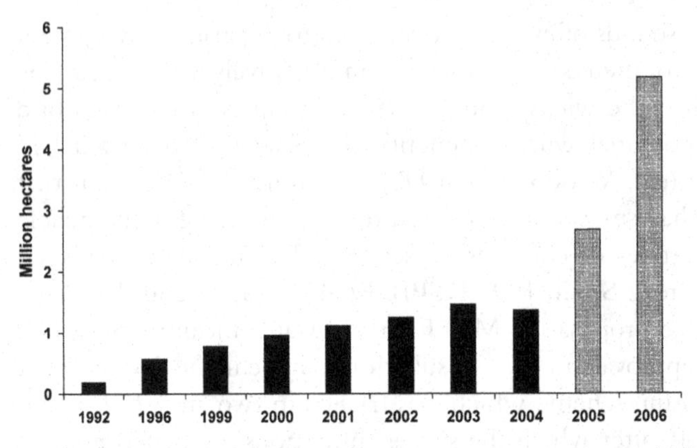

Fig. 10.1 Area of land under agri-environment management (including organic farming schemes) in England since 1992. Data for 2005 and 2006 are shaded grey because these totals are inflated by land under Environmental Stewardship, in which the whole farm area is recorded rather than the area under direct management. Data from Defra (www.defra.gov.uk).

policies which seek to conserve wildlife, in general, in the wider countryside. Birds were chosen for three reasons: first, because of the unparalleled information that is available on annual population changes (Chapter 5); second, for their ability to act as indicators of the general suitability of the farmed environment for wildlife (Gregory *et al.* 2005); and third, for their popularity, meaning that they can be used as 'flagship' species with which to engage farmers and draw attention to society's need for a sustainably managed countryside that is rich in wildlife. In 2001, the UK Government reorganised its departments dealing with environmental matters and created a new body, the Department for Environment, Food and Rural Affairs (Defra), whose stated remit was the pursuit of sustainable development in the UK. Defra brought together the responsibility for all aspects of environment, rural, farming and food production policy under one body, a clear signal of the Government's desire to integrate agriculture more fully with its wider environmental and socio-economic priorities. Indeed, it was the first time since 1889 that the Government department responsible for the farming industry did not have 'agriculture' in its title. These rapid changes towards a more integrated view of management of the countryside for agricultural production and other services justified the view espoused by Buckwell & Armstrong-Brown (2004) that a third generation of agriculture might be emerging.

In 2001, an outbreak of the highly contagious viral disease foot-and-mouth, which is not fatal to livestock but wipes out their commercial value, resulted in the greatest mass slaughter of farm livestock yet seen in Britain. Farmers were compensated individually for losses, but although the overall effects on the agricultural economy were great, they were far exceeded by losses in the recreation and tourism industries as huge swathes of the countryside were effectively closed to public access for fear of contributing to disease spread. The epidemic brought into sharp focus the need for a radical reconsideration of the future role of agriculture in Britain. Indeed, the nation's farming industry was widely perceived to be in crisis with declining farm incomes despite agricultural subsidies costing UK taxpayers annually some £3 billion, and unsustainable practices causing environmental degradation and human health scares, also at high public cost. This prompted the establishment, in August 2001, of a policy commission on the future of farming and food under the chairmanship of Sir Donald Curry. Whilst the Commission's remit was confined to England, many of the issues and recommendations that appear in its report *Farming and Food: A Sustainable Future* (Curry 2002) also have great resonance elsewhere. They considered that farming had become detached from the rest of the economy, society and the environment, and so 'reconnection' with these areas formed a central theme of the report's 105 recommendations. Most importantly for farmland wildlife, the Commission considered that the scope and commitment of public funds to pay farmers who protect and enhance the environment, notably through agri-environment schemes, should increase, funded by the further redirection of agricultural subsidies. They also recommended the piloting of a new 'entry-level' agri-environment scheme, as proposed by English Nature, RSPB and the Game Conservancy Trust, which would pay for relatively simple, low-cost environmental management over a large proportion of England's farmed land. The Commission also highlighted the need for effective farm audit and planning to get the most from environmental payments, the potential environmental benefits of organic farming and the need to keep an open mind over the effects of new technological developments, such as biotechnology and genetic modification of crops.

The Government responded quickly to some of the Commission's recommendations, announcing the introduction of a pilot Entry Level Stewardship scheme in England in July 2002. In March 2005, this was joined by a Higher Level Stewardship scheme consisting of more targeted, complex and costly land management options. Together known as Environmental Stewardship, this scheme replaced England's ESAs and Countryside Stewardship schemes to create an all-England scheme comprising two elements – a lower tier open to all with farmers choosing their options from a menu, and higher tier prescriptions available on a discretionary basis, targeted on areas of high wildlife, landscape or historic interest. A similar combination of entry-level (Tir Cynnal) and higher-level (Tir Gofal) schemes is available in Wales, and has emerged in Scotland in 2008 as the 'Land Managers' Options' and 'Rural Priorities' tiers of Rural Development Contracts. All of the schemes now operate in conjunction with the farmers' receipt of a Single Payment, decoupled from production but dependent on farming to basic standards of animal and plant health, environmental condition and animal welfare. In this context, we are in a period in which agri-environmental management throughout Britain can be seen as operating over a three-tier 'pyramid': 'good agricultural and environmental condition' with which all farmers must comply if they are to receive their main subsidy payment (cross-compliance), entry-level agri-environment available to all, and higher-level

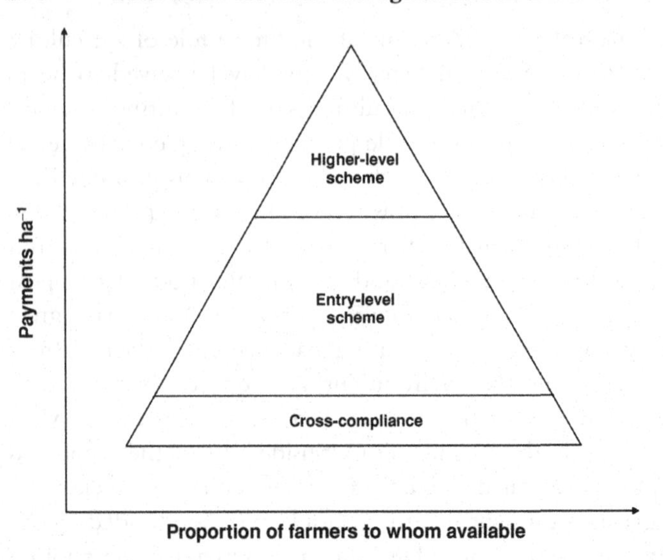

Fig. 10.2 Diagrammatic representation of agri-environment payment schemes as they currently apply in Britain.

agri-environment based on competitive application for more targeted and costly management options (Fig. 10.2).

The only fly in the ointment, but it is a significant one for the foreseeable future, is that overall rural development funding for the period 2007–13 was substantially reduced by the EU (Barnett 2007). Given that the revised EU Rural Development Regulation of 2005 requires monies to be spent across three so-called 'axes' or 'pillars' (I: to develop competitiveness, II: for environmental improvement, and III: for business diversification), with only 25% required to be spent on Axis II, resources may be spread more thinly through in the upper tiers of the agri-environment pyramid than had been hoped.

Agri-environment schemes: a success in reversing wildlife declines on farmland?

It may seem odd to begin this section with a question mark. After all, the history of development of agri-environment schemes – in Britain at least – has been characterised by a close relationship between research studies to design individual measures directed towards specific biodiversity goals, and the adoption of these measures within schemes (e.g. Grice *et al.* 2004). If we take as an example the list of measures currently available within England's Entry Level Stewardship scheme, and primarily available to deliver biodiversity benefits, then for almost every measure there is strong evidence of its benefit for the targeted species (Table 10.1).

Table 10.1 *Management measures available to deliver biodiversity benefits in the Entry Level tier of England's Environmental Stewardship scheme (see www.defra.gov.uk), and the evidence base underpinning the detailed management prescriptions*

Measure	Target(s)	Evidence base
Hedgerow management	Higher, wider hedges that are not cut annually to benefit wildlife in general, including birds, small mammals, invertebrates and plants	Relationship between hedgerow management and bird use for nesting and foraging is well studied, and reviewed by Hinsley & Bellamy (2000)
Ditch management	Varied bankside and aquatic vegetation and undisturbed general wildlife habitat adjacent to ditches	Ditches are a key foraging habitat for birds which rely on aquatic invertebrates or soft ground for accessing soil invertebrates (Peach *et al.* 2002; Brickle & Peach 2004; Duriez *et al.* 2005)
Grass margin buffer strips	Creating new habitat for small mammals, invertebrates and nesting and foraging birds; protecting habitats from agrochemicals and cultivation	Uncropped field margins adjacent to arable fields provide important foraging habitats for several bird species; reviewed by Vickery *et al.* (2002). Few studies of similar strips adjacent to grass fields, but increases in invertebrate diversity and abundance suggest benefits to birds are likely to accrue (Haysom *et al.* 2004; Cole *et al.* 2007; IGER/BTO/CAER 2007)
Wild bird seed mixes	Providing a year-round seed supply for granivorous birds by planting crop mixes in blocks or as strips along field margins. Can also provide a source of invertebrates	Successfully established and seeding crop mixes support high densities of foraging birds relative to conventional crops both in winter and summer (e.g. Henderson *et al.* 2004b; Parish & Sotherton 2004; Stoate *et al.* 2004)
Pollen and nectar flower mixture	Boosting numbers of nectar and pollen-feeding insects, including butterflies and bees	Recent evidence that this measure can provide an attractive foraging resource for these insects (Pywell *et al.* 2006; Carvell *et al.* 2007) but more work needed to assess long-term population effects.
Over-winter stubbles	Providing an important winter foraging habitats for granivorous birds through spilt grain and the seeds of broadleaved weeds	Good evidence from many studies that seed-rich stubbles can attract high densities of granivorous birds (e.g. Wilson *et al.* 1996b; Buckingham *et al.* 1999) and some evidence of subsequent positive effect on breeding abundance (Gillings *et al.* 2005b)

(*cont.*)

Table 10.1 (*cont.*)

Measure	Target(s)	Evidence base
Beetle banks	Providing habitat for insects, especially predators of cereal pests such as aphids, as well as foraging habitat and cover for birds and small mammals	Beetle banks can support high densities of predatory arthropods which may also reduce densities of cereal pests, as well as providing insect-rich habitat for foraging gamebirds (Thomas *et al.* 2001; Collins *et al.* 2002; MacLeod *et al.* 2004)
Skylark plots	Provides undrilled patches in winter cereal fields to increase suitability for nesting skylarks	This measure is based on experimental evidence that winter cereal fields with undrilled patches support higher breeding densities of Skylarks for more of the breeding season than conventional fields and that birds nesting in such fields have enhanced productivity (Morris *et al.* 2004, 2007)
Uncropped, cultivated margins on arable land	Providing habitat for arable plants and insects, and foraging habitat for birds	See under grass margins (above) for benefits to birds and insects. Recent evidence that uncultivated or fallowed margins are better for arable flora diversity than reduced-input or conventionally cropped headlands (Walker *et al.* 2007)
Conservation headlands	Reducing pesticide use in cereal field headlands to increase populations of arable plants and insects, and provide foraging opportunities for birds	Good evidence of beneficial effects on invertebrate communities, recently reviewed by Frampton & Dorne (2007). Benefits to arable plants only realised when fertiliser inputs are also reduced (Walker *et al.* 2007). Benefits to birds little studied beyond value in providing chick food for grey partridges (e.g. Sotherton 1992)
Fodder crops or whole crop silage followed by over-winter stubbles	Providing seed sources for farmland birds	Relatively low-input management of fodder crops makes the subsequent crop stubble especially rich as a seed source for granivorous birds (e.g. Hancock & Wilson 2003)
Undersown spring cereals	Encourages crop diversity (mix of arable and grassland), and grass understorey to cereal crop allows reduced agrochemical input to benefit invertebrate populations	Some correlative evidence that mixed farming systems support richer bird communities (e.g. Siriwardena *et al.* 2000d; Atkinson *et al.* 2002). The lack of subsequent ploughing of undersown cereals is known to benefit populations of sawflies (Barker *et al.* 1999). Other than this, evidence of wildlife benefits of undersowing is limited

Low-input grassland	Low agrochemical inputs designed to sustain higher numbers and diversity of plants and insects	Some evidence the biodiversity in intensive grass swards may begin to recover after time under agri-environment management (e.g. Swetnam *et al.* 2004), and studies of pasture headlands suggest that lower intensity management can favour invertebrate populations (Cole *et al.* 2007; Woodcock *et al.* 2007). More work needed to identify the best management regimes for birds
Management of rush pastures	Creating optimal rush cover and sward structure in damp pastures to benefit breeding waders	Previous studies of habitat associations of breeding waders in marginal grassland suggest that intermediate levels of rush cover are optimal for waders such as snipe, redshank and curlew (O'Brien 2001)
Mixed stocking	Encouraging a diverse sward structure on land grazed by both sheep and cattle to benefit breeding birds	Best evidence comes from experimental studies in upland grasslands where mixed, low-intensity grazing increased abundance of meadow pipits (Evans *et al.* 2006b)

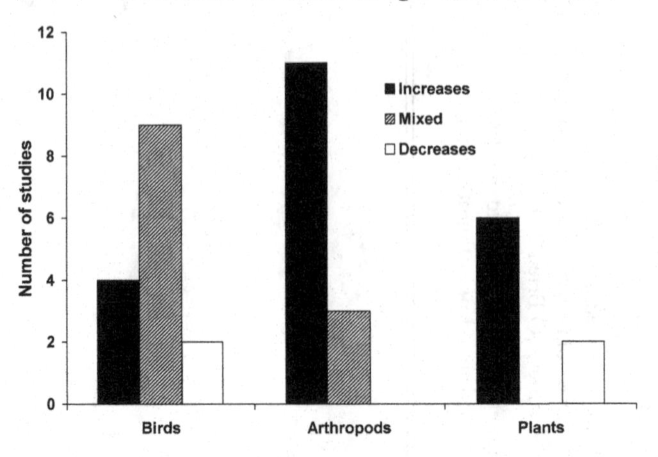

Fig. 10.3 Number of studies of birds, arthropods and plants reviewed by Kleijn & Sutherland (2003) with increases, decreases or a mixture of trends in response to agri-environment schemes.

Nonetheless, for those measures directed at bird populations, Table 10.1 reveals good evidence only of their success in attracting foraging birds or locally increasing nesting densities or reproductive success, rather than understanding whether scheme implementation has been able to reverse population declines at larger scales. Such studies of the overall effectiveness of agri-environment schemes remain scarce. For example, by 2003 the EU had spent approximately €24.3 billion on agri-environment schemes, a small but increasing proportion of the estimated *annual* €16.9 billion cost of the CAP. Yet in a review of what these schemes had achieved for biodiversity conservation, Kleijn & Sutherland (2003) found just 62 evaluation studies from only five EU countries (plus Switzerland), with the majority of these coming from the UK and the Netherlands. Overall, the results were very mixed. Whilst evidence that agri-environment measures were associated with increases in species richness or abundance outweighed the converse outcome, in many cases results were not clear-cut (Fig. 10.3). Where statistical analyses had been carried out, it appeared that agri-environment measures were more likely to have had beneficial effects on arthropod communities than on plants or birds.

Most tellingly of all, Kleijn & Sutherland (see also Berendse *et al.* 2004) concluded that the implementation of agri-environment programmes across Europe had not, in general, been accompanied by robust evaluation studies (i.e. involving collection of baseline data, random allocation of treatment and control sites, and replication) to allow assessment of the effectiveness of schemes and the individual management measures contained within them. Meeting these requirements is by no means straightforward since the EU does not cover the costs of monitoring in the way that it does the implementation of the schemes, and resources devoted to monitoring are often those that would otherwise be spent on delivering management (Carey 2001). Nonetheless, perhaps in part because of Kleijn & Sutherland's admonition, the subsequent increase in published studies of the effectiveness of agri-environment scheme implementation for biodiversity has been marked (Whittingham 2007). Studies of arable and grassland flora, arthropod communities in general and pollinators in particular, and small mammals have all been conducted widely across Europe and, strongly suggest that agri-environment measures have proved beneficial (Table 10.2).

For birds in Britain, there are three cases where there is convincing evidence that agri-environment, coupled with other conservation measures, have reversed national population trends. These are for the corncrake, stone curlew and cirl bunting (Fig. 10.4).

The recent histories of these three species have been described in detail in Chapter 8. In all three cases, population recovery has depended on three key factors (Evans & Green 2007):

(i) thorough diagnosis of the reasons for population decline,
(ii) design and testing of management for population recovery, from biological, practical and economic perspectives, and
(iii) targeted deployment of the tested measures in the right locations and at the right scale.

It is of course easier to target management, and usually easier to secure resources for intensive delivery of bespoke measures when a species has become rare and localised. The recovery of each of these species has depended on this approach, with specific recovery measures supplementing those available in the 'standard' agri-environment schemes (Aebischer *et al.* 2000; Evans & Green 2007). The same is also true for species that are more geographically widespread but restricted by habitat requirements. For example, Ausden & Hirons (2002) found that for agri-environment measures to deliver population recovery of breeding waders (lapwing, redshank and snipe) on lowland wet grassland in the English Broads ESA, the higher-tier management options had to be adopted. These required farmers to maintain high groundwater levels in meadows during the wader breeding season to ensure damp soils and good foraging conditions. Lower-tier measures which simply prevented farmers from cultivating or reseeding meadows and restricted fertiliser usage and timing of field operations were unable to slow population declines. Five years later, A. M. Wilson *et al.* (2007) examined change in breeding density of the same three species more broadly across lowland wet grassland in southern England over a 20-year period (1982–2002) and found very similar results. Even the higher tiers of ESA management had only halted or reversed declines of lapwing and redshank. Catastrophic declines of snipe had continued everywhere (this species was effectively extinct outside higher-tier ESA or nature reserve management), and all three species had declined where only lower-tier ESA management was practised, and in the wider countryside (Fig. 10.5).

Very similar evidence comes from studies of breeding waders and agri-environment schemes in the Netherlands, where monitoring studies initially showed little positive impact of agri-environment management (e.g. Kleijn *et al.* 2001, 2004;

Table 10.2 *Summary of findings of recent (post Kleijn & Sutherland 2003) studies examining the effectiveness of agri-environment schemes (AES) for biodiversity (other than birds)*

Study	Location	AES measures /taxonomic groups studied	Key findings
Askew *et al.* (2007)	UK	AES grasslands/small mammals	In all habitats except farm woodlands, small mammal numbers were positively associated with taller swards characteristic of agri-environment management
Aviron *et al.* (2007a)	Switzerland	Wild flower strips on arable farmland/arthropods	Strips had more arthropod species than conventional crops but a similar or lower activity-density of epigeic arthropods
Aviron *et al.* (2007b)	Switzerland	AES grasslands/butterflies	AES grasslands had a butterfly assemblage more characteristic of higher plant species richness; however, benefits to overall species diversity depended on local site conditions and availability of semi-natural habitats nearby
Carvell *et al.* (2007)	UK	Arable field margins/bumblebees	Margin management designed to provide nectar and pollen can quickly provide an attractive forage resource, but *seasonal flowering phenology and longevity of the seed mixture need to be improved*
Critchley *et al.* (2003)	UK	Semi-natural grassland*/vascular plants	Over 38 sites, 22 showed no significant change in botanical composition, 9 showed some evidence of rehabilitation and 7 deteriorated. *Greater use of site-specific targets and prescriptions would enhance future scheme performance*
Kleijn *et al.* (2004)	Netherlands	*/vascular plants, birds, bees, hoverflies	Over 78 field pairs, neither plant species richness nor meadow bird abundance was higher on fields in AES management. However, species richness of bees and hoverflies was enhanced. *Simple conservation measures may not be sufficient to counteract the impact of factors controlled at the landscape scale*

Kleijn *et al.* (2006)	Germany, Spain, UK, Switzerland, Netherlands	*/ vascular plants, bees, Orthoptera, birds, spiders	Across countries, AES had marginal to moderately positive effects. However, uncommon or rare species rarely benefited. *AES objectives may need to differentiate between common species requiring relatively simple modifications to farming practice and endangered species requiring more elaborate conservation measures*
Knop *et al.* (2006)	Switzerland	Hay meadows/vascular plants, Orthoptera, bees, spiders	Species richness of vascular plants, grasshoppers and bees was significantly higher on AES hay meadows than conventional meadows, but that of spiders did not differ. *Organisms that particularly depend on vegetation structure (e.g. spiders) should be targeted with additional management restrictions*
Kohler *et al.* (2007)	Switzerland, Netherlands	Grasslands/bees and insect-pollinated plants	Species richness of bees and plants, and abundance of bees (other than bumblebees) was higher on AES than non-AES fields in Switzerland. The same effect was not found in the Netherlands where bee diversity and abundance is much lower. *Additional measures are needed that focus specifically on needs of bees and other pollinating arthropods*
Maes *et al.* (2008)	Netherlands	Farmland ditches/amphibions	All native species had higher adult abundance in AES ditches than in non-AES ditches
Marshall *et al.* (2006)	UK	Arable field margins/vascular pants, birds, bees, spiders, Orthoptera, carabid beetles	Diversity and/or abundance of vascular plants, bees and Orthoptera were higher in 6 m sown grass margins than conventional crop margins. There were no negative effects
Pywell *et al.* (2006)	UK	Arable field margins/bumblebees	Bumblebee abundance was higher in pollen and nectar margins than wildflower margins (2x), mature grass margins (14x), recently sown grass margins (10x) and crop margins (430x). Species richness was higher in the first two of these margin types than the others. *Research is needed to determine the quantity and location of foraging habitat required to restore bumblebee populations at the landscape scale*

(*cont.*)

Table 10.2 (*cont.*)

Study	Location	AES measures /taxonomic groups studied	Key findings
Swetnam *et al.* (2004)	UK	Grasslands*/vascular plants	Some evidence when range and abundance of 18 plant species were compared before (1980) and after (1997) ESA designation, of an increase in species number and range
Walker *et al.* (2007)	UK	Arable field margins/vascular plants	Plant species diversity was highest in uncultivated margins, followed by spring-fallowed margins, and cropped but unfertilised conservation headlands. Cropped and fertilised conservation headlands were no different to conventional cereal crop controls. *More precise geographical targeting, improved control of competitive species and research on requirements of rare species will improve scheme efficacy*

* indicates studies that made comparisons at the scheme level rather than comparing individual measures within schemes. All studies compare sample points under AES and control management at a single point in time, except Critchley *et al.* (2003) and Swetnam *et al.* (2004), which are both longitudinal (i.e. before–after) comparisons. Italics indicate recommendations made by the authors for further research or improvement of agri-environment measures.

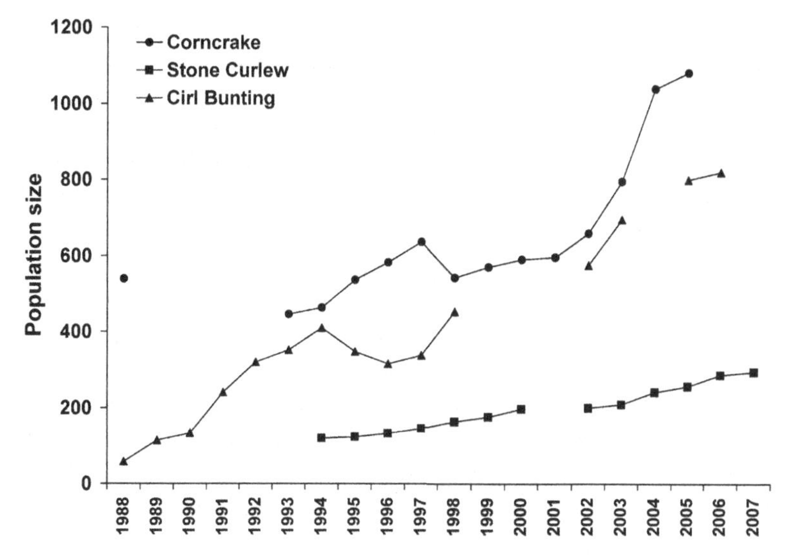

Fig. 10.4 Trends in population size of corncrakes (calling males in core survey areas), stone curlews and cirl buntings (total breeding pairs) in Britain since 1988.

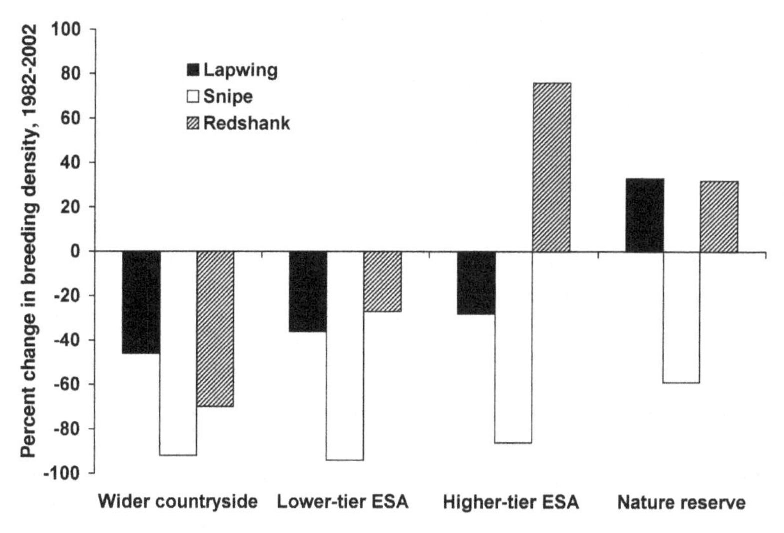

Fig. 10.5 Changes in densities of breeding waders (lapwing, redshank and snipe) on lowland wet grassland across southern England, 1982–2002, as a function of agri-environment and nature reserve management. Data from A. M. Wilson *et al.* (2007).

Kleijn & van Zuijlen 2004). More recent studies suggest that current schemes might be improved for breeding waders if measures are added to raise groundwater levels in meadows (Verhulst *et al.* 2007); effectively the equivalent of the higher-tier measures in the ESA schemes in England.

Broader evaluations of agri-environment scheme performance for more widespread farmland birds remain few. The best example is the monitoring of the short-term response (two and five years after implementation) of breeding bird populations to the pilot Arable Stewardship Scheme introduced in 1998 in two areas of England (Bradbury & Allen 2003; Bradbury *et al.* 2004; Stevens & Bradbury 2006). However, although some positive responses at the field or field boundary scale were evident and accounted for by behavioural responses to increased availability of food resources and nest sites offered by scheme options, there was very little evidence of a farm-level population response over these timescales. Indeed, two recent studies of the response of breeding abundance of seed-eating birds to the availability of seed-rich habitats such as stubbles – now widely available through agri-environment management – have provided only very inconsistent evidence of any positive response (Gillings *et al.* 2005b; Roberts & Pullin 2007; Siriwardena *et al.* 2007; and see Fig. 9.1). Defra's composite farmland bird indicator for England reached its lowest ever level in 2007, despite the increasing areas of land under agri-environment management. As Vickery *et al.* (2004) point out, although entry-level or lower-tier agri-environment measures may, in principle, be delivered very widely across the agricultural countryside, further research-based testing, improvement and targeting of individual measures may yet be needed for these schemes to reach their full potential. It may also be that they need to be deployed over much larger areas to be effective.

Perhaps then it is no coincidence that despite heavy investment in diagnostic research, the population declines in a wide range of more widespread declining farmland birds in Britain have not yet been turned into national population recoveries. Contrast this with the successful delivery at a population level of targeted recovery initiatives built upon higher-level agri-environment schemes for range-restricted species (Fig. 10.4). There seems to be general agreement amongst students of agri-environment management for birds (e.g. Vickery *et al.* 2004; Evans & Green 2007; Whittingham *et al.* 2007) that if the long-term declines of widespread species are to be reversed then much remains to be done in four respects:

(i) adapt and improve individual measures in the light of experience,
(ii) better understand how to target particular measures within agricultural landscapes,
(iii) quantify and plan the appropriate scale and spatial dispersion of measures across landscapes, and
(iv) determine the extent to which the required measures for individual species might vary across their geographical range.

These themes are well reflected in the conclusions reached by studies of responses of other taxonomic groups to agri-environment management (Table 10.2). Several of these studies argue for agri-environment management that is better targeted, both spatially and in terms of the design of individual management measures. Two note that successful agri-environment management may depend on understanding specific ecological requirements of individual species well enough to design bespoke measures for their conservation. Another three note the importance of factors operating at the landscape scale and the need for research at this scale if we are to understand the quantity and spatial distribution of key habitats that will need to be provided if populations are to be restored over large geographical areas. Vigorous research continues to provide the evidence to improve agri-environment management in all these respects, and we now consider some examples focused on birds.

In arable landscapes, Butler *et al.* (2005a) and Whittingham *et al.* (2006) have demonstrated how manipulation of the height and density of post-harvest crop stubbles could be important in determining the extent to which those stubbles are used by seed-eating birds, and Henderson *et al.* (2007b) have shown how regular management to open up the vegetation structure of grass margins of arable fields, for example by scarification, cutting or use of selective graminicides, can enhance their value for insects, arable plants and foraging birds. In grassland landscapes, recent studies by Buckingham & Peach (2006) and Mortimer *et al.* (2007) have shown how leaving either final-cut grass silage or the stubbles of whole-crop cereal silages in situ over winter may be of value in restoring populations of seed-eating birds in grassland landscapes.

Studies of whole-crop silages also provide a simple example of the importance of appropriate targeting of measures within agricultural systems. In intensive lowland grassland systems such as the dairying areas of Cheshire, Shropshire and Staffordshire, the landscape is dominated by grass pastures and silage meadows, with maize – the only arable crop grown over a substantial area – providing little in the way of nesting or foraging resources for farmland birds. Here, growing spring barley as an alternative form of whole-crop silage to maize or ryegrass provided very valuable and otherwise scarce seed resources for farmland birds if the stubbles were left after harvest (Mortimer *et al.* 2007). In sharp contrast, a recent decline of the corn bunting population on the Western Isles of Scotland has been caused by the replacement of traditional reaper–binding of ripe cereals as a cattle fodder (a method that ensured a rich supply of grain over winter for corn buntings as the cereals were fed out to cattle) – by early harvesting of the cereal as a whole crop silage (Wilson *et al.* 2007b). In this case, the grain itself is the critical food resource, and early harvesting of the cereal as whole-crop silage, usually before the grain has formed, removes a key winter food source from the system.

Recent studies have begun to address the challenging task of understanding what might be the optimum spatial and temporal dispersion of seed resources in arable landscapes to benefit seed-eating birds (Calladine *et al.* 2006; Siriwardena *et al.* 2006, 2008; Perkins *et al.* 2008a). For example, should two patches of seed of the same food value be placed far apart so that each benefits a different population of birds, or should they be placed closer together to benefit the same population of birds, for example by providing alternative feeding sites when one is made unavailable due to disturbance or presence of a predator? These spatial studies have not yielded unequivocal conclusions, because of the twin difficulties of assessing the ranging behaviour of individual birds using colour-ringing or radio-tracking, and the great difficulty of imposing an experimental array of seed patches in landscapes where other such patches are available in the normal course of agricultural management. Nonetheless, Siriwardena *et al.* (2006) tentatively conclude that birds of a range of species tend to share seed patches within 500 m of each other, and that patches at least 1 km apart may be a cost-effective minimum recommendation to inform any planning of agri-environment management at the landscape scale.

Lastly, recent research on declining populations of corn buntings in parts of Britain as widely dispersed as Sussex, Aberdeenshire and the Western Isles illustrates the extent to which well-targeted agri-environment management for a single species may differ markedly across its geographical range (Brickle 1998; Wilson *et al.* 2007a, b; Perkins *et al.* 2008b). In Sussex, intensive arable management with autumn-sown cereals, heavy pesticide inputs and a lack of over-winter seed sources limits both winter and breeding-season food supplies for corn buntings, causes nest losses through harvesting of cereal crops and limits opportunities for re-nesting. Here, agri-environment management needs to provide a variety of resources if declines are to be halted (see corn bunting case study). Conversely, on the Western Isles, where the birds nest mainly in dune grassland in a very low-intensity crofting system, agri-environment management that secures an over-winter supply of grain is probably all that is needed to reverse the current population decline. In Aberdeenshire, the situation is intermediate. Here, a mix of autumn- and spring-sown cereals and retention of some areas of stubble over winter offer more propitious conditions for corn buntings than in many areas of lowland England. However, the later springs at this latitude mean that silage is harvested much later than in southern England (late May to late June) and this encourages many corn buntings to nest in silage fields where they are then subject to high losses in much the same way as corncrakes (see case study) when the silage is harvested. In this population, then, an essential piece of the agri-environment jigsaw will be measures to delay silage cutting in fields occupied by nesting corn buntings, and research is now taking place to assess how effective (for corn buntings) and practicable (for farmers) such a measure might be.

A look to the future: birds and agriculture in Britain at a time of global change

The current global human population of 6 billion is expected to grow by 50% over the next 50 years, with rapidly increasing per capita rates of consumption (Lutz *et al.* 2001; Myers & Kent 2003). A combination of increasing yields in the developed world and the conversion of huge new areas of land to agricultural and energy production will feed a predicted two- or threefold increase in demand for food (Tilman *et al.* 2002; Jenkins 2003). However, the loss of pristine habitats, especially in tropical areas, two- to threefold increases in global pesticide use and agricultural eutrophication, and the growing impacts of global climate warming are predicted to commit many species to extinction (C. D. Thomas *et al.* 2004). As Tilman *et al.* (2001) note, over a third of terrestrial and a half of freshwater primary production is already appropriated by human food chains, and this fraction is bound to increase, with an inevitable reduction in the energy available to support non-crop biodiversity worldwide (Krebs *et al.* 1999; Firbank 2005). Set against this daunting cloud, current agricultural management in Europe is a small but important silver lining. Europe has a very long history of human settlement and agricultural management of a high proportion of the land surface, and much wildlife, including many bird species, is now adapted to landscapes characterised by early successional vegetation conditions that are created and maintained by grazing and cultivation. Conservation effort very often involves maintaining these landscapes but doing so in ways that accommodate agricultural production. This is exactly what the agri-environment programmes funded by the EU and its national governments seek to achieve. As Sutherland (2004) notes, the approach contrasts quite starkly with more recently human-modified landscapes, such as those of North America, where the focus of conservation has historically been more on the protection and restoration of natural habitats rather than the more limited wildlife values of agriculturally productive land. This is not surprising as the agricultural ecosystem in Britain has evolved over 5000 years, whilst widespread agricultural modification of landscapes is a much more recent phenomenon in North America.

So, in Europe we have a long history of coexistence of agriculture and wildlife, we have current national and European policy commitments to deliver a wider range of public goods – including environmental gain – from agriculture, and we have some evidence that agri-environment programmes are beginning to reverse some of the wildlife losses driven by post-war agricultural intensification. From this position, the time is ripe to consider how we manage our agricultural landscapes in future so that we can continue to accommodate rich wildlife communities alongside the ability to maintain and probably, in future, increase food production and to do so in the context of the greater demands that climate change is already bringing (see also Firbank 2005). To achieve these goals, the research agenda will need to address at least three key questions.

(1) How do we integrate the conservation of birds and other biodiversity with agricultural production and delivery of an increasing range of other 'ecosystem services' (e.g. flood mitigation, water quality, carbon stewardship and recreational access) (Sutherland 2004; Robertson & Swinton 2005; Tscharntke *et al.* 2005)?

Here we have a head start in that the development of agri-environment schemes has been founded upon an increasing understanding of how to reconcile agricultural production with the delivery of a range of environmental goods and services, including thriving bird populations. Recent research has begun to explore how agri-environment measures might deliver both wildlife benefits and the protection of resources such as water and soil quality. For example, Walter *et al.* (2003), Cunningham *et al.* (2004), Bradbury & Kirby (2006) and Falloon *et al.* (2004) consider how measures such as non-inversion tillage, uncultivated field margins, hedgerows and small-scale wetland features might simultaneously reduce diffuse pollution or provide opportunities for carbon sequestration whilst also supporting farmland wildlife. More specifically, field studies by Cunningham *et al.* (2005) showed that seed-eating birds such as gamebirds and skylarks were more likely to be found on winter cereal fields established by non-inversion tillage than those established after conventional ploughing. Similarly, in the uplands, whilst the effects of existing agricultural and sporting management practices such as sheep grazing, grouse shooting and forestry on habitats and bird communities are increasingly well known (e.g. Avery & Leslie 1990; Tharme *et al.* 2001; Evans *et al.* 2006a; Pearce-Higgins & Grant 2006; Yallop *et al.* 2006), there is now growing interest in understanding how these same management practices influence upland hydrology and the security of carbon stocks in organic soils (e.g. Holden *et al.* 2007; Worrall *et al.* 2007).

Much remains to be done to understand to what extent management of agricultural land for environmental services as diverse as wildlife conservation, soil and water quality and carbon sequestration is synergistic or will require trade-offs to be understood and made. For example, in semi-natural grazed systems such as upland heathlands and peatlands, one might propose that simply refraining from management activities such as drainage, grazing, burning and cutting (which retard vegetation succession) is a good starting point for the long-term stewardship of carbon and water, and the reduction of flood risk further downstream. What would the consequences of such 're-wilding' be for wildlife, including many bird populations whose conservation is strongly associated with the low-intensity agricultural systems and vegetation structures maintained by such management (see case studies)? As the RSPB's (2007) think-piece *The Uplands: Time to Change?* notes, 'uplands gather more than 70% of our drinking water, store billions of tons of carbon in peat and soils and are home to some of our most special wildlife'. Yet with declining employment and marginal agricultural conditions, much of the upland area is designated 'Less Favoured' with heavy dependence of continuing agricultural management on subsidies and rural development grant support. We face big scientific, socio-economic and policy challenges in deciding how to manage our uplands for the diverse range of ecosystem services that they offer – from renewable energy and agriculture to wildlife, recreation and water and carbon storage – whilst maintaining thriving local communities and enterprise (Orr *et al.* 2008).

At the other end of the spectrum, in very intensively managed lowland agricultural landscapes, management to reduce diffuse pollution, retain water for longer in more naturally functioning floodplains to reduce flood risk, and better protect tilled agricultural soils from erosion and carbon loss all seem likely to be broadly consistent with the delivery of wildlife conservation benefits (e.g. Falloon *et al.* 2004; Ratcliffe *et al.* 2005; Bradbury & Kirby 2006). The task here will be to identify the right management measures in the right places in order to come as close as possible to genuine 'win–wins' (Smith 2004) in delivering a wider range of environmental benefits from more generally sustainable environmental management of the countryside. One noteworthy avian study is Field & Anderson's (2004)

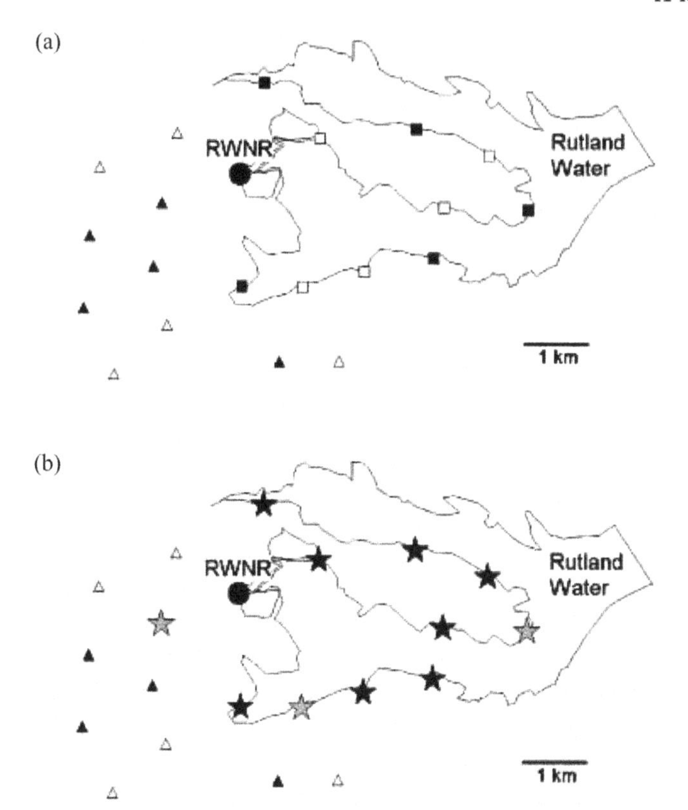

Fig. 10.6 (a) Layout of nestbox groups relative to the central tree sparrow colony at Rutland Water Nature Reserve. Triangles indicate groups on farmland, and squares indicate those next to the reservoir edge. Filled symbols indicate those with supplementary feeding, and open symbols indicate those without. (b) Colonisation of nestbox groups by breeding tree sparrows after three years. Black stars indicate colonisation in at least one year by colour-ringed birds from the central colony. Shaded stars indicate colonization from elsewhere. By the third year, all nestbox groups next to the reservoir edge supported at least one pair of tree sparrows. Reproduced from Field & Anderson (2004) with kind permission.

experimental study of the relative role of the rich invertebrate resources associated with wetland edge habitats – critical for feeding nestlings – and the availability of seed resources for full-grown birds, in determining the spatial dispersion of tree sparrow colonies in the agricultural landscape. Field & Anderson established 10 clusters of nest boxes close to aquatic habitat (a reservoir edge), half with associated seed provision and half without, and another 10 in the dry, agricultural hinterland (again half with and half without seed). The average distance of clusters from a central established breeding colony in each treatment group was the same. Over the course of three years, all wetland-edge nest box clusters were occupied by nesting tree sparrows whilst only one farmland cluster was occupied (Fig. 10.6). These striking results strongly suggest that whilst the availability of seed resources may determine tree sparrow presence and density it probably does so primarily at the landscape scale due to the mobility of the species during the non-breeding season. However, the availability of a sufficiently rich invertebrate resource to support provisioning of young proved critical to local settlement, and in this case was provided by wetland habitats in an otherwise dry agricultural landscape. Here then is one example which suggests that management to retain water for longer within catchments may have benefits not only for the obvious candidate wetland bird species, but also for birds much less obviously associated with such habitats.

(2) At what spatial scales should we seek to integrate and to separate agricultural production from the delivery of environmental services (e.g. Sutherland 2004; Firbank 2005; Mattison & Norris 2005; Tscharntke *et al.* 2005)?

Ask farmers which of a menu of agri-environment measures they find most attractive, and (aside from the payment rate), most will point to those measures that can be applied around field margins, along hedgerows and woodland edges, or on their least productive ground; anywhere in fact where the measures interfere least with what they see as their primary role – to grow food. This response illustrates a long-standing conundrum in agri-environment management. What is the best spatial grain at which to separate land allocated to production from that allocated to delivery of other environmental goods? One end of the spectrum is perhaps best represented by organic farming. Here the withdrawal of synthetic agrochemical inputs and greater reliance on organic manures and crop rotations for nutrition and pest control provide the opportunity to integrate production, wildlife conservation and (for example) carbon sequestration within individual cropped fields (e.g. Bengtsson *et al.* 2005; Hole *et al.* 2005; Smith *et al.* 2005). At the other extreme might be the abrupt distinctions between land dedicated solely to agriculture, and pristine habitats (e.g. National Parks) at the landscape scale in North America (Sutherland 2004). In Europe, and certainly in Britain, the long history of integration between agriculture and conservation and the recent growth of agri-environment management create landscapes where production and environmental management are separated at intermediate scales: hedgerows and field margins, ponds and woodland plantings in field corners, occasional fields of fallow/set-aside, and small nature reserves set in wider agricultural landscapes.

The reality is that it is almost certainly necessary to create greater ecological heterogeneity in agricultural systems by making space (and time) for the life cycles of non-crop organisms at all spatial scales (Benton *et al.* 2003; Firbank 2005; Tscharntke *et al.* 2005), and to do this both globally and locally. Agricultural intensification has tended to create homogeneity at all scales from simplified crop rotations to the structure of vegetation within individual fields (Benton *et al.* 2003; J. D. Wilson *et al.* 2005). For example, in Britain Tallowin *et al.* (2005) note 'the ubiquity of grass-dominated, species-poor and structurally uniform grasslands' and that 'the consequences of this spatial uniformity on grassland biodiversity are likely to be profound'.

Globally, the fact that the world's rapidly growing human population in developing countries is being fed by massive conversion of pristine habitats to agriculturally productive land means that encouraging intensive agricultural production there and thus allowing more land to be spared for wild nature (Balmford *et al.* 2005; Green *et al.* 2005; Matson & Vitousek 2006) may reduce loss of pristine habitat and species extinctions in the long run. At home, such large-scale allocation of land either to production or to conservation does not sit comfortably with the long history and cultural importance of our more intimate farmed landscapes. Nonetheless, as we saw in Chapter 1, many of our most important semi-natural grazed systems, especially in the lowlands, have been reduced to fragmented remnants of their former extents. Large-scale restoration of these habitats, alongside continued refinement of finer-grained agri-environment approaches in the wider countryside, may be critical to give these habitats and their wildlife greater resilience to the effects of climate change, pollution and fragmentation. They also offer opportunities to deliver other ecosystem services, including softening the impacts of sea-level rise, reducing flood risks and reversing the soil carbon losses caused by conversion of grassland to cultivated farmland (Spurgeon 1998; van der Ploeg *et al.* 1999; RSPB 2001; Sutherland 2002). In Britain, the recent large-scale restoration of wetlands from agricultural or brownfield land, such as the RSPB's reserves at Otmoor, Lakenheath and in the Dearne Valley, shows what can be done. Even more ambitious projects are planned. For example, the Needingworth Wetland Project plans to convert over 700 hectares of land used for gravel extraction to a wetland that will support 40% of the UK's Biodiversity Action Plan target for reedbed, whilst the nearby Great Fen Project aims to connect the existing fenland fragments of Holme Fen and Woodwalton Fen by conversion of arable farmland to create a single 3700-hectare fenland (www.greatfen.org.uk).

There may also be utilitarian benefits to agriculture of integrating ecological heterogeneity at all spatial scales that have yet to be fully explored. Tscharntke *et al.* (2005) provide numerous examples of the role of biodiversity as an insurance that guarantees resilience of production to disturbance and uncertainty in agricultural systems. A neat example is the fact that to achieve efficient control of cereal aphids across different years and landscape types, a wide range of parasitoid and predator natural enemies is required (Thies *et al.* 2005). Indeed, Tscharntke *et al.* (2005) conclude that 'only high diversity agroecosystems connected with a diversity of habitats in complex landscapes may have the capacity to provide resilient ecosystems and a sustainable multi-functional agriculture', and that seeking to maintain as rich a biodiversity as possible in agricultural systems is the only reliable option to maintain sustainable land-use over the long term. The scientific challenges here will be to understand how best to integrate agri-environmental management at field, farm and landscape scales with restoration of larger areas of semi-natural habitats to meet the combined needs of agricultural production, wildlife conservation, recreation and the delivery of ecosystem services.

(3) How do we do all of the above in ways that accommodate the increasing impacts of climate change?

The possible impacts of climate change on agriculture and bird populations loom over this chapter. Will agricultural land change not at all, a little or out of all recognition? How will the breeding ranges of bird species have changed? How fast will change happen? Will everything we have learned about how to reconcile agricultural production with conservation in the agricultural systems of the twentieth century be of any relevance by the end of the twenty-first?

Addressing these questions is largely an exercise in the use of predictive models (and their uncertainties), whether they be models of greenhouse gas emissions, of climate change itself, or of changes in land-use or other environmental conditions under specified scenarios of climate change. What is certain is that although climatic change has happened throughout Earth's history, the size and rate of changes expected during this century as a result of human activity are equal to or greater than the largest natural changes in recent geological history. These changes will result in a global climate unlike any in recent geological history (e.g. Houghton *et al.* 2001). In Britain, the UK Climate Impacts Programme (www.ukcip.org.uk) predicts an overall warming of between 1 and 5 °C during this century, especially in the south and east, with hotter and drier summers (a longer growing season, but with reduced soil moisture and up to a 50% reduction in precipitation), milder, wetter winters and reduced snowfall. The frequency of extreme weather events is also likely to change with a greater frequency of summer heatwave conditions, fewer 'cold snaps' and a greater frequency of stormy weather with intense rainfall.

At a European scale, Huntley *et al.* (2007) have recently published *A Climatic Atlas of European Breeding Birds*. This atlas was made possible by the fact that the current distribution of many species across Europe, when viewed at a 50-km grid square resolution (Hagemeijer & Blair 1997), is well explained by association with just three bioclimate variables: winter cold (the mean temperature of the coldest month of the year), warmth over the growing season (annual temperature sum above 5 °C) and moisture availability (the ratio of actual to potential evapotranspiration). These same variables are also known to be important in determining global vegetation patterns (Prentice *et al.* 1992). When these simple 'climate envelope' models are used in conjunction with a leading global climate model and a 'middle-of-the-road' emissions scenario for this century, the potential future breeding ranges of these species can be predicted for the end of the century (2071–90). The results suggest a general north-eastwards shift of breeding ranges with many species having much smaller future ranges that overlap little or not at all with their current distribution. These are perhaps the species most at risk from the impacts of climate change over the next century. If we add the even more uncertain effects of demographic and phenological change, impacts on migration and wintering conditions for migratory species and habitat change, then the extent to which species ultimately occupy these

Table 10.3 *Predicted change in British breeding range by 2071–90 of 15 birds characteristic of agricultural habitats*

Species	Description of predicted range change in Britain
Sparrowhawk	No change
Black grouse	Most of current range (except N Scotland) becomes unsuitable
Grey partridge	No change
Corncrake	Uncertain
Stone curlew	Spread of range in southern England
Lapwing	Lowland habitats in England predicted to be unsuitable
Golden plover	Most of current range (except N Scotland) becomes unsuitable
Woodpigeon	No change
Turtle dove	Spread northwards
Skylark	No change
Song thrush	No change
Chough	Slight range expansion predicted
Linnet	No change
Cirl bunting	Spread northwards
Corn bunting	No change

Source: Derived from inspection of maps in *A Climatic Atlas of European Breeding Birds* (Huntley *et al.* 2007).

predicted future ranges becomes even more uncertain. The consequences of climate change may be yet more severe.

If we consider species of agricultural habitats in Britain, then, examining just the species for which we undertook detailed case studies in Chapter 8, predicted future ranges vary considerably (Table 10.3). A few species currently restricted to southern areas (e.g. stone curlew, turtle dove and cirl bunting) are predicted to find larger areas of lowland Britain climatically suitable, whilst the prognosis is at first sight grim for a range of species associated with upland heathland and grassland, especially in England, Wales and southern Scotland, including black grouse and golden plover. Lapwings may retreat primarily to upland habitats as much of lowland England becomes climatically unsuitable.

As Huntley *et al.* (2007) suggest, a fundamental change in conservation thinking will be required in order to accommodate the needs of species whose distributions are likely to change, perhaps rapidly, over the coming decades. This is likely to require a rethink of the role played by existing networks of protected areas and could involve increasing the extent of existing site networks to increase their resilience to climate change. Moreover, the wider countryside matrix between these important patches of semi-natural habitat may need to be more 'permeable' to species whose ranges are changing. This matrix is very often agricultural, thus illustrating a key role that agri-environment management is likely to play in future. For example, Vickery *et al.* (2004) and Donald & Evans (2006) illustrate a range of management measures currently available in England's Environmental Stewardship scheme and the

Table 10.4 *Environmental Stewardship measures and some of the resources they offer that could serve to soften intensively managed agricultural landscapes*

	Resource provided					
Management measure	Flying insects	Epigeal/foliar invertebrates	Aquatic invertebrates	Pollen/ nectar	Seeds and fruit	Refuge habitat
Hedge planting or restoration	×	×			×	×
Ditch management or restoration	×		×			×
Pond creation or restoration	×		×			×
Water-level management			×			
Field margin creation	×	×		×	×	×
Reduced agrochemical inputs	×	×	×	×	×	
Wild bird seed provision through planting or stubbles		×			×	
Planting for pollen/nectar	×	×		×		
Summer fallow	×	×		×	×	

Source: Modified from Vickery *et al.* (2004) and Donald & Evans (2006).

resources they offer that could help in softening intensively managed landscapes for exploitation by a wider variety of species (Table 10.4). Their effectiveness may be all the greater in this respect if future development of agri-environment scheme implementation takes us from the current farm-by-farm approach to a more integrated planning of agri-environment delivery across landscapes and catchments.

For birds at least, this management of the wider countryside need not necessarily require the establishment of physical corridors of semi-natural habitat. As Bailey (2007) notes in the context of woodlands, the belief in biodiversity gain from increasing connectivity to create 'habitat networks' (Peterken 2000) is currently rather greater than the evidence to support it. Rather, one might think of management of the intervening agricultural countryside as 'softening the matrix' to allow easier movement of species between patches of other habitat types, to limit the effect of intensive management on protected areas, or to provide 'stepping stones' to enable a species to breed successfully, and ease the spread of its population across the landscape.

Although everything we have said so far depends upon testing and interpreting the implications for birds of possible future worlds, our understanding is growing rapidly, not least because evidence of the real impact of climate change on birds is accumulating fast. Climate envelope models are able to 'retrodict' recent changes in bird distribution and abundance (Green *et al.* 2008). For example, the fact that the 2004 survey of the increasing Dartford warbler population in Britain found them nesting on upland heathlands in south-west England and south Wales may well be accounted for by climate warming. There is good evidence of the effect of warming on the migratory and nesting phenology of European breeding birds (Both *et al.* 2004; Crick 2004) and the potential impact of this in generating mismatches between the timing of migration, breeding seasons, and availability of key food supplies (e.g. Visser *et al.* 1998; Ahola *et al.* 2004; Pearce-Higgins *et al.* 2005; Both *et al.* 2006; Laaksonen 2006), and increasing interspecific competition (Ahola *et al.* 2007). What are now needed are more studies that begin to unravel the mechanisms through which climate change influences population dynamics. If we can do this, then not only can we consider conservation measures to accommodate changes in range and population in response to climate change, but we can also undertake management to ensure that species remain robust for as long as possible to the effects of climate change within their current ranges. This buys crucial time for both mitigation of climate change itself, and adaptive management responses to changing climate. In Finland, for example, Ludwig *et al.* (2006) found that black grouse had responded to spring warming by laying and hatching eggs earlier. However, temperatures at the time of hatching have not warmed to the same extent, chicks are suffering higher mortality, and populations have declined severely because of these cold post-hatching conditions. In Scotland, Beale *et al.* (2006) found that warmer, drier late summers on the breeding grounds could account for observed population declines of ring ouzels, and further work is now under way to understand the mechanisms through which warm, dry weather in the late breeding season might have adverse impacts on adult or juvenile birds. In both of these cases, management to mitigate the key climate effect – for example by increasing food availability or accessibility for black grouse chicks in cold weather – may be critical in mitigating climate change effects.

Although studies of the impact of climate change on bird populations are in their infancy, predicting the impacts of climate change on the agricultural land-uses which shape bird habitats on farmland is even more daunting. Here, a whole range of social, economic, political and environmental uncertainties come into play to render it very difficult to predict future land-use. Recently Rounsevell *et al.* (2005) attempted to project possible changes in areas of cropland and grassland across Europe using (amongst others) the same climate and emissions models employed by Huntley *et al.* (2007). They showed that across much of Europe the areas of arable and agricultural grassland required would decline substantially (by as much as 50%) if current rates of technological development in crop production continue as they have done over the past 50 years. As the authors themselves recognise, if food or bioenergy demand increases, if current agri-environmental policies do not continue to limit maximisation of production, or if there is any return to future policy support for overproduction then these reductions will not be seen. However, what this exploration of future scenarios does show is that continued technological advance in agriculture – for example through biotechnological research to develop crops which assist in the delivery of ecosystem services rather than solely delivering increased production efficiency (Krebs *et al.* 1999; Ervin *et al.* 2003; Lal 2007) – may help to provide some breathing space. This breathing space will help us to meet the challenges of integrating wildlife conservation with the increasing list of demands that will be placed on our agricultural ecosystems and to do so in ways that provide resilience to climate change whilst accommodating its inevitable impacts.

In ending this section it is worth remembering that all of the above challenges would be more easily met if we understood how bird distributions and populations would respond to a range of possible futures. Twenty years ago any attempt at prediction for birds of agricultural systems in Britain would have been frustrated by lack of data for almost any species except perhaps, grey partridge. However, the last 20 years of research which Chapters 6–9 of this volume have attempted to synthesise, have provided rich data on the abundance, distribution and demographic rates of many species as a function of measures of agricultural land use and practice, across the agricultural systems of Britain and Europe. These data provide real opportunities to apply a range of modelling techniques to predict and to visualise the impacts of possible future agricultural worlds on our bird populations. Methods include multivariate regression models of environmental associations at a variety of scales (e.g. Swetnam *et al.* 2005; Huntley *et al.* 2007), spatially explicit population dynamics models (e.g. Rushton *et al.* 1997), depletion models extrapolating population consequences from individual behaviour (e.g. Bradbury *et al.* 2001; Stephens *et al.* 2003), and simple risk assessment frameworks which use existing knowledge of species' population responses to key components of past change to predict their response to future change (e.g. Butler *et al.* 2007). All of these approaches have strengths and weaknesses, adherents and critics, and varying track records in assisting decision-making. Suffice it to say that, as Stephens *et al.* (2003) point out, field trials, monitoring and experiments are time-consuming, expensive and limited in their spatial and temporal scope. The use of predictive modelling tools such as those described above to test ranges of future land-use and management scenarios, backed up by field validation, will be a critical strategy in the face of time and resource constraints.

Conclusions

In concluding a major review of farming and birds in Britain 20 years ago, O'Connor & Shrubb (1986) expressed concerns about the advancing impact of agricultural technology allied to a production-orientated farming philosophy and policy. Thus:

it is well to remember that farming has a very long tradition of striving to improve the productivity of the land in the only way it understands (raising crop yields) and of this being regarded by all as an entirely admirable goal. It is essentially difficult for farmers to grasp modern attitudes towards this or to move towards the ecologist's concept of productivity. In the view of conservation bodies, amenity and conservation are the main alternatives to food production in land use but the increasing spread of alternative crops and crop uses being studied and developed makes it certain that this is not so; we really must not underestimate the speed with which new technology will be applied.

In so many ways, the concerns expressed by the authors have proven well founded as the catastrophic losses of biodiversity from intensive agricultural systems, described in this book as they have happened in Britain, have come to pass. Almost as *Farming and Birds* was published, however, the first agri-environment schemes (ESAs) appeared in Britain. Since then, we have seen the widespread adoption and funding of such schemes across Europe and the gradual reform of the CAP to direct subsidy support to environmental goods in addition to food and fibre. This supports Buckwell & Armstrong-Brown's (2004) more optimistic view that at the start of the twenty-first century we are moving towards a genuinely more sustainable agriculture which combines production with stewardship of the natural environment. The scientific, agricultural, economic and political experience gained during this emergence of agri-environmental management now needs to be seen as the starting point in meeting the much greater challenges now faced. These include: climate change, continuing human population growth and the growing need to see the management of agricultural landscapes as being directed towards the stewardship of a wide range of resources, from food, fibre and renewable energy, to water, carbon and wildlife, not to mention the recreational and health benefits likely to be gained from access to a diverse and vibrant countryside. In a recent collation of 100 ecological questions of high policy relevance in the UK, and 25 novel threats and opportunities facing UK biodiversity, at least half relate to the issues discussed in this chapter (Sutherland *et al.* 2006, 2008).

Meeting these challenges is a long-term objective, and although it remains an exciting time to view the prospects for agriculture and wildlife, and to engage in finding the solutions to the challenges outlined above, we must not be blinded to more immediate problems. Today, bird conservation, and indeed the conservation of biodiversity generally, risks being swamped by the rush to respond to global warming. Finding ways to mitigate the effect of greenhouse gas emissions, develop renewable energy sources and adapt ecosystem management to the climate change to which we are already committed are indeed daunting prospects, requiring urgent attention. However, they can lead decision-makers to consider biodiversity conservation as of secondary importance, or to emphasise only the utilitarian value of biodiversity in terms of ecosystem function (e.g. Kremen 2005), or even to suggest that the protection of species and habitats and the site networks which support them are anachronistic in times of global change (e.g. Stokstad 2005). Nonetheless, as Sekercioğlu *et al.* (2004) have shown, the global extinction of bird species could

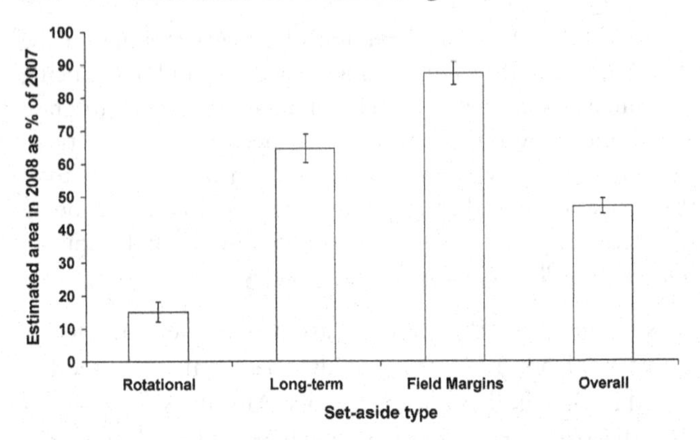

Fig. 10.7 Estimates of area of land uncropped in England in 2008 as a percentage of the equivalent set-aside area in 2007. Data from Langton (2008).

have marked impacts on ecosystem process such as decomposition, pollination and seed dispersal, so the utilitarian significance of bird conservation should not be underestimated. Equally, the conservation measures associated with the network of SPAs for birds designated by individual countries in response to the EU's Birds Directive are associated with positive bird population responses (Donald *et al.* 2007). As Huntley *et al.* (2007) conclude, far from being anachronisms, these protected areas are the critical nodes from which the conservation response to climate change must begin.

In Britain, immediate challenges to bird conservation in agricultural systems include the fact that England's Farmland Bird Indicator reached a record low in 2007 and that, as we write in 2008, the set-aside rate has been set to zero for the second cropping season running, and will probably be abolished altogether by the EU very soon. With production surpluses a thing of the past, and grain prices rising, set-aside has served its primary purpose. However, as Chapter 6 showed, its environmental benefits, alongside agri-environment measures and the rise of organic farming, may have been considerable. Will agri-environment schemes, and uptake of these by farmers expand to compensate the probable loss of set-aside land? As yet, we do not know. However, no government in the UK had the foresight to act pre-emptively to safeguard the environmental benefits brought by set-aside, and the scramble to return rotational set-aside to production in late 2007 (Fig. 10.7) increases the probability that the Farmland Bird Indicator will fall further. Retaining the environmental benefits brought by set-aside will be a key test of farmers in their role as custodians of farmland wildlife.

Results from a recent study by Vickery *et al.* (2007) confirm a weak but statistically significant correlation between the area of set-aside and the between-year changes in the Farmland Bird Index (Fig. 10.8). This apparent national-scale relationship is even more striking in view of the marked variability (in both quantity and quality) of set-aside as a bird habitat. It is perhaps no surprise that only two of the 19 species

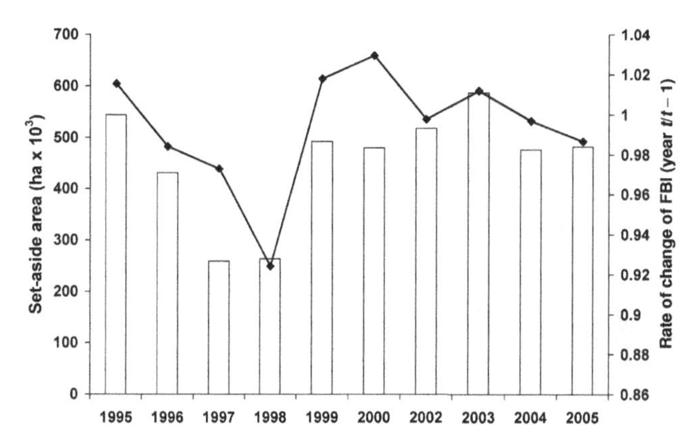

Fig. 10.8 Covariation in the availability of set-aside land and the rate of change in the Farmland Bird Index (FBI) in England, 1994–2005. Data from Vickery *et al.* (2007). Note that although the required set-aside rate between 1999 and 2005 remained constant at 10%, the total area fluctuated due to additional voluntary set-aside land. Mean winter temperature was incorporated into the analysis to control for any fluctuation in the FBI attributable to effects of severe winter weather. Partial correlation between set-aside area and rate of change of FBI = 0.685 (p = 0.04). Data for 2001 were disregarded because the outbreak of foot-and-mouth disease in that year and the subsequent restrictions regarding access to the countryside meant that measures of bird population trends had to be interpolated from data in previous and subsequent years.

that make up the index showed a significant correlation based on their individual trends, although a further 11 species showed non-significant positive associations with the amount of set-aside. Taken together with results of the field-scale studies discussed above, this work suggests that there is a high risk that bird populations on lowland farmland will fall significantly if there is no implementation of mitigation measures to compensate for the loss of set-aside.

On the other side of the coin, where CAP reforms have brought to an end the worst production excesses on intensively managed land, those same reforms, through the loss of headage payments, and the subsequent impacts of BSE and foot-and-mouth disease have further marginalised the agricultural viability of sheep and cattle enterprises in the uplands. The arrival in Britain in 2007 of bluetongue disease, carried by biting midges whose range is expanding northwards with warming climate, risks further economic impacts on livestock enterprises. So, where once conservationists worried about the 'overgrazing' of the uplands by increasing stock densities (e.g. Fuller & Gough 1999), now abandonment by grazing animals is a serious concern in some areas for both birds and other wildlife dependent on low-intensity agricultural systems and vegetation communities maintained by grazing (e.g. Bignal & McCracken 1996).

We also need to know more about the more immediate impacts of climate change on agricultural land-use and conditions for farmland birds. As we have seen, lowland England is predicted to experience progressively warmer, drier summer conditions. Much of this landscape is already heavily agriculturally drained, and we know that

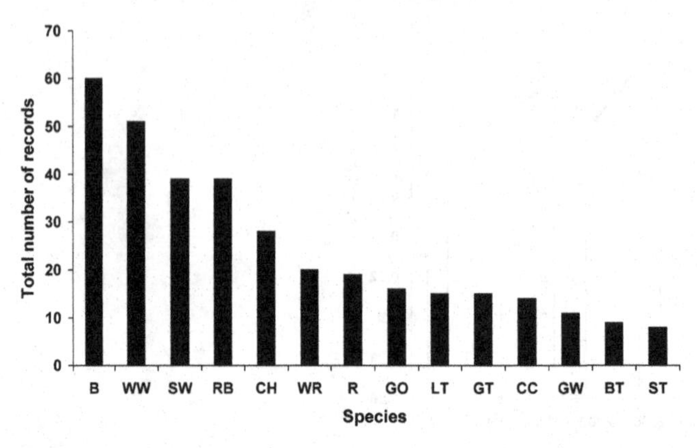

Fig. 10.9 Total number of songbirds recorded from 147 spring point counts in 49 plots of short-rotation willow and poplar coppice across the UK. Data from Sage & Robertson (1996). B, blackbird; WW, willow warbler; SW, sedge warbler; RB, reed bunting; CH, chaffinch; WR, wren; R, robin; GO, goldfinch; LT, long-tailed tit; GT, great tit; CC, chiffchaff; GW, garden warbler; BT, blue tit; ST, song thrush. A further 16 species were recorded on fewer than eight occasions.

lack of water is already limiting the distribution and abundance of a surprising range of farmland bird species, from the obvious cases such as breeding waders to passerines as diverse as yellow wagtails, song thrushes, starlings and tree sparrows (Chapters 6 and 7 and Fig. 10.6). In this context, developing the pioneering work of Bradbury & Kirby (2006) to better understand how small-scale management for wildlife and water resources on dry farmland can best go hand in hand should be a priority. Even in the uplands, the condition of blanket bogs has deteriorated following drainage and burning to improve the quality of upland grazing. Blocking moorland 'grips' (drains) to restore hydrological function and condition of blanket bogs is now commonplace (e.g. www.blanketbogs.org), although the biodiversity benefits of this measure are only now being studied in detail, as at RSPB's peatland reserve at Forsinard in Caithness.

As well as lowland summers becoming warmer and drier, growing seasons are predicted to become longer across much of Britain. Given that vegetation structure is already known to be a key limiting factor for nesting and foraging farmland birds (J. D. Wilson *et al.* 2005), we may need more innovations like that of that of unsown patches in cereal fields designed, and now successfully implemented, to extend the breeding season of skylarks (Morris *et al.* 2004). Lastly, it seems likely that there will be an increase in the use of agricultural land to grow biomass or biofuel crops (e.g. Andersen *et al.* 2005). Where these crops are familiar varieties such as oilseed rape, there may be little new to learn from a bird conservation perspective. However, the biodiversity implications of large-scale growing of crops such as short-rotation coppice willow or poplar have received little attention so far (see Sage & Robertson 1996; Sage 1998; Anderson & Fergusson 2006). There seems to be some potential for these crops to support a range of bird species (Fig. 10.9), especially where use of insecticides and herbicides in management is limited (Coates & Say 1999). Kennedy & Southwood (1984) long ago noted that species of willow could support a rich diversity of invertebrates, especially Lepidoptera, relative to many other tree species. The value of novel crops such as the perennial grass *Miscanthus* remains almost unknown (Semere & Slater 2007).

In the uplands, renewable energy is increasingly generated by the development of windfarms, and adverse impacts on upland bird populations through collision, behavioural avoidance of turbines, barrier effects and habitat loss require continued study in order to understand the scale of any population effects and to improve guidance on location and mitigation measures (Drewitt & Langston 2006).

So, both in the immediate and long term, the conservation of bird populations in agricultural landscapes must remain central to our concerns. When Rachel Carson published *Silent Spring* in the early 1960s and alerted the public to the toxic environmental effects of the early generations of pesticides that fuelled the post-war agricultural revolution, it was the idea of landscapes lacking birds and bird song that held the attention. Since then, bird populations have been central to developing our understanding of the environmental impacts of agricultural intensification, the dependence of a rich wildlife on agriculture in many low-intensity, grazed systems and the effectiveness of management to reverse those effects and restore wildlife to intensively managed agricultural landscapes. We have no doubt that the conservation of birds will continue to be central to reconciling agricultural production with delivery of the environmental goods in ways that will make for sustainable management of our countryside in a future of global change. We remain confident that a silent spring in Britain will remain as unacceptable a prospect in this century as Rachel Carson helped to make it in the last.

Plate 10.1 Uncultivated field margins, provided through agri-environment schemes, can provide important nesting and foraging habitats for farmland birds. © RSPB.

Plate 10.2 A 'skylark plot' in an winter wheat field in Cambridgeshire. Leaving a scatter of small patches unsown in this manner allows fields which would otherwise become unsuitable for nesting and foraging skylarks due to crop growth, to be used throughout the summer. This relatively new agri-environment option has yet to be adopted on a sufficiently large scale to judge whether it could assist in reversing the national population decline of skylarks. © RSPB.

Plate 10.3 An experimental grazing plot in Glen Finglas, Scotland. This study contrasted a series of different moorland sheep-grazing regimes, and found that a low-intensity, mixed grazing regime using cattle and sheep created more heterogeneous sward structures, a richer invertebrate fauna and increases in nesting densities of meadow pipits. © Peter Dennis.

Plate 10.4 Amberley Wild Brooks was saved from drainage for agricultural improvement by a 1978 public inquiry which demonstrated that agricultural production did not have universal priority over conservation interests. To what extent might we see more river floodplains restored in future as a means of protecting human settlements against increasing flood risk that can bring huge biodiversity benefits? © RSPB.

Plate 10.5 In the 1960s and 1970s ornithologists questioned whether Britain's population of the Dartford warbler, confined to lowland heaths along the south coast of England, could withstand a succession of severe winters. Today, the species breeds in unprecedented numbers and pairs have been found on upland heaths on Dartmoor and in the Brecon Beacons. How much further will its range extend as the climate warms? © RSPB.

Plate 10.6 Short-rotation willow coppice. What will the implications be for farmland birds of increasing acreages of this crop, grown to produce biofuel? At present studies of the use of this crop by birds, and the opportunities for management to increase its conservation value are few. © Guy Anderson.

Appendix 1

Habitat associations and population status of birds of agricultural habitats in Britain

Broad habitat usage of bird species that nest or feed in farmland or grazed semi-natural habitats in Britain. **(A)** Species for which a high proportion of the British breeding or wintering population is associated with these habitats. Those of European conservation concern and Red- or Amber-listed in the UK due to population decline or range contraction in these habitats are listed first in bold. **(B)** Generalist species or species whose breeding or wintering populations are concentrated in other habitats. Habitats: A, arable; B, buildings; E, field edges and marginal habitats; G, grassland; H, lowland and upland heathland; M, machair; S, saltmarsh; W, woodland. Wetland species associated only with saltmarsh habitats, and marine species only associated with coastal heathlands through nesting colonies are not included. * indicates species probably now extinct as annually breeding species in Britain. ** indicates species extinct in Britain, but currently subject of a re-introduction attempt. Status column is divided into three parts. The first lists UK conservation status (R, Red list; A, Amber list; G, Green list – Gregory *et al.* 2002). The second lists European conservation status from BirdLife International (2004) (1, species of global conservation concern; 2, species concentrated in Europe and with an unfavourable conservation status; 3, species not concentrated in Europe, but with an unfavourable conservation status; E, species concentrated in Europe and with a favourable conservation status; 0, species not of European conservation concern and not concentrated in Europe; E, endangered; V, vulnerable; D, declining; R, rare; H, depleted; L, localised; S, secure). The third lists the population status of the species in Britain (R, resident; S, summer migrant breeder; W, winter visitor; M, passage migrant). Where two status codes are given, the first represents the main status of the species in Britain. INT, introduced species not considered by Gregory *et al.* (2002). Taxonomic order follows Dudley *et al.* (2006).

English name[a]	Scientific name	Nesting habitats (nb = not breeding in Britain)	Feeding habitats	Status
(A)				
(Greater) White-fronted goose	*Anser albifrons*	**nb**	**A,G**	**A/0S/W**
Red grouse	*Lagopus lagopus*	**H**	**G,H**	**A/0S/R**
Black grouse	*Tetrao tetrix*	**G,H,W**	**A,G,H,W**	**R/3H/R**
Grey partridge	*Perdix perdix*	**A,E,G,H**	**A,E,G,H**	**R/3V/R**
(Common) Quail	*Coturnix coturnix*	**A,G**	**A,E,G**	**R/3H/S**
Red kite	*Milvus milvus*	**W**	**A,G,W**	**A/2D/R**
Hen harrier	*Circus cyaneus*	**G,H**	**A,E,G,H,M,S**	**R/3H/RW**
Montagu's harrier	*Circus pygargus*	**A**	**A,E,G,H**	**A/ES/S**
(Common) Kestrel	*Falco tinnunculus*	**B,E,W**	**A,B,E,G,H,M,W**	**A/3D/R**
Merlin	*Falco columbarius*	**G,H**	**A,E,G,H,M,S**	**A/0S/RW**
Corncrake	*Crex crex*	**E,G,M**	**E,G,M**	**R/1H/S**
Great bustard**	*Otis tarda*	**A,G,H**	**A,G,H**	**–/3D/(R)**
(Eurasian Thick-knee) Stone curlew	*Burhinus oedicnemus*	**A,G,H**	**A,G,H**	**R/3V/S**
(Northern) lapwing	*Vanellus vanellus*	**A,G,H,M,S**	**A,G,H,M,S**	**A/2V/RW**
(Common) Snipe	*Gallinago gallinago*	**H,G,M,S**	**H,G,M,S**	**A/3D/RW**
Black-tailed godwit	*Limosa limosa*	**A,G**	**A,G**	**R/2V/SW**
(Common) Redshank	*Tringa totanus*	**G,H,M, S**	**G,H,M, S**	**A/2D/RW**
(Common) Black-headed gull	*Larus ridibundus*	**G,H**	**A,G,H,M,S**	**A/ES/RW**
(Mew gull) Common gull	*Larus canus*	**G,H**	**A,G,H,M,S**	**A/2H/RW**
(European) Turtle dove	*Streptopelia turtur*	**E,W**	**A,E,G,H,W**	**R/3D/S**
(Common) Cuckoo	*Cuculus canorus*	**(E,G,H)**	**E,G,H,M,S,W**	**A/0S/S**
Barn owl	*Tyto alba*	**B,E,W**	**A,B,E,G,H,S,W**	**A/3D/R**
(Eurasian) Nightjar	*Caprimulgus europaeus*	**H,W**	**G,H,W**	**R/2H/S**
(Wood lark) Woodlark	*Lullula arborea*	**A,H,W**	**A,H,W**	**R/2H/R**
(Eurasian) Skylark	*Alauda arvensis*	**A,E,G,H,M,S**	**A,E,G,H,M,S**	**R/3H/RW**
Tree pipit	*Anthus trivialis*	**H,W**	**H,W**	**A/0S/S**
Meadow pipit	*Anthus pratensis*	**G,H,M,S**	**A,E,G,H,M,S**	**A/ES/RW**
Yellow wagtail	*Motacilla flava*	**A,E,G,S**	**A,E,G,S**	**A/0S/S**
(Hedge accentor) Dunnock	*Prunella modularis*	**B,E,H,W**	**A,B,E,G,H,W**	**A/ES/R**
Ring ouzel	*Turdus torquatus*	**G,H**	**G,H**	**R/ES/S**

Song thrush	*Turdus philomelos*	**B,E,H,W**	**A,B,E,G,H,W**	**R/ES/RW**
Mistle thrush	*Turdus viscivorus*	**E,H,W**	**A,E,G,H,W**	**A/ES/R**
(Common) Grasshopper warbler	*Locustella naevia*	**E,G,H,W**	**E,G,H,W**	**R/ES/S**
Dartford warbler	*Sylvia undata*	**H**	**H**	**A/2H/R**
Spotted flycatcher	*Muscicapa striata*	**B,E,W**	**B,E,W**	**R/3H/S**
Red-backed shrike*	*Lanius collurio*	**E,G,H**	**E,G,H**	**R/3H/(S)**
(Red-billed) Chough	*Pyrrhocorax pyrrhocorax*	**B**	**A,E,G,H**	**A/3D/R**
(Common) Starling	*Sturnus vulgaris*	**B,E,M,W**	**A,B,E,G,H,M,S,W**	**R/3D/RW**
House sparrow	*Passer domesticus*	**B,E**	**A,B,E,G,M**	**R/3D/R**
(Eurasian) Tree sparrow	*Passer montanus*	**B,E,W**	**A,B,E,G**	**R/3D/R**
(Eurasian) Linnet	*Carduelis cannabina*	**E,H,W**	**A,B,E,G,H,S**	**R/2D/RS**
Twite	*Carduelis flavirostris*	**H,G,M**	**A,E,G,H,M,S**	**R/0S/RW**
(Eurasian) Bullfinch	*Pyrrhula pyrrhula*	**E,H,W**	**A,E,G,H,W**	**R/0S/R**
Yellowhammer	*Emberiza citrinella*	**E,H,W**	**A,B,E,G,H**	**R/ES/R**
Cirl bunting	*Emberiza cirlus*	**E**	**A,E,G**	**R/ES/R**
Reed bunting	*Emberiza shoeniclus*	**A,E,H,M,S**	**A,B,E,G,H,M,S**	**R/0S/R**
Corn bunting	*Emberiza calandra*	**A,E,G,M**	**A,B,E,G,M**	**R/2D/R**
Mute swan	*Cygnus olor*	S	A,G,M,S	A/ES/R
(Tundra swan) Bewick's swan	*Cygnus columbianus*	nb	A,G,S	A/3V/W
Whooper swan	*Cygnus cygnus*	nb	A,G,M,S	A/ES/W
Bean goose	*Anser fabalis*	nb	A,G	A/ES/W
Pink-footed goose	*Anser brachyrhynchus*	nb	A,G,S	A/ES/W
Greylag goose	*Anser anser*	G,M,S	A,G,M,S	A/0S/R
Barnacle goose	*Branta leucopsis*	nb	A,G,S	A/ES/W
Brent goose	*Branta bernicla*	nb	A,G,S	A/3V/W
(Eurasian) Wigeon	*Anas penelope*	G,H	G,H,M,S	A/ES/RW
(Common) Teal	*Anas crecca*	G,H,M,S	G,H,M,S	A/0S/RW
Mallard	*Anas platyrhynchos*	E,G,M,S	A,B,E,G,M,S	G/0S/RW
Red-legged partridge	*Alectoris rufa*	A,E,G,H	A,E,G,H	INT/2D/R
Pheasant	*Phasianus colchicus*	B,E,G,H,W	A,E,G,H,W	INT/0S/R
(Eurasian) Sparrowhawk	*Accipiter nisus*	E,W	A,E,G,H,S,W	G/0S/R
(Common) Buzzard	*Buteo buteo*	H,M,W	A,E,G,H,M,S,W	G/0S/R
Golden eagle	*Aquila chrysaetos*	G,H	G,H	A/3R/R

(cont.)

English name[a]	Scientific name	Nesting habitats (nb = not breeding in Britain)	Feeding habitats	Status
(Eurasian) Hobby	*Falco subbuteo*	E,H,W	A,G,H,S	G/0S/S
(Common) Moorhen	*Gallinula chloropus*	E,G,S	A,B,E,G,S	G/0S/R
(Eurasian) Oystercatcher	*Haematopus ostralegus*	A,G,M,S	A,G,M,S	A/ES/R
(Common) Ringed plover	*Charadrius hiaticula*	A,G,H,M,S	A,G,H,M,S	A/ES/RW
(Eurasian) Golden plover	*Pluvialis apricaria*	G,H	A,G,H,M,S	G/ES/RW
Dunlin	*Calidris alpina*	G,H,M,S	G,H,M,S	A/3H/RW
Ruff	*Philomachus pugnax*	G	G,S	A/2D/SM
Jack snipe	*Lymnocryptes minimus*	nb	G,S	G/3D/WM
(Eurasian) Curlew	*Numenius arquata*	G,H	A,G,H,M,S	A/2D/RW
Lesser black-backed gull	*Larus fuscus*	marine	A,G,M,S	A/ES/RS
Herring gull	*Larus argentatus*	marine	A,G,M,S	A/ES/RW
(Rock pigeon) Rock dove	*Columba livia*	B	A,B,G,H,M,S	G/0S/R
(Stock pigeon) Stock dove	*Columba oenas*	B,E,W	A,B,E,G,H,W	A/ES/R
(Common) Woodpigeon	*Columba palumbus*	B,E,W	A,B,E,G,H,W	G/ES/R
(Eurasian) Collared dove	*Streptopelia decaocto*	B,E,W	A,B,E,G,H,W	G/0S/R
Little owl	*Athene noctua*	B,E,W	A,B,E,G,H,W	INT/3D/R
Long-eared owl	*Asio otus*	H,W	E,G,H,S,W	G/0S/RW
Short-eared owl	*Asio flammeus*	G,H	E,G,H,M,S	A/3H/RW
(Eurasian) Green woodpecker	*Picus viridis*	E,W	E,G,H,W	A/2H/R
Barn swallow	*Hirundo rustica*	B	A,B,E,G,H,S,W	A/3H/S
White/pied wagtail	*Motacilla alba alba/yarrellii*	B,E	A,B,E,G,H,M,S	G/0S/RM
(Winter) Wren	*Troglodytes troglodytes*	B,E,G,H,M,S,W	B,E,G,H,M,S,W	G/0S/R
(European) Robin	*Erithacus rubecula*	B,E,H,W	B,E,G,H,W	G/ES/R
Whinchat	*Saxicola rubetra*	G,H	G,H,M	G/ES/S
(Common) Stonechat	*Saxicola torquata*	G,H,M	A,G,H,M,S	A/0S/R
(Northern) Wheatear	*Oenanthe oenanthe*	G,H,M	A,G,H,M,S	G/3D/SM
(Eurasian) Blackbird	*Turdus merula*	B,E,H,M,W	A,B,E,G,H,M,W	G/ES/RW
Fieldfare	*Turdus pilaris*	W	A,E,G,H,M,W	A/ES/W
Redwing	*Turdus iliacus*	W	A,E,G,H,M,W	A/ES/WS
Lesser whitethroat	*Sylvia curruca*	E,H,W	E,H,W	G/0S/S

(Common) Whitethroat	*Sylvia communis*	A,E,H,W	A,E,H,W	G/ES/S
Great tit	*Parus major*	B,E,W	B,E,H,W	G/0S/R
Blue tit	*Cyanistes caeruleus*	B,E,W	B,E,H,W	G/ES/R
Great grey shrike	*Lanius excubitor*	nb	G,H	–/3H/W
(Black-billed) magpie	*Pica pica*	E,W	A,B,E,G,H,S,W	G/0S/R
(Eurasian) Jackdaw	*Corvus monedula*	B,E,W	A,B,E,G,H,S,W	G/ES/R
Rook	*Corvus frugilegus*	W	A,B,E,G,H,S,W	G/0S/R
Carrion crow	*Corvus corone*	E,G,H,M,W	A,B,E,G,H,M,S,W	G/0S/R
Hooded crow	*Corvus cornix*	E,G,H,M,W	A,B,E,G,H,M,S,W	G/0S/R
(Common) Raven	*Corvus corax*	G,H	G,H,M	G/0S/R
Chaffinch	*Fringilla coelebs*	E,W	A,B,E,G,H,S,W	G/ES/RW
(European) Greenfinch	*Carduelis chloris*	E,W	A,B,E,G,H,S,W	G/ES/R
(European) Goldfinch	*Carduelis carduelis*	G,H,W	A,B,E,G,H,S,W	G/0S/RS
(B)				
Canada goose	*Branta canadensis*	S	A,G,S	INT/0S/R
(Common) Shelduck	*Tadorna tadorna*	S	A,G,S	A/0S/R
Grey heron	*Ardea cinerea*	S,W	G,S	G/0S/R
(Western) Marsh harrier	*Circus aeruginosus*	A	A,E,G,S	A/0S/R
Peregrine falcon	*Falco peregrinus*	G,H	A,E,G,H,M,S	A/3R/R
Water rail	*Rallus aquaticus*	S	G,S	A/0S/RW
(Common) Coot	*Fulica atra*	S	A,E,G,S	G/0S/R
(Eurasian) Dotterel	*Charadrius morinellus*	montane	A,G	A/0S/SM
(Eurasian) Woodcock	*Scolopax rusticola*	W	E,G,W	A/3D/RW
Great black-backed gull	*Larus marinus*	marine	A,G,M	G/ES/RW
Common tern	*Sterna hirundo*	M,S		G/0S/S
Arctic tern	*Sterna paradisaea*	M		A/0S/S
Little tern	*Sterna albifrons*	M		A/3D/S
Tawny owl	*Strix aluco*	E,W	A,E,G,H,W	G/ES/R
(Common) Swift	*Apus apus*	B	A,E,G,H,S,W	G/0S/S
(Eurasian) Wryneck*	*Jynx torquilla*	H,W	E,G,H,W	R/3D/M(S)
Great spotted woodpecker	*Dendrocopos major*	E,W	B,E,W	G/0S/R
Lesser spotted woodpecker	*Dendrocopos minor*	E,W	E,W	R/0S/R

(*cont.*)

English name[a]	Scientific name	Nesting habitats (nb = not breeding in Britain)	Feeding habitats	Status
(Horned lark) Shore lark	*Eremophila alpestris*	nb	A,G,S	–/0S/W
Sand martin	*Riparia riparia*	riparian	A,E,G,H,S	A/3H/S
(Northern) House martin	*Delichon urbicum*	B	A,B,E,G,H,S	A/3D/S
Rock pipit	*Anthus petrosus*	G,H	G,H,M,S	G/ES/R
Water pipit	*Anthus spinoletta*	nb	G,S	G/0S/W
Grey wagtail	*Motacilla cinerea*	riparian	B,G,S	A/0S/R
(Bohemian) Waxwing	*Bombycilla garrulus*	nb	B,E	G/0S/W
(Common) Nightingale	*Luscinia megarhynchos*	E,W	E,W	A/ES/S
Cetti's warbler	*Cettia cetti*	E	E	G/0S/R
Sedge warbler	*Acrocephalus schoenobaenus*	A,E,G,M,S	A,E,G,M,S	G/ES/S
(Common) Reed warbler	*Acrocephalus scirpaceus*	E	E	G/ES/S
Garden warbler	*Sylvia borin*	E,W	E,W	G/ES/S
Blackcap	*Sylvia atricapilla*	E,W	B,E,W	G/ES/SW
(Common) Chiffchaff	*Phylloscopus collybita*	W	B,E,H,W	G/0S/S
Willow warbler	*Phylloscopus trochilus*	H,W	B,E,H,W	A/0S/S
Goldcrest	*Regulus regulus*	W	B,E,H,W	A/ES/R
Long-tailed tit	*Aegithalos caudatus*	B,E,W	B,E,H,W	G/0S/R
Marsh tit	*Poecile palustris*	W	E,H,W	R/3D/R
Willow tit	*Poecile montanus*	W	E,H,W	R/0S/R
Coal tit	*Periparus ater*	B,E,W	B,E,H,W	G/0S/R
(Eurasian) Treecreeper	*Certhia familiaris*	E,W	E,W	G/0S/R
(Eurasian) Jay	*Garrulus glandarius*	W	E,H,W	G/0S/R
Brambling	*Fringilla montifringilla*	nb	B,E,W	G/0S/W
(Eurasian) Siskin	*Carduelis spinus*	W	B,E,W	G/ES/RW
Lesser redpoll	*Carduelis cabaret*	W	A,E,H,S,W	A/0S/R
Common redpoll	*Carduelis flammea*	W	A,E,H,S,W	G/0S/W
Snow bunting	*Plectrophenax nivalis*	montane	A,G,M,S	G/0S/WR
Lapland longspur	*Calcarius lapponicus*	nb	A,G,S	G/0S/W

[a] For ease of reading, given that this book has a British focus, those parts of formal English names in brackets will not be used in the main text. For example, Northern lapwings will simply be lapwings.

Appendix 2

Scientific names of non-avian species mentioned in the text

Micro-organisms

Take-all *Ophiobulus graminis* (fungus)
Labyrinthula zosterae (protozoan)

Plants (non-flowering)

Woolly fringe-moss *Racomitrium lanuginosum*
Bracken *Pteridium aquilinum*

Plants (flowering)

Field crops

Hop *Humulus lupulus*
Sugar beet *Beta vulgaris*
Oilseed rape/swede *Brassica napus*
Cabbage, kale *B. oleracea*
Turnip *B. rapa*
Black mustard *B. nigra*
Sainfoin *Onobrychis viciifolia*
Common vetch *Vicia sativa*
Broad bean *V. faba*
Common bean *Phaseolus vulgaris*
Pea *Pisum sativum*
Red clover *Trifolium pratense*
White clover *T. repens*
Linseed *Linum usitatissimum*
Carrot *Daucus carota*
Potato *Solanum tuberosum*
Sunflower *Helianthus annuus*
Italian ryegrass *Lolium multiflorum*
Perennial ryegrass *L. perenne*
Oat *Avena sativa*

Barley *Hordeum vulgare*
Rye *Secale cereale*
Wheat *Triticum aestivum*
Maize *Zea mays*

Other species

Norway spruce *Picea abies*
Sitka spruce *Picea sitchensis*
Larch *Larix decidua*
Scots pine *Pinus sylvestris*
Lodgepole pine *P. contorta*
Juniper *Juniperus communis*
Yew *Taxus baccata*
Pasque flower *Pulsatilla vulgaris*
Meadow buttercup *Ranunculus acris*
Corn buttercup *R. arvensis*
Poppy *Papaver rhoeas*
Wych elm *Ulmus glabra*
Bog myrtle *Myrica gale*
Beech *Fagus sylvatica*
Oak *Quercus robur/petraea*
Silver birch *Betula pendula*
Hazel *Corylus avellana*
Fat-hen *Chenopodium album*
Sea purslane *Atriplex portulacoides*
Glasswort *Salicornia europaea*
Annual seablite *Suaeda maritima*
Common chickweed *Stellaria media*
Corncockle *Agrostemma githago*
Knotgrass *Polygonum aviculare*
Black bindweed *Fallopia convolvulus*
Common sorrel *Rumex acetosa*
Curled dock *R. crispus*
Broad-leaved dock *R. obtusifolius*
Sea lavender *Limonium vulgare*
Thrift *Armeria maritima*
Small-leaved lime *Tilia cordata*
Willow *Salix* spp.
Charlock *Sinapis arvensis*
Crowberry *Empetrum nigrum*
Heather *Calluna vulgaris*
Dorset heath *Erica ciliaris*

Cross-leaved heath *E. tetralix*
Bell heather *E. cinerea*
Cornish heath *E. vagans*
Bilberry *Vaccinium myrtillus*
Meadowsweet *Filipendula ulmaria*
Bramble *Rubus fruticosus* agg.
Great burnet *Sanguisorba officinalis*
Dog rose *Rosa canina*
Blackthorn *Prunus spinosa*
Rowan *Sorbus aucuparia*
Hawthorn *Crataegus monogyna*
Horseshoe vetch *Hippocrepis comosa*
Gorse *Ulex* spp.
Box *Buxus sempervirens*
Buckthorn *Rhamnus cathartica*
Milkwort *Polygala* spp.
Field maple *Acer campestre*
Wood cranesbill *Geranium sylvaticum*
Ivy *Hedera helix*
Pignut *Conopodium majus*
Hogweed *Heracleum sphondylium*
Felwort *Gentiana amarella*
Wild thyme *Thymus serpyllum*
Hoary plantain *Plantago media*
Privet *Ligustrum vulgare*
Elder *Sambucus nigra*
Wayfaring tree *Viburnum lantana*
Spear thistle *Cirsium vulgare*
Creeping thistle *C. arvense*
Greater knapweed *Centaurea scabiosa*
Sow-thistle *Sonchus* spp.
Dandelion *Taraxacum officinale* agg.
Sea aster *Aster tripolium*
Corn marigold *Chrysanthemum segetum*
Eel-grass *Zostera marina*
Sea rush *Juncus maritimus*
Soft rush *J. effusus*
Cotton-grass *Eriophorum* spp.
Mat grass *Nardus stricta*
Meadow fescue *Festuca pratensis*
Tall fescue *F. arundinacea*
Red fescue *F. rubra*
Crested dogstail *Cynosurus cristatus*

Saltmarsh grass *Puccinellia maritima*
Annual meadow-grass *Poa annua*
Cocksfoot *Dactylis glomerata*
False oat-grass *Arrhenatherum elatius*
Wild oat *Avena fatua*
Wavy hair-grass *Deschampsia flexuosa*
Sweet vernal grass *Anthoxanthum odoratum*
Common bent *Agrostis capillaris*
Marram *Ammophila arenaria*
Meadow foxtail *Alopecurus pratensis*
Black grass *A. myosuroides*
Timothy *Phleum pratense*
Purple moor-grass *Molinia caerulea*
Common reed *Phragmites australis*
Cord-grass *Spartina anglica*
Yellow iris *Iris pseudacorus*

Invertebrates

Earthworm *Lumbricus* spp.
Sandhopper *Talitrus saltator*
Knotgrass beetle *Gastrophysa polygoni*
Wheat bulb fly *Delia coarctata*
Kelp-fly *Coelopa frigida*
Cranefly *Tipula* spp.
Earwig *Forficula* spp.

Mammals

Hedgehog *Erinaceus europaeus*
Common/pygmy shrew *Sorex araneus/minutus*
Pipistrelle *Pipistrellus pipistrellus*
Rabbit *Oryctolagus cuniculus*
Brown hare *Lepus europaeus*
Mountain hare *Lepus timidus*
Wild boar *Sus scrofa*
Grey squirrel *Sciurus carolinensis*
Field vole *Microtus agrestis*
Orkney vole *Microtus arvalis orcadensis*
Wood mouse *Apodemus sylvaticus*
Harvest mouse *Micromys minutus*
Common rat *Rattus norvegicus*
Red fox *Vulpes vulpes*

Stoat *Mustela erminea*
Weasel *Mustela nivalis*
Badger *Meles meles*
Moose *Alces alces*
Red deer *Cervus elaphus*
Bison *Bison bonasus*
Sheep *Ovis aries*

Appendix 3

Commonly used abbreviations

Organisations

BTO British Trust for Ornithology
CPRE Council for the Preservation of Rural England
Defra Department for Environment, Food and Rural Affairs
EEC European Economic Community
EU European Union
GWCT Game and Wildlife Conservation Trust
JNCC Joint Nature Conservation Committee
MAFF Ministry for Agriculture, Fisheries and Food
NCC Nature Conservancy Council
RSPB Royal Society for the Protection of Birds
SNH Scottish Natural Heritage
WWT Wildfowl and Wetlands Trust

Surveys

BBS Breeding Bird Survey
CBC Common Bird Census
WeBS Wetland Bird Survey

Schemes and policies

CAP Common Agricultural Policy
CPS Countryside Premium Scheme
CSS Countryside Stewardship Scheme
RSS Rural Stewardship Scheme
ESA Environmentally Sensitive Area

Designations

SPA Special Protection Area
SSSI Site of Special Scientific Interest

References

Aebischer, NJ 1991. Twenty years of monitoring invertebrates and weeds in cereal fields in Sussex. In Firbank, LG, Carter, N, Darbyshire, JF & Potts, GR (eds) *The Ecology of Temperate Cereal Fields*, pp 305–331. Oxford: Blackwell Scientific Publications.

Aebischer, NJ 1997. Gamebirds: management of the Grey Partridge in Britain. In Bolton, M (ed) *Conservation and the Use of Wildlife Resources*, pp 131–151. London: Chapman & Hall.

Aebischer, NJ 1999. Multi-way comparisons and generalized linear models of nest success: extensions of the Mayfield method. *Bird Study* 46 (Suppl.): S22–S31.

Aebischer, NJ 2002. European Turtle Dove. In Wernham, CV, Toms, MP, Marchant, JH, Clark, JA, Siriwardena, GM & Baillie, SR (eds) *The Migration Atlas: Movements of the Birds of Britain and Ireland*, pp 420–422. London: T. & A. D. Poyser.

Aebischer, NJ & Ewald, JA 2004. Managing the UK Grey Partridge *Perdix perdix* recovery: population change, reproduction, habitat and shooting. *Ibis* 146 (Suppl. 2): 181–191.

Aebischer, NJ & Potts, GR 1998. Spatial changes in Grey Partridge (*Perdix perdix*) distribution in relation to 25 years of changing agriculture in Sussex, UK. *Gibier Faune Sauvage* 15: 293–308.

Aebischer, NJ, Green, RE & Evans, AD 2000. From science to recovery: four case studies of how research has been translated into conservation action in the UK. In Aebsicher, NJ, Evans, AD, Grice, PV & Vickery, JA (eds) *Ecology and Conservation of Lowland Farmland Birds*, pp 43–54. Tring: British Ornithologists' Union.

Ahola, M, Laaksonen, T, Sippola, K, Eeva, T, Rainio, K & Lehikoinen, E 2004. Variation in climate warming along the migration route uncouples arrival and breeding dates. *Global Change Biology* 10: 1610–1617.

Ahola, MP, Laaksonen, T, Eeva, T & Lehikoinen, E 2007. Climate change can alter competitive relationships between resident and migratory birds. *Journal of Animal Ecology* 76: 1045–1052.

Alexander, HG 1914. A report on the Land-Rail inquiry. *British Birds* 8: 83–92.

Alexander, WB & Lack, D 1944. Changes in status among British breeding birds. *British Birds* 38: 42–45, 62–69, 82–88.

Allen, DS 1995. Habitat selection by Whinchats: a case for bracken in the uplands? In Thompson, DBA, Hester, AJ & Usher, MB (eds) *Heaths and Moorland: Cultural Landscapes*, pp 43–50. Edinburgh: Her Majesty's Stationery Office.

Amar, A & Redpath, SM 2005. Habitat use by Hen Harriers *Circus cyaneus* on Orkney: implications of land-use change for this declining population. *Ibis* 147: 37–47.

Amar, A, Thirgood, S, Pearce-Higgins, J & Redpath, S 2008. The impact of raptors on the abundance of upland passerines and waders, *Oikos* 117: 1143–1152.

Ambrosini, R, Bolzern, AM, Canova, L, Arieni, S, Møller, AP & Saino, N 2002. The distribution and colony size of barn swallows in relation to agricultural land use. *Journal of Applied Ecology* 39: 524–534.

Andersen, PS, Towers, W & Smith, P 2005. Assessing the potential for biomass energy to contribute to Scotland's renewable energy resource. *Biomass and Bioenergy* 29: 73–82.

Anderson, DR & Yalden, DW 1981. Increased sheep numbers and the loss of heather moorland in the Peak District, England. *Biological Conservation* 20: 195–213.

Anderson, GQA & Fergusson, MJ 2006. Energy from biomass in the UK: sources, processes and biodiversity implications. *Ibis* 148: 180–183.

Anderson, RM, Donnelly, CA, Ferguson, NM, Woolhouse, MEJ, Watt, CJ, Udy, HJ, Mawhinney, S, Dunstan, SP, Southwood, TRE, Wilesmith, JW, Ryan, JBM, Hoinville, LJ, Hillerton, JE, Austin, AR & Wells, GAH 1996. Transmission dyamics and epidemiology of BSE in British cattle. *Nature* 382: 779–788.

Andreasen, C, Stryhn, H & Streibig, JC 1996. Decline of the flora in Danish arable fields. *Journal of Applied Ecology* 33: 619–626.

Andrén, H 1992. Corvid density and nest predation in relation to forest fragmentation: a landscape perspective. *Ecology* 73: 794–804.

Andrén, H & Angelstam, P 1988. Elevated predation rates as an edge effect in habitat islands: experimental evidence. *Ecology* 69: 544–547.

Andrzejewska, L 1976. The effect of mineral fertilization of a meadow on the Auchenorrhyncha (Homoptera) fauna. *Polish Ecological Studies* 2: 111–127.

Anon. 1995. *Biodiversity: The UK Steering Group Report, Volume 2: Action Plans.* London: Her Majesty's Stationery Office.

Anon. 1998–99. *UK Biodiversity Group Tranche 2 Action Plans. Volume I–VI.* Peterborough: English Nature.

Appleton, GF 2002. Northern Lapwing *Vanellus vanellus*. In Wernham, CV, Toms, MP, Marchant, JH, Clark, JA, Siriwardena, GM & Baillie, SR (eds) *The Migration Atlas: Movements of the Birds of Britain and Ireland*, pp 290–292. London: T. & A. D. Poyser.

Appleyard, I 1994. *Ring Ouzels of the Yorkshire Dales.* Leeds: Maney & Son.

Armstrong, HM, Gordon, IJ, Hutchings, NJ, Illius, AW, Milne, JA & Sibbald, AR 1997. A model of the grazing of hill vegetation by sheep in the UK. 2. The prediction of offtake by sheep. *Journal of Applied Ecology* 34: 186–207.

Arnold, GW 1983. The influence of ditch and hedgerow structure, length of hedgerows, and area of woodland and garden on bird numbers on farmland. *Journal of Applied Ecology* 20: 731–750.

Arroyo, B, Garcia, JT & Bretagnolle, V 2002. Conservation of Montagu's Harrier (*Circus pygargus*) in agricultural areas. *Animal Conservation* 5: 283–290.

Asher, J, Warren, M, Fox, R, Harding, P, Jeffcoate, G & Jeffcoate, S 2001. *The Millennium Atlas of Butterflies in Britain and Ireland.* Oxford University Press.

Askew, NP, Searle, JB & Moore, NP 2007. Agri-environment schemes and foraging of barn owls *Tyto alba*. *Agriculture, Ecosystems and Environment* 118: 109–114.

Atkinson, PW 1998. The wintering ecology of Twite *Carduelis flavirostris* and the consequences of habitat loss. Unpublished PhD thesis, University of East Anglia.

Atkinson, PW, Fuller, RJ & Vickery, JA 2002. Large scale patterns of summer and winter bird distribution in relation to farmland type in England and Wales. *Ecography* 25: 466–480.

Atkinson, PW, Crooks, S, Drewitt, A, Grant, A, Rehfisch, MM, Sharpe, J & Tyas, CJ 2004. Managed realignment in the UK: the first 5 years of colonization by birds. *Ibis* 146 (Suppl. 1): S101–S110.

Atkinson, PW, Fuller, RJ, Vickery, JA, Conway, GJ, Tallowin, JRB, Smith, REN, Haysom, KA, Ings, TC, Asteraki, EJ & Brown, VK 2005. Influence of agricultural management, sward structure and food resources on grassland use by birds in lowland England. *Journal of Applied Ecology* 42: 932–942.

Ausden, M & Hirons, GJM 2002. Grassland nature reserves for breeding wading birds in England and the implications for the ESA agri-environment scheme. *Biological Conservation* 106: 279–291.

Ausden, M, Sutherland, WJ & James, R 2001. The effects of flooding lowland wet grassland on soil macroinvertebrate prey of breeding wading birds. *Journal of Applied Ecology* 38: 320–338.

Ausden, M, Rowlands, A, Sutherland, WJ & James, R 2003. Diet of breeding Lapwing *Vanellus vanellus* and Redshank *Tringa totanus* on coastal grazing marsh and implications for habitat management. *Bird Study* 50: 285–293.

Avery, MI 1989. Effects of upland afforestation on some birds of the adjacent moorlands. *Journal of Applied Ecology* 26: 957–966.

Avery, MI & Haines-Young, RH 1990. Population estimates for the dunlin *Calidris alpina* derived from remotely sensed satellite imagery of the Flow Country of northern Scotland. *Nature* 344: 860–862.

Avery, M & Leslie, R 1990. *Birds and Forestry*. London: T. & A. D. Poyser.

Aviron, S, Herzog, F, Klaus, I, Luka, H, Pfiffner, L, Schüpbach, B & Jeanneret, P 2007a. Effects of Swiss agri-environmental measures on arthropod biodiversity in arable landscapes. *Aspects of Applied Biology* 81: 101–109.

Aviron, S, Jeanneret, P, Schupbach, B & Herzog, F 2007b. Effects of agri-environmental measures, site and landscape conditions on butterfly diversity of Swiss grassland. *Agriculture, Ecosystems and Environment* 122: 295–304.

Bailey, S 2007. Increasing connectivity in fragmented landscapes: an investigation of evidence for biodiversity gain in woodlands. *Forest Ecology and Management* 238: 7–23.

Baillie, SR 1990. Integrated population monitoring of breeding birds in Britain and Ireland. *Ibis* 132: 151–166.

Baillie, SR & Green, RE 1987. The importance of variation in recovery rates when estimating survival rates from ringing recoveries. *Acta Ornithologica* 23: 41–60.

Baillie, SR & Peach, WJ 1992. Population limitation in Palearctic–African migrant passerines. *Ibis* 134 (Suppl. 1): 120–132.

Baillie, SR, Marchant, JH, Crick, HQP, Noble, DG, Balmer, DE, Barimore, C, Coombes, RH, Downie, IS, Freeman, SN, Joys, AC, Leech, DI, Raven, MJ, Robinson, RA & Thewlis, RM 2007. *Breeding Birds in the Wider Countryside: Their Conservation Status 2007*, BTO Research Report No. 487. Thetford: British Trust for Ornithology. (www.bto.org/birdtrends)

Baines, D 1988. The effects of improvement of upland, marginal grasslands on the distribution and density of breeding wading birds (Charadriiformes) in Northern England. *Biological Conservation* 45: 221–236.

Baines, D 1989. The effects of improvement of upland, marginal grasslands on the breeding success of Lapwings *Vanellus vanellus* and other waders. *Ibis* 131: 497–506.

Baines, D 1990. The roles of predation, food and agricultural practice in determining the breeding success of the lapwing (*Vanellus vanellus*) on upland grasslands. *Journal of Animal Ecology* 59: 915–929.

Baines, D 1994. Seasonal differences in habitat selection by Black Grouse *Tetrao tetrix* in the northern Pennines, England. *Ibis* 136: 39–43.

Baines, D 1996. The implications of grazing and predator management on the habitats and breeding success of black grouse *Tetrao tetrix*. *Journal of Applied Ecology* 33: 54–62.

Baines, D 2003. *The Black Grouse*, a special report and management guide by the Game Conservancy Trust. Fordingbridge: Game Conservancy Trust.

Baines, D & Andrew, M 2003. Marking of deer fences to reduce frequency of collisions by woodland grouse. *Biological Conservation* 110: 169–176.

Baines, D & Hudson, PJ 1995. The decline of Black Grouse in Scotland and northern England. *Bird Study* 42: 122–131.

Baines, D & Summers, RW 1997. Assessment of bird collisions with deer fences. *Journal of Applied Ecology* 34: 941–948.

Baines, D, Wilson, IA & Beeley, G 1996. Timing of breeding in Black Grouse *Tetrao tetrix* and Capercaillie *Tetrao urogallus* and distribution of insect food for chicks. *Ibis* 138: 181–187.

Baines, D, Warren, P & Richardson, M 2007. Variation in vital rates of black grouse *Tetrao tetrix* in the United Kingdom. *Wildlife Biology* 13: 109–116.

Baker, H, Stroud, DA, Aebischer, NJ, Cranswick, PA, Gregory, RD, McSorley, CA, Noble, DG & Rehfisch, MM 2006. Population estimates of birds in Great Britain and the United Kingdom. *British Birds* 99: 25–44.

Balmford, A, Green, RE & Scharlemann, JPW 2005. Sparing land for nature: exploring the potential impact of changes in agricultural yield on the area needed for crop production. *Global Change Biology* 11: 1594–1605.

Banks, AN, Collier, MP, Austin, GE, Hearn, RD & Musgrove, AJ 2006. *Waterbirds in the UK 2004/05: The Wetland Bird Survey.* Thetford: British Trust for Ornithology, Wildfowl and Wetlands Trust, Royal Society for the Protection of Birds and Joint Nature Conservation Committee.

Barber, D (ed) 1970. *Farming and Wildlife: A Study in Compromise.* Sandy: Royal Society for the Protection of Birds.

Barker, AM, Brown, NJ & Reynolds, CJM 1999. Do host-plant requirements and mortality from soil cultivation determine the distribution of graminivorous sawflies on farmland? *Journal of Applied Ecology* 36: 271–282.

Barnard, C & Thompson, DBA 1985. *Gulls and Plovers.* London: Croom Helm.

Barnett, A 2007. Agri-environmental policy: a European overview. *Aspects of Applied Biology* 81: 1–6.

Barnett, PR, Whittingham, MJ, Bradbury, RB & Wilson, JD 2004. Use of unimproved and improved lowland grassland by wintering birds in the UK. *Agriculture, Ecosystems and Environment* 102: 49–60.

Batten, LA, Bibby, CJ, Clement, P, Elliott, GD & Porter, RF 1990. *Red Data Birds in Britain: Action for Rare, Threatened and Important Species.* London: T. & A. D. Poyser.

Baxter, EV & Rintoul, LJ 1953. *The Birds of Scotland.* Edinburgh: Oliver & Boyd.

Beale, CM, Burfield, IJ, Sim, IMW, Rebecca, GW, Pearce-Higgins, JW & Grant, MC 2006. Climate change may account for the decline in British ring ouzels *Turdus torquatus*. *Journal of Animal Ecology* 75: 826–835.

Bealey, CE, Green, RE, Robson, R, Taylor, CR & Winspear, R 1999. Factors affecting the numbers and breeding success of Stone Curlews *Burhinus oedicnemus* at Porton Down, Wiltshire. *Bird Study* 46: 145–156.

Beecher, NA, Johnson, RJ, Brandle, JR, Case, RM & Young, LJ 2002. Agroecology of birds in organic and nonorganic farmland. *Conservation Biology* 16: 1620–1631.

Beintema, AJ & Müskens, JDM 1987. Nesting success of birds breeding in Dutch agricultural grasslands. *Journal of Applied Ecology* 24: 743–758.

Beintema, AJ, Thissen, JB, Tensen, D & Visser, GH 1991. Feeding ecology of charadriiform chicks in agricultural grassland. *Ardea* 79: 31–44.

Bengtsson, J, Ahnstrom, J & Weibull, A-C 2005. The effects of organic agriculture on biodiversity and abundance: a meta-analysis. *Journal of Applied Ecology* 42: 261–269.

Benton, TG, Bryant, DM, Cole, L & Crick, HQP 2002. Linking agricultural practice to insect and bird populations: a historical study over three decades. *Journal of Applied Ecology* 39: 673–687.

Benton, TG, Vickery, JA & Wilson, JD 2003. Farmland biodiversity: is habitat heterogeneity the key? *Trends in Ecology and Evolution* 18: 182–188.

Berendse, F, Chamberlain, D, Kleijn, D & Schekkerman, H 2004. Declining biodiversity in agricultural landscapes and the effectiveness of agri-environment schemes. *Ambio* 33: 499–502.

Berg, A 1996. Predation on artificial, solitary and aggregated wader nests on farmland. *Oecologia* 107: 343–346.

Berg, A & Gustafson, T 2007. Meadow management and occurrence of corncrake *Crex crex*. *Agriculture, Ecosystems and Environment* 120: 139–144.

Berg, A, Lindberg, T & Källebrink, K 1992. Hatching success of Lapwings on farmland: differences between habitats and colonies of different sizes. *Journal of Animal Ecology* 61: 469–476.

van den Berg, LJL, Bullock, JM, Clarke, RT, Langston, RHW & Rose, RJ 2001. Territory selection by the Dartford Warbler (*Sylvia undata*) in Dorset, England: the role of vegetation type, habitat fragmentation and population size. *Biological Conservation* 101: 217–228.

Berthold, P, Helbig, AJ, Mohr, G & Querner, U 1992. Rapid microevolution of migratory behaviour in a wild bird species. *Nature* 360: 668–670.

Bibby, CJ 1978. A heathland bird census. *Bird Study* 25: 87–96.

Bibby, CJ 1979. Food of the Dartford Warbler *Sylvia undata* on southern English heathland. *Journal of Zoology, London* 188: 557–576.

Bibby, CJ & Buckland, ST 1987. Bias of bird census results due to detectability varying with habitat. *Acta Oecologia* 8: 103–112.

Biesmeijer, JC, Roberts, SPM, Reemer, M, Ohlemüller, R, Edwards, M, Peeters, T, Schaffers, AP, Potts, SG, Kleukers, R, Thomas, CD, Settele, J & Kunin, WE 2006. Parallel declines in pollinators and insect-pollinated plants in Britain and the Netherlands. *Science* 313: 351–354.

Bignal, EM & McCracken, DI 1996. Low-intensity farming systems in the conservation of the countryside. *Journal of Applied Ecology* 33: 413–424.

Bignal, E, Bignal, S & McCracken, D 1997. The social life of the Chough. *British Wildlife* 8: 373–383.

BirdLife International 2000. *Threatened Birds of the World*. Barcelona and Cambridge, UK: Lynx Editions and BirdLife International.

BirdLife International 2004. *Birds in Europe: Population Estimates, Trends and Conservation Status*, BirdLife Conservation Series No. 12. Cambridge, UK: BirdLife International.

Birks, HJB 1988. Long-term ecological change in the British uplands. In Usher, MB & Thompson, DBA (eds) *Ecological Change in the Uplands*, pp 37–56. Oxford: Blackwell Scientific Publications.

Blackstock, TH, Stevens, JP, Howe, EA & Stevens, DP 1995. Changes in the extent and fragmentation of heathland and other semi-natural habitats between 1920–22 and 1987–88 in the Llyn peninsula, Wales, UK. *Biological Conservation* 72: 33–44.

Blake, S, Foster, GN, Eyre, MD & Luff, ML 1994. Effects of habitat type and grassland management practices on the body size and distribution of carabid beetles. *Pedobiologia* 38: 502–512.

Blake, S, Foster, GN, Fisher, GEJ & Ligertwood, GEL 1996. Effects of management practices on the carabid faunas of newly-established wildflower meadows in southern Scotland. *Annales Zoologici Fennici* 33: 139–147.

Blanco, G, Tella, JL & Torre, I 1998. Traditional farming and key foraging habitats for chough *Pyrrhocorax pyrrhocorax* conservation in a Spanish pseudosteppe landscape. *Journal of Applied Ecology* 35: 232–239.

Boatman, ND, Brickle, NW, Hart, JD, Milsom, TP, Morris, AJ, Murray, AWA, Murray, KA & Robertson, PA 2004. Evidence for the indirect effects of pesticides on farmland birds. *Ibis* 146 (Suppl. 2): 131–143.

Bohan, DA, Boffey, CWH, Brooks, DR, Clark, SJ, Dewar, AM, Firbank, LG, Haughton, AJ, Hawes, C, Heard, MS, May, MJ, Osborne, JL, Perry, JN, Rothery, P, Roy, DB, Scott, RJ, Squire, GR, Woiwod, IP & Champion, GT 2005. Effects on weed and invertebrate abundance and diversity of herbicide management in genetically-modified herbicide-tolerant winter-sown oilseed rape. *Proceedings of the Royal Society B* 272: 463–474.

Bolton, M, Tyler, G, Smith, K & Bamford, R 2007. The impact of predator control on lapwing *Vanellus vanellus* breeding success on wet grassland nature reserves. *Journal of Applied Ecology* 44: 434–544.

Borg, C & Toft, S 1999. Value of the aphid (*Rhopalosiphum padi*) as food for grey partridge chicks. *Wildlife Biology* 5: 55–59.

Borg, C & Toft, S 2000. Importance of insect prey quality for grey partridge chicks *Perdix perdix*: a self-selection experiment. *Journal of Applied Ecology* 37: 557–563.

Both, C, Artemyev, AV, Blaauw, B, Cowie, RJ, Dekhuijzen, AJ, Eeva, T, Enemar, A, Gustaffson, L, Ivankina, EV, Jarvinen, A, Metcalfe, NB, Nyholm, NEI, Potti, J, Ravussin, PA, Sanz, JJ, Silverin, B, Slater, FM, Sokolov, LV, Torok, J, Winkel, W, Wright, J, Zang, H & Visser, ME 2004. Large-scale geographical variation confirms that climate change causes birds to lay earlier. *Proceedings of the Royal Society of London B* 271: 1657–1662.

Both, C, Bouwhuis, S, Lessells, CM & Visser, ME 2006. Climate change and population declines in a long-distance migratory bird. *Nature* 441: 81–83.

Boutin, J-M 2001. Elements for a Turtle Dove (*Streptopelia turtur*) management plan. *Game and Wildlife Science* 18: 87–112.

Bowden, CGR 1990. Selection of foraging habitats by Woodlarks (*Lullula arborea*) nesting in pine plantations. *Journal of Applied Ecology* 27: 410–419.

Bradbury, RB & Allen, DS 2003. Evaluation of the impact of the pilot UK Arable Stewardship Scheme on breeding and wintering birds. *Bird Study* 50: 131–141.

Bradbury, RB & Bradter, U 2004. Habitat associations of Yellow Wagtails *Motacilla flava flavissima* on lowland wet grassland. *Ibis* 146: 241–246.

Bradbury, RB & Kirby, WB 2006. Farmland birds and resource protection in the UK: cross-cutting solutions for multi-functional farming? *Biological Conservation* 129: 530–542.

Bradbury, RB, Kyrkos, A, Morris, AJ, Clark, SC, Perkins, AJ & Wilson, JD 2000. Habitat associations and breeding success of yellowhammers on lowland farmland. *Journal of Applied Ecology* 37: 789–805.

Bradbury, RB, Payne, RJH, Wilson, JD & Krebs, JR 2001. Predicting population responses to resource management. *Trends in Ecology and Evolution* 16: 440–445.

Bradbury, RB, Browne, SJ, Stevens, DK & Aebsicher, NJ 2004. Five-year evaluation of the impact of the Arable Stewardship Pilot Scheme on birds. *Ibis* 146 (Suppl. 2): 171–180.

Bradbury, RB, Bailey, CM, Wright, D & Evans, AD 2008. Wintering cirl buntings *Emberiza cirlus* in southwest England select cereal stubbles that follow a low-input herbicide regime. *Bird Study* 55: 23–31.

Brambilla, M, Rubolini, D & Guidali, F 2007. Between land abandonment and agricultural intensification: habitat preferences of Red-backed Shrikes *Lanius collurio* in low-intensity farming conditions. *Bird Study* 54: 160–167.

Brickle, NW 1998. The effect of agricultural intensification on the decline of the Corn Bunting, *Miliaria calandra*. Unpublished DPhil thesis, University of Sussex.

Brickle, NW & Harper, DGC 1999. Diet of nestling Corn Buntings *Miliaria calandra* in southern England examined by compositional analysis of faeces. *Bird Study* 46: 319–329.

Brickle, NW & Harper, DGC 2000. Habitat use by Corn Buntings *Miliaria calandra* in winter and summer. In Aebsicher, NJ, Evans, AD, Grice, PV & Vickery, JA (eds) *Ecology and Conservation of Lowland Farmland Birds*, pp 156–164. Tring: British Ornithologists' Union.

Brickle, NW & Harper, DGC 2002. Agricultural intensification and the timing of breeding of Corn Buntings *Miliaria calandra*. *Bird Study* 49: 219–228.

Brickle, NW & Peach, WJ 2004. The breeding ecology of Reed Buntings *Emberiza schoeniclus* in farmland and wetland habitats in lowland England. *Ibis* 146 (Suppl. 2): 69–77.

Brickle, NW, Harper, DGC, Aebischer, NJ & Cockayne, SH 2000. Effects of agricultural intensification on the breeding success of corn buntings *Miliaria calandra*. *Journal of Applied Ecology* 37: 742–755.

BRIG (2007). *Report on the Species and Habitats Review*. Peterborough: Joint Nature Conservation Committee.

Briggs, D & Courtenay, F 1985. *Agriculture and Environment: The Physical Geography of Temperate Farming Systems*. Harlow: Longman.

Brindley, E, Norris, K, Cook, T, Babbs, S, Forster Brown, C, Massey, P, Thompson, R & Yaxley, R 1998. The abundance and conservation status of redshank *Tringa totanus* nesting on saltmarshes in Great Britain. *Biological Conservation* 86: 289–297.

British Trust for Ornithology 2002. *The Effects of Different Crop Stubbles and straw disposal methods on wintering birds and arable plants*, Report to Defra of Project BD1610. Thetford: British Trust for Ornithology. (www.defra.gov.uk)

Britschgi, A, Spaar, R & Arlettaz, R 2006. Impact of grassland farming intensification on the breeding ecology of an indicator insectivorous passerine, the Whinchat *Saxicola rubetra*: lessons for overall Alpine meadowland management. *Biological Conservation* 130: 193–205.

Bro, E, Sarrazin, F, Clobert, J & Reitz, F 2000. Demography and the decline of the grey partridge *Perdix perdix* in France. *Journal of Applied Ecology* 37: 432–448.

Bro, E, Reitz, F, Clobert, J, Migot, P & Massot, M 2001. Diagnosing the environmental causes of the decline in grey partridge survival in France. *Ibis* 143: 120–132.

Bro, E, Mayot, P, Corda, E & Reitz, F 2004. Impact of habitat management on grey partridge populations: assessing wildlife cover using a multisite BACI experiment. *Journal of Applied Ecology* 41: 846–857.

Brooks, DR, Bohan, DA, Champion, GT, Haughton, AJ, Hawes, C, Heard, MS, Clark, SJ, Dewar, AM, Firbank, LG, Perry, JN, Rothery, P, Scott, RJ, Woiwod, IP, Birchall, C, Skellern, MP, Walker, JH, Baker, P, Bell, D, Browne, EL, Dewar, AJG, Fairfax, CM, Garner, BH, Haylock, LA, Horne, SL, Hulmes, SE, Mason, NS, Norton, LR, Nuttall, P, Randle, Z, Rossall, MJ, Sands, RJN, Singer, EJ & Walker, MJ 2003. Invertebrate responses to the management of genetically modified herbicide-tolerant and conventional spring crops. 1. Soil-surface-active invertebrates. *Philosophical Transactions of the Royal Society of London B* 358: 1847–1862.

Brown, AF & Atkinson, PW 1996. Habitat associations of coastal wintering passerines. *Bird Study* 43: 188–200.

Brown, A & Grice, P 2005. *Birds in England*. London: T. & A. D. Poyser.

Brown, AF & Stillman, RA 1993. Bird-habitat associations in the eastern Highlands of Scotland. *Journal of Applied Ecology* 33: 413–424.

Brown, AF, Crick, HQP & Stillman, RA 1995. The distribution, numbers and breeding ecology of Twite *Acanthis flavirostris* in the south Pennines of England. *Bird Study* 42: 107–121.

Brown, JH 1984. On the relationship between abundance and distribution of species. *American Naturalist* 124: 255–279.

Browne, SJ & Aebischer, NJ 2003a. Temporal changes in the migration phenology of turtle doves *Streptopelia turtur* in Britain, based on sightings from coastal bird observatories. *Journal of Avian Biology* 34: 65–71.

Browne, SJ & Aebischer, NJ 2003b. Habitat use, foraging ecology and diet of Turtle Doves *Streptopelia turtur* in Britain. *Ibis* 145: 572–582.

Browne, SJ & Aebischer, NJ 2004. Temporal changes in the breeding ecology of European Turtle Doves *Streptopelia turtur* in Britain, and implications for conservation. *Ibis* 146: 125–137.

Browne, S, Vickery, J & Chamberlain, D 2000. Densities and population estimates of breeding Skylarks *Alauda arvensis* in Britain in 1997. *Bird Study* 47: 52–65.

Browne, SJ, Aebischer, NJ & Crick, HQP 2005. Breeding ecology of Turtle Doves *Streptopelia turtur* in Britain during the period 1941–2000: an analysis of BTO nest record cards. *Bird Study* 52: 1–9.

Brownie, C, Anderson, DR, Burnham, KP & Robson, DS 1985. *Statistical Inference from Band Recovery: A Handbook*. Washington, DC: US Department of the Interior, Fish & Wildlife Service.

Bruun, M & Smith, HG 2003. Landscape composition affects habitat use and foraging flight distances in breeding European starlings. *Biological Conservation* 114: 179–187.

Bryson, B 2000. *The English Landscape.* London: Profile Books.

Buchanan, GM, Pearce-Higgins, JW, Wotton, SR, Grant, MC & Whitfield, DP 2003. Correlates of change in Ring Ouzel *Turdus torquatus* abundance in Scotland from 1988–91 to 1999. *Bird Study* 50: 97–105.

Buckingham, DL 2001. Within-field habitat selection by wintering skylarks *Alauda arvensis* in southwest England. In Donald, PF & Vickery, JA (eds) *The Ecology and Conservation of Skylarks* Alauda arvensis, pp 149–158. Sandy: Royal Society for the Protection of Birds/British Trust for Ornithology.

Buckingham, DL & Peach, WJ 2006. Leaving final-cut grass silage *in situ* overwinter as a seed resource for declining farmland birds. *Biodiversity and Conservation* 15: 3827–3845.

Buckingham, DL, Evans, AD, Morris, AJ, Orsman, CJ & Yaxley, R 1999. Use of set-aside in winter by declining farmland bird species in the UK. *Bird Study* 46: 157–169.

Buckingham, DL, Peach, WJ & Fox, DS 2006. Effects of agricultural management on the use of lowland grassland by foraging birds. *Agriculture, Ecosystems and Environment* 112: 21–40.

Buckland, ST, Anderson, DR, Burnham, KP, Laake, JL, Borchers, DL & Thomas, L 2001. *Introduction to Distance Sampling: Estimating Abundance of Biological Populations.* Oxford University Press.

Buckwell, A & Armstrong-Brown, S 2004. Changes in farming and future prospects: technology and policy. *Ibis* 146 (Suppl. 2): 14–21.

Bullock, ID, Drewitt, DR & Mickleburgh, SP 1983. The Chough in Britain and Ireland. *British Birds* 76: 377–401.

Burfield, IJ 2002. The breeding ecology and conservation of the Ring Ouzel *Turdus torquatus* in Britain. Unpublished PhD thesis, University of Cambridge.

Burn, AJ 2000. Pesticides and their effects on lowland farmland birds. In Aebsicher, NJ, Evans, AD, Grice, PV & Vickery, JA (eds) *Ecology and Conservation of Lowland Farmland Birds*, pp 89–104. Tring: British Ornithologists' Union.

Burton, NHK, Watts, PN, Crick, HQP & Edwards, PJ 1999. The effects of preharvesting operations on Reed Buntings *Emberiza shoeniclus* nesting in Oilseed Rape *Brassica napus. Bird Study* 46: 369–172.

von Busche, G. 1989. The decline of the Corn Bunting population in Schleswig-Holstein, FRG. *Die Vogelwarte* 35: 11–20.

Butler, SJ & Gillings, S 2004. Quantifying the effects of habitat structure on prey detectability and accessibility to farmland birds. *Ibis* 146 (Suppl 2.): 123–130.

Butler, SJ, Bradbury, RB & Whittingham, MJ 2005a. Stubble height affects the use of stubble fields by farmland birds. *Journal of Applied Ecology* 42: 469–476.

Butler, SJ, Whittingham, MJ, Quinn, JL & Cresswell, W 2005b. Quantifying the interaction between food density and habitat structure in determining patch selection. *Animal Behaviour* 69: 337–343.

Butler, SJ, Vickery, JA & Norris, K 2007. Farmland biodiversity and the footprint of agriculture. *Science* 315: 381–384.

Byrkjedal, I & Thompson, DBA 1998. *Tundra Plovers: the Eurasian, Pacific and American Golden Plovers and Grey Plover.* London: T. & A. D. Poyser.

Cadbury, CJ 1980. The status and habitats of the corncrake in Britain 1978/79. *Bird Study* 27: 203–218.

Calladine, J, Baines, D & Warren, P 2002. Effects of reduced grazing on population density and breeding success of black grouse in northern England. *Journal of Applied Ecology* 39: 772–780.

Calladine, J, Robertson, D & Wenham, CV 2006. The movements of some granivorous passerines in winter on farmland. *Ibis* 148: 169–173.

Campbell, LH, Avery, MI, Donald, PF, Evans, AD, Green, RE & Wilson, JD 1997. *A Review of the Indirect Effects of Pesticides on Birds*, JNCC Report No. 227. Peterborough: Joint Nature Conservation Committee.

Carey, PD 2001. Schemes are monitored and effective in the UK. *Nature* 414: 687.

Carson, R 1963. *Silent Spring*. London: Hamish Hamilton.

Carter, I, Cross, AV, Douse, A, Duffy, K, Etheridge, B, Grice, PV, Newberry, P, Orr-Ewing, DC, O'Toole, L, Simpson, D & Snell, N 2003. Re-introduction and conservation of the red kite (*Milvus milvus*) in Britain: current threats and prospects for future range expansion. In Thompson, DBA, Redpath, SM, Fielding, AH, Marquiss, M & Galbraith, CA (eds) *Birds of Prey in a Changing Environment*, pp 407–416. Edinburgh: The Stationery Office.

Carvell, C, Meek, WR, Pywell, RF, Goulson, D & Nowakowski, M 2007. Comparing the efficacy of agri-environment schemes to enhance bumble bee abundance and diversity on arable field margins. *Journal of Applied Ecology* 44: 29–40.

Catchpole, CK & Phillips, JF 1992. Territory quality and reproductive success in the Dartford Warbler *Sylvia undata* in Dorset, England. *Biological Conservation* 61: 209–215.

Catchpole, EA, Morgan, BJT, Freeman, SN & Peach, WJ 1999. Modelling the survival of British lapwings using ring-recovery data and weather covariates. *Bird Study* 46: S5–S13.

Cayford, J & Hope-Jones, P 1989. Black Grouse in Wales. *RSPB Conservation Review* 3: 79–81.

Chalfoun, AD, Thompson, FR & Ratnaswamy, MJ 2002. Nest predators and fragmentation: a review and meta-analysis. *Conservation Biology* 16: 306–318.

Chamberlain, DE 2001. Habitat associations and trends in reproductive performance of skylarks *Alauda arvensis* breeding in the uplands of the UK. In Donald, PF & Vickery, JA (eds) *The Ecology and Conservation of Skylarks* Alauda arvensis, pp 25–39. Sandy: Royal Society for the Protection of Birds/British Trust for Ornithology.

Chamberlain, DE & Crick, HQP 1999. Population declines and reproductive performance of skylarks *Alauda arvensis* in different regions and habitats of the United Kingdom. *Ibis* 141: 38–51.

Chamberlain, DE & Fuller, RJ 2000. Local extinctions and changes in species richness of lowland farmland birds in England and Wales in relation to recent changes in agricultural land-use. *Agriculture, Ecosystems and Environment* 78: 1–17.

Chamberlain, DE & Fuller, RJ 2001. Contrasting patterns of change in the distribution and abundance of farmland birds in relation to farming system in lowland Britain. *Global Ecology and Biogeography* 10: 399–409.

Chamberlain, DE & Gregory, RD 1999. Coarse and fine scale habitat associations of breeding Skylarks *Alauda arvensis* in the UK. *Bird Study* 46: 34–47.

Chamberlain, DE & Wilson, JD 2000. The contribution of hedgerow structure to the value of organic farms to birds. In Aebsicher, NJ, Evans, AD, Grice, PV & Vickery, JA (eds) *Ecology and Conservation of Lowland Farmland Birds*, pp 57–68. Tring: British Ornithologists' Union.

Chamberlain, DE, Glue, DE & Toms, MP 2008a. Sparrowhawk *Accipiter nisus* presence and winter bird abundance. *Journal of Ornithology* in press.

Chamberlain, D, Gough, S, Vickery, J, Anderson, G, Grice, P & Cooke, A 2008b. The plot thickens. *BTO News* 278: 24.

Chamberlain, DE, Hatchwell, BJ & Perrins, CM 1995. Spaced out nests and predators: an experiment to test the effect of habitat structure. *Journal of Avian Biology* 26: 346–349.

Chamberlain, DE, Wilson, JD & Fuller, RJ 1999. A comparison of bird populations on organic and conventional farm systems in southern Britain. *Biological Conservation* 88: 307–320.

Chamberlain, DE, Fuller, RJ, Bunce, RGH, Duckworth, JC & Shrubb, M 2000. Changes in the abundance of farmland birds in relation to the timing of agricultural intensification in England and Wales. *Journal of Applied Ecology* 37: 771–788.

Chamberlain, DE, Vickery, JA, Glue, DE, Robinson, RA, Conway, GJ, Woodburn, RJW & Cannon, AR 2005. Annual and seasonal trends in the use of garden feeders by birds in winter. *Ibis* 147: 563–575.

Channell, R & Lomolino, MV 2000. Trajectories to extinction: spatial dynamics of the contraction of geographic ranges. *Journal of Biogeography* 27: 169–180.

Christensen, KD, Jacobsen, EM & Nohr, H 1996. A comparative study of bird faunas in conventionally and organically farmed areas. *Dansk Ornitologisk Forenings Tidsskrift* 90: 21–28.

Clarke, J 1992. *Set-Aside*, British Crop Protection Council Monograph No. 50. Farnham: British Crop Protection Council.

Clarke, R, Combridge, M & Combridge, P 1997. A comparison of the feeding ecology of wintering Hen Harriers *Circus cyaneus* centred on two heathland areas in England. *Ibis* 139: 4–18.

Clergeau, P & Burel, F 1997. The role of spatio-temporal patch connectivity at the landscape level: an example in a bird distribution. *Landscape and Urban Planning* 38: 37–43.

Coates, A & Say, A 1999. *Ecological Assessment of Short Rotation Coppice*, ETSU B/W5/00216/REP/1. Harwell: Energy Technology Support Unit.

Cole, LJ, McCracken, DI, Baker, L & Parish, D 2007. Grassland conservation headlands: their impact on invertebrate assemblages in intensively managed grassland. *Agriculture, Ecosystems and Environment* 122: 252–258.

Collins, KL, Boatman, ND & Chaney, K 2002. Influence of beetle banks on cereal aphid predation in winter wheat. *Agriculture, Ecosystems and Environment* 93: 337–350.

Colquhoun, MK & Morley, A 1941. The density of downland birds. *Journal of Animal Ecology* 10: 35–46.

Conrad, KF, Woiwod, IP, Parsons, M, Fox, R & Warren, MS 2004. Long-term population trends in widespread British moths. *Journal of Insect Conservation* 8: 119–136.

Conrad, KF, Warren, MS, Fox, R, Parsons, MS & Woiwod, IP 2006. Rapid declines of common, widespread British moths provide evidence of an insect biodiversity crisis. *Biological Conservation* 132: 279–291.

Conway, G, Wotton, S, Henderson, I, Langston, R, Drewitt, A & Currie, F 2007. Status and distribution of European Nightjars *Caprimulgus europaeus* in the UK in 2004. *Bird Study* 54: 98–111.

Cook, AS, Grant, MC, McKay, CR & Peacock, MA 2001. Status, distribution and breeding success of the Red-billed Chough in Scotland in 1998. *Scottish Birds* 22: 82–91.

Cooke, AS 1973. Shell thinning in avian eggs by environmental pollutants. *Environmental Pollution* 4: 85–152.

Cooke, AS 1979. Changes in eggshell characteristics of the Sparrowhawk (*Accipiter nisus*) and Peregrine (*Falco peregrinus*) associated with exposure to environmental pollutants during recent decades. *Journal of Zoology, London* 187: 245–263.

Cope, DR, Pettifor, RA, Griffin, LR & Rowcliffe, JM 2003. Integrating farming and wildlife conservation: the Barnacle Goose Management Scheme. *Biological Conservation* 110: 113–122.

Cotton, PA, Kacelnik, A & Wright, J 1996. Chick begging as a signal: are signals honest? *Behavioural Ecology* 7: 178–182.

Coulson, JC & Butterfield, JEL 1985. The invertebrate communities of peat and upland grasslands in the north of England and some conservation implications. *Biological Conservation* 34: 197–225.

Coulson, JC & Whittaker, JB 1978. The ecology of moorland animals. In Heal, OW & Perkins, DF (eds) *Production Ecology of British Moors and Mountain Grasslands*, pp 52–93. Berlin: Springer-Verlag.

Court, I, Baker, D, Cleasby, I, Gibson, M, Smith, J, Straker, C & Thom, TJ 2001. *A Survey of Yellow Wagtails in the Yorkshire Dales National Park in 2000 and a Review of their Historical Population Status*. Grassington: Yorkshire Dales National Park Authority.

Cowley, MJR, Thomas, CD, Thomas, JA & Warren, MS 1999. Flight areas of British butterflies: assessing species status and decline. *Proceedings of the Royal Society of London B* 266: 1587–1592.

Cramp, S, Conder, PJ & Ash, J 1962. *Deaths of Birds and Mammals from Toxic Chemicals*, 2nd report of the Joint Committee of the British Trust for Ornithology, the Royal Society for the Protection of Birds and the Game Research Association.

Crick, HQP 1992. A bird-habitat coding system for use in Britain and Ireland incorporating aspects of land management and human activity. *Bird Study* 39: 1–12.

Crick, HQP 2004. The impact of climate change on birds. *Ibis* 146 (Suppl. 1): 48–56.

Crick, HQP & Baillie, SR 1996. *A Review of BTO's Nest Record Scheme: Its Value to the Joint Nature Conservation Committee, and Country Agencies, and its Methodology*, BTO Research Report No. 159. Thetford: British Trust for Ornithology.

Crick, HQP & Siriwardena, GM 2002. National trends in the breeding performance of House Sparrows *Passer domesticus*. In Crick, HQP, Robinson, RA, Appleton, GF, Clark, NA & Rickard, AD (eds) *Investigations into the Causes of Decline of Starlings and House Sparrows in Great Britain*, pp 163–192. Bristol: Defra.

Crick, HQP, Dudley, C, Evans, AD & Smith, KW 1994. Causes of nest failure among buntings in the UK. *Bird Study* 41: 88–94.

Crick, HQP, Baillie, SR & Leech, DI 2003. The UK nest record scheme: its value for science and conservation. *Bird Study* 50: 254–270.

Critchley, CNR, Burke, MJW & Stevens, DP 2003. Conservation of lowland semi-natural grasslands in the UK: a review of botanical monitoring results from agri-environment schemes. *Biological Conservation* 115: 263–278.

Cunningham, HM, Chaney, K, Bradbury, RB & Wilcox, A 2004. Non-inversion tillage and farmland birds: a review with special reference to the UK and Europe. *Ibis* 146 (Suppl. 2): 192–202.

Cunningham, HM, Bradbury, RB, Chaney, K & Wilcox, A 2005. Effect of non-inversion tillage on field usage by UK farmland birds in winter. *Bird Study* 52: 173–179.

Curry, D 2002. *Farming and Food: A Sustainable Future*, Report on the Policy Commission on the Future of Farming and Food. London: Defra.

Curry, JP & Tuohy, CF 1978. Studies on the epigeal microarthropod fauna of grassland swards managed for silage production. *Journal of Applied Ecology* 15: 727–741.

D'Arcy-Burt, S & Blackshaw, RP 1991. Bibionids (Diptera: Bibionidae) in agricultural land: a review of damage, benefits, natural enemies and control. *Annals of Applied Biology* 118: 695–708.

Dearborn, DC 1999. Brown-headed Cowbird nestling vocalisation and the risk of predation. *Auk* 116: 448–457.

Delany, S & Scott, DA 2002. *Waterbird Population Estimates*, 3rd edn, Wetlands International Global Series no. 12. Wageningen: Wetlands International.

Dennis, P, Redpath, S, McCracken, D & Grant, M 2005. *Effects of Grazing Management on Upland Bird Populations: Disentangling Habitat Structure and Arthropod Food Supply at Appropriate Spatial Scales*. Edinburgh: Scottish Executive Environment and Rural Affairs Department.

Dennis, P, Skartveit, J, McCracken, DI, Pakeman, RJ, Beaton, K, Kunaver, A & Evans, DM 2008. The effects of livestock grazing on foliar arthropods associated with bird diet in upland grasslands of Scotland. *Journal of Applied Ecology* 45: 279–287.

Devereux, CL, McKeever, CU, Benton, TG & Whittingham, MJ 2004. The effect of sward height and drainage on Common Starlings *Sturnus vulgaris* and Northern Lapwings *Vanellus vanellus* foraging in grassland habitats. *Ibis* 146 (Suppl. 2): 115–122.

Devereux, CL, Whittingham, MJ, Krebs, JR, Fernández-Juricic, E & Vickery, JA 2006. What attracts birds to newly mown pastures? Decoupling the action of mowing from the provision of short swards. *Ibis* 148: 302–306.

Dierschke, J & Bairlein, F 2004. Habitat selection of wintering passerines in salt marshes of the German Wadden Sea. *Journal of Ornithology* 145: 48–58.

Dobinson, HM & Richards, AJ 1964. The effects of the severe winter of 1962/63 on birds in Britain. *British Birds* 59: 373–434.

Dolton, CS & Brooke, M de L 1999. Changes in the biomass of birds breeding in Great Britain, 1968–88. *Bird Study* 46: 274–278.

Donald, PF 1997. The corn bunting *Miliaria calandra* in Britain: a review of current status, patterns of decline and possible causes. In Donald, PF & Aebischer, NJ (eds) *The Ecology and Conservation of Corn Buntings* Miliaria calandra, UK Nature Conservation No. 13, pp 11–26. Peterborough: Joint Nature Conservation Committee.

Donald, PF 2004. *The Skylark*. London: T. & A. D. Poyser.

Donald, PF & Aebischer, NJ (eds) 1997. *The Ecology and Conservation of Corn Buntings* Miliaria calandra, UK Nature Conservation No. 13. Peterborough: Joint Nature Conservation Committee.

Donald, PF & Evans, AD 1994. Habitat selection by Corn Buntings *Miliaria calandra* in winter. *Bird Study* 41: 199–210.

Donald, PF & Evans, AD 1995. Habitat selection and population size of corn buntings *Miliaria calandra* breeding in Britain in 1993. *Bird Study* 42: 190–204.

Donald, PF & Evans, AD 2006. Habitat connectivity and matrix restoration: the wider implications of agri-environment schemes. *Journal of Applied Ecology* 43: 209–218.

Donald, PF & Greenwood, JJD 2001. Spatial patterns of range contraction in British breeding birds. *Ibis* 143: 593–601.

Donald, PF, Wilson, JD & Shepherd, M 1994. The decline of the Corn Bunting. *British Birds* 87: 106–132.

Donald, PF, Buckingham, DL, Moorcroft, D, Muirhead, LB, Evans, AD & Kirby, WB 2001a. Habitat use and diet of skylarks *Alauda arvensis* wintering on lowland farmland in southern England. *Journal of Applied Ecology* 38: 536–547.

Donald, PF, Buckingham, DL, Muirhead, LB, Evans, AD, Kirby, WB & Schmitt, SIA 2001b. Factors affecting clutch size, hatching rates and partial brood losses in skylark *Alauda arvensis* nests on lowland farmland. In Donald, PF & Vickery, JA (eds) *The Ecology and Conservation of Skylarks* Alauda arvensis, pp 63–78. Sandy: Royal Society for the Protection of Birds.

Donald, PF, Evans, AD, Buckingham, DL, Muirhead, LB & Wilson, JD 2001c. Factors affecting the territory distribution of Skylarks *Alauda arvensis* breeding on lowland farmland. *Bird Study* 48: 271–278.

Donald, PF, Green, RE & Heath, MF 2001d. Agricultural intensification and the collapse of Europe's farmland bird populations. *Proceedings of the Royal Society of London B* 268: 25–29.

Donald, PF, Muirhead, LB, Buckingham, DL, Evans, AD, Kirby, WB & Gruar, DJ 2001e. Body condition, growth rates and diet of Skylark *Alauda arvensis* nestlings on lowland farmland. *Ibis* 143: 658–669.

Donald, PF, Evans, AD, Muirhead, LB, Buckingham, DL, Kirby, WB & Schmitt, SIA 2002a. Survival rates, causes of failure and productivity of Skylark *Alauda arvensis* nests on lowland farmland. *Ibis* 144: 652–664.

Donald, PF, Pisano, G, Rayment, M & Pain, DJ 2002b. The Common Agricultural Policy, EU enlargement and the conservation of Europe's farmland birds. *Agriculture, Ecosystems and Environment* 89: 167–182.

Donald, PF, Sanderson, FJ, Burfield, IJ & van Bommel, FPJ 2006. Further evidence of continent-wide impacts of agricultural intensification on European farmland birds, 1990–2000. *Agriculture, Ecosystems and Environment* 116: 189–196.

Donald, PF, Sanderson, FJ, Burfield, IJ, Bierman, SM, Gregory, RD & Waliczky, Z 2007. International conservation policy delivers benefits for birds in Europe. *Science* 317: 810–813.

Dougall, TW 1996. Movement and mortality of British-ringed Skylarks *Alauda arvensis*. *Ringing and Migration* 17: 81–92.

Drachmann, J, Komdeur, J & Boomsma, JJ 2000. Mate guarding in the Linnet *Carduelis cannabina*. *Bird Study* 47: 238–241.

Draycott, RAH, Hoodless, AN & Sage, RB 2008. Effects of pheasant management on vegetation and birds in lowland woods. *Journal of Applied Ecology* 45: 334–341.

Drewitt, AL & Langston, RHW 2006. Assessing the impacts of wind farms on birds. *Ibis* 148 (Suppl. 1): 29–42.

Dudley, SP, Gee, M, Kehoe, C, Melling, TM & British Ornithologists' Union Records Committee 2006. The British List: A Checklist of Birds of Britain, 7th edn. *Ibis* 148: 526–563.

Dunning, RA 1974. Bird damage to sugar beet. *Annals of Applied Biology* 76: 325–366.

Durant, D, Tichit, M, Fritz, H & Kernéïs, E 2008a. Field occupancy by breeding lapwings *Vanellus vanellus* and redshanks Tringa *totanus* in agricultural wet grasslands. *Agriculture, Ecosystems and Environment* 128: 146–150.

Durant, D, Tichit, M, Kernéïs, E & Fritz, H 2008b. Management of agricultural wet grasslands for breeding waders: integrating ecological and livestock system perspectives – a review. *Biodiversity and Conservation* 17: 2275–2295.

Durell, SEA Le V dit & Clarke, RT 2004. The buffer effect of non-breeding birds and the timing of farmland bird declines. *Biological Conservation* 120: 375–382.

Duriez, O, Ferrand, Y, Binet, F, Corda, E, Gossmann, F & Fritz, H 2005. Habitat selection of the Eurasian Woodcock in winter in relation to earthworm availability. *Biological Conservation* 122: 479–490.

Easterbrook, TG 1999. Population trends of wintering birds around Banbury, Oxfordshire, 1975–96. *Bird Study* 46: 16–24.

Eaton, MA, Stoate, C, Whittingham, MJ & Bradbury, RB 2002. Determinants of Whitethroat *Sylvia communis* distribution in different agricultural landscapes. In Chamberlain, D & Wilson, A (eds) *Avian Landscape Ecology: Pure and Applied Issues in the Large-Scale Ecology of Birds*, Proceedings of 11th conference of International Association for Landscape Ecology (UK), pp 300–304.

Eaton, MA, Dillon, IA, Stirling-Aird, PK & Whitfield, DP 2007. Status of Golden Eagle *Aquila chrysaetos* in Britain in 2003. *Bird Study* 54: 212–220.

Edwards, CA & Lofty, JR 1975. The influence of soil cultivation on soil animal populations. In Vanek, J (ed) *Progress in Soil Zoology*, pp 399–407. The Hague: Dr W. Junk.

Edwards, CA & Lofty, JR 1982. Nitrogenous fertilizers and earthworm populations in agricultural soils. *Soil Biology and Biochemistry* 14: 515–521.

Edwards, PJ 1977. 'Re-invasion' by some farmland bird species following capture and removal. *Polish Ecological Studies* 3: 53–70.

Eglington, SM, Gill, JA, Bolton, M, Smart, MA, Sutherland, WJ & Watkinson, AR 2008. Restoration of wet features for breeding waders on lowland wet grassland. *Journal of Applied Ecology* 45: 305–314.

Eislöffel, F 1997. The corn bunting *Miliaria calandra* in south-west Germany: population decline and habitat requirements. In Donald, PF & Aebischer, NJ (eds) *The Ecology and Conservation of Corn Buntings* Miliaria calandra, UK Nature Conservation No. 13, pp. 170–173. Peterborough: Joint Nature Conservation Committee.

Elliot, RD 1985. The exclusion of avian predators from aggregations of nesting Lapwings (*Vanellus vanellus*). *Animal Behaviour* 33: 308–314.

Elmegaard, N, Andersen, PN, Odderskær, P & Prang, A 1999. Food supply and breeding activity of skylarks in fields with different pesticide treatments. In Adams, NJ & Slotow, RH (eds) *Proceedings of the 22nd International Ornithological Congress, Durban*, pp 1058–1069. Johannesburg: BirdLife South Africa.

Eraud, C & Boutin, J-M 2002. Density and productivity of breeding skylarks *Alauda arvensis* in relation to crop type on agricultural lands in western France. *Bird Study* 49: 287–296.

Ervin, DE, Welsh, R, Batie, SS & Carpentier, CL 2003. Towards an ecological systems approach in public research for environmental regulation of transgenic crops. *Agriculture, Ecosystems and Environment* 99: 1–14.

Etheridge, B, Summers, RW & Green, RE 1997. The effects of illegal killing and destruction of nests by humans on the population dynamics of the hen harrier *Circus cyaneus* in Scotland. *Journal of Applied Ecology* 34: 1081–1105.

Evans, A, Vickery, J & Shrubb, M 2004. Importance of overwintered stubble for farmland bird recovery: a reply to Potts. *Bird Study* 51: 94–96.

Evans, AD 1992. The numbers and distribution of Cirl Buntings breeding in Britain in 1989. *Bird Study* 39: 17–22.

Evans, AD 1997. The importance of mixed farming for seed-eating birds in the UK. In Pain, DJ & Pienkowski, MW (eds) *Farming and Birds in Europe: The Common Agricultural Policy and its Implications for Bird Conservation*, pp 331–357. London: Academic Press.

Evans, AD & Green, RE 2007. An example of a two-tiered agri-environment scheme designed to deliver effectively the ecological requirements of both localised and widespread bird species in England. *Journal of Ornithology* 148: S279–S286.

Evans, AD & Smith, KW 1994. Habitat selection of Cirl Buntings wintering in Britain. *Bird Study* 41: 81–87.

Evans, AD, Curtoys, J, Kew, J, Lea, A & Rayment, M 1997a. Set-aside: conservation by accident or design? *RSPB Conservation Review* 11: 59–66.

Evans, AD, Smith, KW, Buckingham, DL & Evans, J 1997b. Seasonal variation in breeding performance and nestling diet of Cirl Buntings in England. *Bird Study* 44: 66–79.

Evans, AD, Armstrong-Brown, S & Grice, PV 2002. The role of research and development in the evolution of a 'smart' agri-environment scheme. *Aspects of Applied Biology* 67: 253–262.

Evans, DM, Redpath, SM, Elston, DA, Evans, SA, Mitchell, RJ & Dennis, P 2006a. To graze or not to graze? Sheep, voles, forestry and nature conservation in the British uplands. *Journal of Applied Ecology* 43: 499–505.

Evans, DM, Redpath, SM, Evans, SA, Elston, DA, Gardner, CJ, Dennis, P & Pakeman, RJ 2006b. Low intensity, mixed livestock grazing improves the breeding abundance of a common insectivorous passerine. *Biology Letters* 2: 636–638.

Evans, KL 2004. The potential for interactions between predation and habitat change to cause population declines of farmland birds. *Ibis* 146: 1–13.

Evans, KL, Bradbury, RB & Wilson, JD 2003a. Selection of hedgerows by Swallows *Hirundo rustica* foraging on farmland: the influence of local habitat and weather. *Bird Study* 50: 8–14.

Evans, KL, Wilson, JD & Bradbury, RB 2003b. Swallow *Hirundo rustica* population trends in England: data from repeated historical surveys. *Bird Study* 50: 178–181.

Evans, KL, Wilson, JD & Bradbury, RB 2007. Effects of crop type and aerial invertebrate abundance on foraging barn swallows *Hirundo rustica. Agriculture, Ecosystems and Environment* 122: 267–273.

Ewald, JA & Aebsicher, NJ 1999. *Pesticide Use, Avian Food Resources and Bird Densities in Sussex*, JNCC Report No. 296. Peterborough: Joint Nature Conservation Committee.

Falloon, P, Powlson, D & Smith, P 2004. Managing field margins for biodiversity and carbon sequestration: a Great Britain case study. *Soil Use and Management* 20 (Suppl. S): 240–247.

Feare, CJ 1994. Changes in numbers of common starlings and farming practice in Lincolnshire. *British Birds* 87: 200–204.

Feber, RE, Smith, H & Macdonald, DW 1996. The effects of management of uncropped edges of arable fields on butterfly abundance. *Journal of Applied Ecology* 33: 1191–1205.

Fenner, M & Palmer, L 1998. Grassland management to promote diversity: creation of a patchy sward by mowing and fertilizer regimes. *Field Studies* 9: 313–324.

Ferguson, NM, Donnelly, CA & Anderson, RM 2001. The foot-and-mouth epidemic in Great Britain: patterns of spread and impact of interventions. *Science* 292: 1155–1160.

Fewster, RM, Buckland, ST, Siriwardena, GM, Baillie, SR & Wilson, JD 2000. Analysis of population trends for farmland birds using generalized additive models. *Ecology* 81: 1970–1984.

Field, RH & Anderson, GQA 2004. Habitat use by breeding Tree Sparrows *Passer montanus*. *Ibis* 146 (Suppl. 2): 60–68.

Firbank, L 2003. The Farm Scale Evaluations of spring-sown genetically modified crops: Introduction. *Philosophical Transactions of the Royal Society of London B* 358: 1777–1778.

Firbank, L 2005. Striking a new balance between agricultural production and biodiversity. *Annals of Applied Biology* 146: 163–175.

Fisher, NM, Davies, DHK & Christel, A 1994. The potential for creating wildlife habitats and amenity grassland from non-rotational set-aside. *Aspects of Applied Biology* 40: 479–487.

Forrester, R, Andrews, IJ, McInerney, CJ, Murray, RD, McGowan, RY, Zonfrillo, B, Betts, MW, Jardine, DC & Grundy, DS (eds) 2007. *The Birds of Scotland*. Aberlady: Scottish Ornithologists' Club.

Fox, AD 2004. Has Danish agriculture maintained farmland bird populations? *Journal of Applied Ecology* 41: 427–439.

Fox, AD & Heldbjerg, H 2008. Which regional features of Danish agriculture favour the corn bunting in the contemporary farming land scape? *Agriculture, Ecosystems and Environment* 126: 261–269.

Frampton, GK & Dorne, JLCM 2007. The effects on terrestrial invertebrates of reducing pesticide inputs in arable crop edges: a meta-analysis. *Journal of Applied Ecology* 44: 362–373.

Frampton, GK, van den Brink, PJ & Gould, PJL 2000. Effects of spring drought and irrigation on farmland arthropods in southern Britain. *Journal of Applied Ecology* 37: 865–883.

Freeman, SN & Crick, HQP 2002. Population dynamics of House Sparrows *Passer domesticus* breeding in Britain: an integrated analysis. In Crick, HQP, Robinson, RA, Appleton, GF, Clark, NA & Rickard, AD (eds) *Investigations into the Causes of Decline of Starlings and House Sparrows in Great Britain*, pp 193–212. Bristol: Defra.

Freeman, SN & Crick, HQP 2003. The decline of the Spotted Flycatcher *Muscicapa striata* in the UK: an integrated population model. *Ibis* 145: 400–412.

Freeman, SN, Robinson, RA, Clark, JA, Griffin, BM & Adams, SY 2002. Population dynamics of Starlings *Sturnus vulgaris* breeding in Britain: an integrated analysis. In Crick, HQP, Robinson, RA, Appleton, GF, Clark, NA & Rickard, AD (eds) *Investigations into the Causes of Decline of Starlings and House Sparrows in Great Britain*, pp 121–143. Bristol: Defra.

Freeman, SN, Robinson, RA, Clark, JA, Griffin, BM & Adams, SY 2007. Changing demography and population decline in the Common Starling *Sturnus vulgaris*: a multisite approach to Integrated Population Monitoring. *Ibis* 149: 587–596.

Freemark, KE & Kirk, DA 2001. Birds on organic and conventional farms in Ontario: partitioning effects of habitat and practices on species composition and abundance. *Biological Conservation* 101: 337–350.

Frey, M 1989. Nahrungsökologie und Raumnutzung einer subalpinen Population des Hänflings *Carduelis cannabina. Der Ornithologische Beobachter* 86: 291–305.

Fuller, RJ 1982. *Bird Habitats in Britain.* Calton: T. & A. D. Poyser.

Fuller, RJ 1995. *Bird Life of Woodland and Forest.* Cambridge University Press.

Fuller, RJ 2000. Relationships between recent changes in lowland British agriculture and farmland bird populations: an overview. In Aebsicher, NJ, Evans, AD, Grice, PV & Vickery, JA (eds) *Ecology and Conservation of Lowland Farmland Birds*, pp 5–16. Tring: British Ornithologists' Union.

Fuller, RJ & Gough, SJ 1999. Changes in sheep numbers in Britain: implications for bird populations. *Biological Conservation* 91: 73–89.

Fuller, RJ & Youngman, RE 1979. The utilisation of farmland by Golden Plovers wintering in southern England. *Bird Study* 26: 37–46.

Fuller, RJ, Marchant, JH & Morgan, RA 1985. How representative of agricultural practice in Britain are Common Birds Census farmland plots? *Bird Study* 32: 56–70.

Fuller, RJ, Reed, TM, Buxton, NE, Webb, A, Williams, TD & Pienkowski, MW 1986. Populations of breeding waders Charadrii and their habitats on the crofting lands of the Outer Hebrides, Scotland. *Biological Conservation* 37: 333–361.

Fuller, RJ, Gregory, RD, Gibbons, DW, Marchant, JH, Wilson, JD, Baillie, SR & Carter, N 1995. Population declines and range contractions among lowland farmland birds in Britain. *Conservation Biology* 9: 1425–1441.

Fuller, RJ, Trevelyan, RJ & Hudson, RW 1997. Landscape composition models for breeding bird populations in lowland English farmland over a 20-year period. *Ecography* 20: 295–307.

Fuller, RJ, Chamberlain, DE, Burton, NHK & Gough, SJ 2001. Distributions of birds in lowland agricultural landscapes of England and Wales: how distinctive are bird communities of hedgerows and woodland? *Agriculture, Ecosystems and Environment* 84: 79–92.

Fuller, RJ, Hinsley, SA & Swetnam, RD 2004. The relevance of non-farmland habitats, uncropped areas and habitat diversity to the conservation of farmland birds. *Ibis* 146 (Suppl. 2): 22–31.

Fuller, RJ, Norton, LR, Feber, RE, Johnson, PJ, Chamberlain, DE, Joys, AC, Mathews, F, Stuart, RC, Townsend, MC, Manley, WJ, Wolfe, MS, Macdonald, DW & Firbank, LG 2005. Benefits of organic farming vary among taxa. *Biology Letters* 1: 431–434.

Fuller, RJ, Atkinson, PW, Garnett, MC, Conway, GJ, Bibby, CJ & Johnstone, IG 2006. Breeding bird communities of the upland margins (ffridd) of Wales in the mid-1980s. *Bird Study* 53: 177–186.

Fuller, RM 1987. The changing extent and conservation interest of lowland grasslands in England and Wales: a review of grassland surveys 1930–1984. *Biological Conservation* 40: 281–300.

Gabriel, D & Tscharntke, T 2007. Insect pollinated plants benefit from organic farming. *Agriculture, Ecosystems and Environment* 118: 43–48.

Galbraith, H 1988a. Effects of agriculture on the breeding ecology of Lapwings *Vanellus vanellus. Journal of Applied Ecology* 25: 487–503.

Galbraith, H 1988b. Adaptation and constraint in the growth pattern of Lapwing *Vanellus vanellus* chicks. *Journal of Zoology, London* 215: 537–548.

Galbraith, H 1988c. Effects of egg size and composition on the size, quality and survival of Lapwing *Vanellus vanellus* chicks. *Journal of Zoology, London* 214: 383–398.

Game Conservancy Trust 2004. *Review of 2003*. Fordingbridge: Game Conservancy Trust.

Garwood, EA 1988. Water deficiency and excess in grassland: the implications for grass production and for the efficient use of N. In Wilkins, RJ (ed) *Nitrogen and Water Use by Grassland*, pp 155–179. North Wyke, Devon: Institute for Grassland and Environmental Research.

Gates, S & Donald, PF 2000. Local extinction of British farmland birds and the prediction of further loss. *Journal of Applied Ecology* 37: 806–820.

Gates, S, Gibbons, DW, Lack, PC & Fuller, RJ 1994. Declining farmland bird species: modelling geographical patterns of abundance in Britain. In Edwards, PJ, May, RM & Webb, N (eds) *Large Scale Ecology and Conservation Biology*, pp 155–179. Oxford: Blackwell Scientific Publications.

Gauthier-Clerc, M, Lebarbenchon, C & Thomas, F 2007. Recent expansion of highly pathogenic avian influenza H5N1: a critical review. *Ibis* 149: 202–214.

Gerard, PW 1995. *Agricultural Practices, Farm Policy and the Conservation of Biological Diversity*, USDI Biological Sciences Report No. 4. Washington, DC: US Department of the Interior.

Ghani, AC, Ferguson, NM, Donnelly, CA, Hagenaars, TJ & Anderson, RM 1998. Epidemiological determinants of the pattern and magnitude of the vCJD epidemic in Great Britain. *Proceedings of the Royal Society of London B* 265: 2443–2452.

Gibbons, DW & Wotton, SR 1996. The Dartford Warbler in the United Kingdom in 1994. *British Birds* 89: 203–212.

Gibbons, DW, Reid, JB & Chapman, RA 1993. *The New Atlas of Breeding Birds in Britain and Ireland: 1988–1991*. London: T. & A. D. Poyser.

Gibbons, DW, Avery, MI, Baillie, SR, Gregory, RD, Kirby, J, Porter, RF, Tucker, GM & Williams, G 1996a. Bird species of conservation concern in the United Kingdom, Channel Islands and Isle of Man: revising the Red Data List. *RSPB Conservation Review* 10: 7–18.

Gibbons, DW, Avery, MI & Brown, AF 1996b. Population trends of breeding birds in the United Kingdom since 1800. *British Birds* 89: 291–305.

Gibbons, DW, Bohan, DA, Rothery, P, Stuart, RC, Haughton, AJ, Scott, RJ, Wilson, JD, Perry, JN, Clark, SJ, Dawson, RJG & Firbank, LG 2006. Weed seed resources for birds in fields with contrasting conventional and genetically modified herbicide-tolerant crops. *Proceedings of the Royal Society of London B* 273: 1921–1928.

Gibbons, DW, Amar, A, Anderson, GQA, Bolton, M, Bradbury, RB, Eaton, MA, Evans, AD, Grant, MC, Gregory, RD, Hilton, GM, Hirons, GJM, Hughes, J, Johnstone, I, Newbery, P, Peach, WJ, Ratcliffe, N, Smith, KW, Summers, RW, Walton, P & Wilson, JD 2007a. *The Predation of Wild Birds in the UK: A Review of Its Conservation Impact and Management*. Sandy: Royal Society for the Protection of Birds.

Gibbons, DW, Donald, PF, Bauer, H-G, Fornasari, L & Dawson, IK 2007b. Mapping avian distributions: the evolution of bird atlases. *Bird Study* 54: 324–334.

Gilbert, OL 1992. Lichen reinvasion with declining air pollution. In Bates, JW & Farmer, AM (eds) *Bryophytes and Lichens in a Changing Environment*, pp 159–177. London: Oxford Science Publications.

Gill, JA, Watkinson, AR & Sutherland, WJ 1997. Cause of the redistribution of Pink-footed Geese *Anser brachyrhynchus* in Britain. *Ibis* 139: 497–503.

Gillings, S 2001. Factors affecting the distribution of skylarks *Alauda arvensis* wintering in Britain and Ireland during the early 1980s. In Donald, PF & Vickery, JA (eds) *The Ecology and Conservation of Skylarks* Alauda arvensis, pp 115–128. Sandy: Royal Society for the Protection of Birds/British Trust for Ornithology.

Gillings, S 2003. Diurnal and nocturnal ecology of Golden Plovers *Pluvialis apricaria* and Lapwings *Vanellus vanellus* wintering on arable farmland. Unpublished PhD thesis, University of East Anglia.

Gillings, S & Dougall, TW 2002. Skylark *Alauda arvensis*. In Wernham, CV, Toms, MP, Marchant, JH, Clark, JA, Siriwardena, GM & Baillie, SR (eds) *The Migration Atlas: Movements of the Birds of Britain and Ireland*, pp 455–457. London: T. & A. D. Poyser.

Gillings, S & Fuller, RJ 1998. Changes in bird populations on sample lowland English farms in relation to loss of hedgerows and other non-crop habitats. *Oecologia* 116: 120–127.

Gillings, S & Fuller, RJ 2001. Habitat selection by Skylarks *Alauda arvensis* wintering in Britain in 1997/98. *Bird Study* 48: 293–307.

Gillings, S, Fuller, RJ & Henderson, ACB 1998. Avian community composition and patterns of bird distribution within birch–heath mosaics in north-east Scotland. *Ornis Fennica* 75: 27–37.

Gillings, S, Fuller, RJ & Sutherland, WJ 2005a. Diurnal studies do not predict nocturnal habitat choice and site selection of European Golden Plovers (*Pluvialis apricaria*) and Northern Lapwings (*Vanellus vanellus*). *Auk* 122: 1249–1260.

Gillings, S, Newson, SE, Noble, DG & Vickery, JA 2005b. Winter availability of cereal stubbles attracts declining farmland birds and positively influences breeding population trends. *Proceedings of the Royal Society of London B* 272: 733–739.

Gillings, S, Austin, GE, Fuller, RJ & Sutherland, WJ 2006. Distribution shifts in wintering Golden Plovers *Pluvialis apricaria* and Lapwings *Vanellus vanellus* in Britain. *Bird Study* 53: 274–284.

Gillings, S, Fuller, RJ & Sutherland, WJ 2007. Winter field use and habitat selection by Eurasian Golden Plovers *Pluvialis apricaria* and Northern Lapwings *Vanellus vanellus* on arable farmland. *Ibis* 149: 509–520.

Gillings, S, Wilson, AM, Conway, GJ, Vickery, JA & Fuller, RJ 2008. Distribution and abundance of birds and their habitats within the lowland farmland of Britain in winter. *Bird Study* 55: 8–22.

Gilroy, JJ 2006. Breeding ecology and conservation of Yellow Wagtails (*Motacilla flava*) in intensive arable farmland. Unpublished PhD thesis, University of East Anglia.

Gooch, S, Baillie, SR & Birkhead, TR 1991. Magpie *Pica pica* and songbird populations: retrospective investigation of trends in population density and breeding success. *Journal of Applied Ecology* 28: 1068–1086.

Gosler, AG, Greenwood, JJD & Perrins, C 1995. Predation risk and the cost of being fat. *Nature* 377: 621–623.

Gotmark, F & Post, P 1996. Prey selection by Sparrowhawks, *Accipiter nisus*: relative predation risk for breeding passerine birds in relation to their size, ecology and behaviour. *Philosophical Transactions of the Royal Society of London B* 351: 1559–1577.

Grant, MC, Orsman, C, Easton, J, Lodge, C, Smith, M, Thompson, G, Rodwell, S & Moore, N 1999. Breeding success and causes of breeding failure of curlew *Numenius arquata* in Northern Ireland. *Journal of Applied Ecology* 36: 59–74.

Green, FHW 1979. *Field Drainage in Europe: A Quantitative Survey*, Institute of Hydrology Report No. 57. Wallingford: Institute of Hydrology.

Green, JO 1982. *A Sample Survey of Grassland in England and Wales 1970–72*. Hurley: Grassland Research Institute.

Green, RE 1978. Factors affecting the diet of farmland Skylarks *Alauda arvensis*. *Journal of Animal Ecology* 47: 913–928.

Green, RE 1980. Food selection by skylarks and grazing damage to sugar beet seedlings. *Journal of Applied Ecology* 17: 613–630.

Green, RE 1984. The feeding ecology and survival of Partridge chicks (*Alectoris rufa* and *Perdix perdix*) on arable farmland in East Anglia. *Journal of Applied Ecology* 21: 817–830.

Green, RE 1988. Effects of environmental factors on the timing and success of breeding common snipe (Aves: Scolopacidae). *Journal of Applied Ecology* 25: 79–93.

Green, RE 1995a. Diagnosing causes of bird population declines. *Ibis* 137: S47–S55.

Green, RE 1995b. The decline of the Corncrake *Crex crex* in Britain continues. *Bird Study* 42: 66–75.

Green, RE 1996. Factors affecting the population density of the Corncrake *Crex crex* in Britain and Ireland. *Journal of Applied Ecology* 33: 237–248.

Green, RE 1999. Applications of large-scale studies of demographic rates to bird conservation. *Bird Study* 46 (Suppl.): S279–S288.

Green, RE 2002. Stone-curlew *Burhinus oedicnemus*. In Wernham, CV, Toms, MP, Marchant, JH, Clark, JA, Siriwardena, GM & Baillie, SR (eds) *The Migration Atlas: Movements of the Birds of Britain and Ireland*, pp 273–275. London: T. & A. D. Poyser.

Green, RE & Etheridge, B 1999. Breeding success of the hen harrier *Circus cyaneus* in relation to the distribution of grouse moors and the red fox *Vulpes vulpes*. *Journal of Applied Ecology* 36: 472–483.

Green, RE & Gibbons, DW 2000. The status of the Corncrake *Crex crex* in Britain in 1998. *Bird Study* 47: 129–137.

Green, RE & Griffiths, GH 1994. Use of preferred nesting habitat by Stone-curlews *Burhinus oedicnemus* in relation to vegetation structure. *Journal of Zoology, London* 233: 457–471.

Green, RE & Robins, MM 1993. The decline in the ornithological importance of the Somerset Levels and Moors, England and changes in management of water levels. *Biological Conservation* 66: 95–106.

Green, RE & Stowe, TJ 1993. The decline of the Corncrake in Britain and Ireland in relation to habitat change. *Journal of Applied Ecology* 30: 689–695.

Green, RE & Taylor, CR 1995. Changes in Stone-curlew *Burhinus oedicnemus* distribution and abundance and vegetation height on chalk grassland at Porton Down, Wiltshire. *Bird Study* 42: 177–181.

Green, RE, Hirons, GJM & Cresswell, BH 1990. Foraging habitats of female common snipe *Gallinago gallinago* during the incubation period. *Journal of Applied Ecology* 27: 325–335.

Green, RE, Osborne, PE & Sears, J 1994. The distribution of passerine birds in hedgerows during the breeding season in relation to characteristics of the hedgerow and adjacent farmland. *Journal of Applied Ecology* 31: 677–692.

Green, RE, Tyler, GA, Stowe, TJ & Newton, AV 1997. A simulation model of the effect of mowing of agricultural grassland on the breeding success of the Corncrake (*Crex crex*). *Journal of Zoology, London* 243: 81–115.

Green, RE, Tyler, GA & Bowden, CGR 2000. Habitat selection, ranging behaviour and diet of the stone curlew (*Burhinus oedicnemus*) in southern England. *Journal of Zoology, London* 250: 161–183.

Green, RE, Cornell, SJ, Scharlemann, JPW & Balmford, A 2005. Farming and the fate of wild nature. *Science* 307: 550–555.

Green, RE, Collingham, YC, Willis, SG, Gregory, RD, Smith, KW & Huntley, B 2008. Performance of climate envelope models in retrodicting recent changes in bird population size from observed climatic change. *Biology Letters* 4: 599–602.

Greenwood, JJD & Baillie, SR 1991. Effects of density dependence and weather on population changes of English passerines using a non-experimental paradigm. *Ibis* 133: 121–133.

Gregory, RD 1999. Broad-scale habitat use of sparrows, finches and buntings in Britain. *Vogelwelt* 120 (Suppl.): 163–173.

Gregory, RD & Baillie, SR 1998. Large-scale habitat use of some declining British birds. *Journal of Applied Ecology* 35: 785–799.

Gregory, RD & Marchant, JH 1996. Population trends of jays, magpies and carrion crows in the United Kingdom. *Bird Study* 43: 28–37.

Gregory, RD, Wilkinson, NI, Noble, DG, Robinson, JA, Brown, AF, Hughes, J, Procter, D, Gibbons, DW & Galbraith, CA 2002. The population status of birds in the United Kingdom, Channel Islands and Isle of Man: an analysis of conservation concern 2002–2007. *British Birds* 95: 410–448.

Gregory, RD, Noble, DG & Custance, J 2004. The state of play of farmland birds: population trends and conservation status of lowland farmland birds in the United Kingdom. *Ibis* 146 (Suppl. 2): 1–13.

Gregory, RD, van Strien, A, Vorisek, P, Meyling, AWG, Noble, DG, Foppen, RPB & Gibbons, DW 2005. Developing indicators for European birds. *Philosophical Transactions of the Royal Society of London B* 360: 269–288.

Grice, P, Evans, A, Osmond, J & Brand-Hardy, R 2004. Science into policy: the role of research in the development of a recovery plan for farmland birds in England. *Ibis* 146 (Suppl. 2): 239–249.

Grigg, D 1989. *English Agriculture: An Historical Perspective.* Oxford: Blackwell.

Groom, DW 1993. Magpie *Pica pica* predation on blackbird *Turdus merula* nests in urban areas. *Bird Study* 40: 55–62.

Gruar, D, Peach, W & Taylor, R 2003. Summer diet and body condition of Song Thrushes *Turdus philomelos* in stable and declining farmland populations. *Ibis* 145: 637–649.

Gruar, D, Barritt, D & Peach, WJ 2006. Summer utilization of Oilseed Rape by Reed Buntings *Emberiza schoeniclus* and other farmland birds. *Bird Study* 53: 47–54.

Grüebler, M, Morand, M & Naef-Daenzer, B 2008. A predictive model of the density of airborne insects in agricultural environments. *Agriculture, Ecosystems and Environment* 123: 75–80.

Hagemeijer, WJM & Blair, MJ (eds) 1997. *The EBCC Atlas of European Breeding Birds: Their Distribution and Abundance.* London: T. & A. D. Poyser.

Haines-Young, R, Barr, CJ, Black, HIJ, Briggs, DJ, Bunce, RGH, Clarke, RT, Cooper, A, Dawson, FH, Firbank, LG, Fuller, RM, Furse, RT, Gillespie, MK, Hill, R, Hornung, M, Howard, DC, McCann, T, Morecroft, MD, Petit, S, Sier, ARJ, Smart, SM, Smith, GM, Stott, A, Stuart, RC & Watkins, JW 2000. *Accounting for Nature: Assessing Habitats in the UK Countryside.* London: Department of the Environment, Transport and the Regions.

Hancock, M & Avery, M 1998. Changes in breeding bird populations in peatlands and young forestry in north east Sutherland and Caithness between 1988 and 1995. *Scottish Birds* 19: 195–205.

Hancock, MH & Wilson, JD 2002. Winter habitat associations of grey partridge *Perdix perdix* in Scotland, 1997–9. *Aspects of Applied Biology* 67: 171–178.

Hancock, MH & Wilson, JD 2003. Winter habitat associations of seed-eating passerines on Scottish farmland. *Bird Study* 50: 116–130.

Hancock, M, Baines, D, Gibbons, D, Etheridge, B & Shepherd, M 1999. Status of male Black Grouse *Tetrao tetrix* in Britain in 1995–96. *Bird Study* 46: 1–15.

Hancock, MH, Grant, MC & Wilson, JD 2009. Associations between distance to forest and spatial and temporal variation in abundance of key peatland breeding bird species. *Bird Study* in press.

Harding, NJ, Green, RE & Summers, RW 1994. *The Effects of Future Changes in Land Use on Upland Birds in Britain.* Sandy: Royal Society for the Protection of Birds.

Harris, S, Morris, P, Wray, S & Yalden, D 1995. *A Review of British Mammals: Population Estimates and Conservation status of British Mammals Other than Cetaceans.* Peterborough: Joint Nature Conservation Committee.

Hart, JD, Milsom, TP, Baxter, A, Kelly, PF & Parkin, WK 2002. The impact of livestock grazing on lapwing (*Vanellus vanellus*) breeding densities and performance on coastal grazing marsh. *Bird Study* 49: 67–78.

Hart, JD, Milsom, TP, Fisher, G, Wilkins, V, Moreby, SJ, Murray, AWA & Robertson, PA 2006. The relationship between yellowhammer breeding performance, arthropod abundance and insecticide applications on arable farmland. *Journal of Applied Ecology* 43: 81–91.

Hartley, IR & Quicke, DJL 1994. The diet of nestling Corn Buntings on North Uist: insects not grain. *Scottish Birds* 17: 169–170.

Hartley, IR & Shepherd, M 1994. Nesting success in relation to timing of breeding in the Corn Bunting on North Uist. *Ardea* 82: 173–184.

Hartley, IR, Shepherd, M & Thompson, DBA 1995. Habitat selection and polygyny in breeding Corn Buntings *Miliaria calandra*. *Ibis* 137: 508–514.

Haskell, DG 1994. Experimental evidence that nestling begging behaviour incurs a cost due to nest predation. *Proceedings of the Royal Society of London B* 257: 161–164.

Hassall, M & Lane, SJ 2001. Effects of varying rates of autumn fertilizer application to pastures in eastern England on feeding site selection by Brent Geese *Branta b. bernicla*. *Agriculture, Ecosystems and Environment* 86: 203–209.

Hassall, M, Riddington, R & Helden, A 2001. Foraging behaviour of brent geese *Branta b. bernicla* on grasslands: effects of sward length and nitrogen content. *Oecologia* 127: 97–104.

Haughton, AJ, Champion, GT, Hawes, C, Heard, MS, Brooks, DR, Bohan, DA, Clark, SJ, Dewar, AM, Firbank, LG, Osborne, JL, Perry, JN, Rothery, P, Roy, DB, Scott, RJ, Woiwod, IP, Birchall, C, Skellern, MP, Walker, JH, Baker, P, Browne, EL, Dewar, AJG, Garner, BH, Haylock, LA, Horne, SL, Mason, NS, Sands, RJN & Walker, MJ 2003. Invertebrate responses to the management of genetically modified herbicide-tolerant and conventional spring crops. 2. Within-field epigeal and aerial arthropods. *Philosophical Transactions of the Royal Society of London B* 358: 1863–1877.

Haworth, PF & Thompson, DBA 1990. Factors associated with the breeding distribution of upland birds in the South Pennines, England. *Journal of Applied Ecology* 27: 562–577.

Haysom, KA, McCracken, DI, Foster, GN & Sotherton, NW 2004. Developing grassland conservation headlands: response of carabid assemblages to different cutting regimes in a silage field edge. *Agriculture, Ecosystems and Environment* 102: 263–277.

Heard, MS, Hawes, C, Champion, GT, Clark, SJ, Firbank, LG, Haughton, AJ, Parish, AM, Perry, JN, Rothery, P, Scott, RJ, Skellern, MP, Squire, GR & Hill, MO 2003a. Weeds in fields with contrasting conventional and genetically modified herbicide-tolerant crops. 1. Effects on abundance and diversity. *Philosophical Transactions of the Royal Society of London B* 358: 1819–1832.

Heard, MS, Hawes, C, Champion, GT, Clark, SJ, Firbank, LG, Haughton, AJ, Parish, AM, Perry, JN, Rothery, P, Scott, RJ, Skellern, MP, Squire, GR & Hill, MO 2003b. Weeds in fields with contrasting conventional and genetically modified herbicide-tolerant crops. 2. Effects on individual species. *Philosophical Transactions of the Royal Society of London B* 358: 1833–1846.

Heath, MF & Evans, MI (eds) 2000. *Important Bird Areas in Europe: Priority Sites for Conservation*, 2 vols, BirdLife Conservation Series No. 8. Cambridge, UK: BirdLife International.

Henderson, IG & Evans, AD 2000. Responses of farmland birds to set-aside and its management. In Aebischer, NJ, Evans, AD, Grice, PV & Vickery, JA (eds) *Ecology and Conservation of Lowland Farmland Birds*, pp 69–76. Tring: British Ornithologists' Union.

Henderson, IG, Cooper, J, Fuller, RJ & Vickery, J 2000a. The relative abundance of birds on set-aside and neighbouring fields in summer. *Journal of Applied Ecology* 37: 335–347.

Henderson, IG, Vickery, JA & Fuller, RJ 2000b. Summer bird abundance and distribution on set-aside fields on intensive arable farms in England. *Ecography* 23: 50–59.

Henderson, IG, Critchley, NR, Cooper, J & Fowbert, JA 2001. Breeding season responses of Skylarks *Alauda arvensis* to vegetation structure in set-aside (fallow arable land). *Ibis* 143: 317–321.

Henderson, IG, Wilson, AM, Steele, D & Vickery, JA 2002. Population estimates, trends and habitat associations of breeding Lapwing *Vanellus vanellus*, Curlew *Numenius arquata* and Snipe *Gallinago gallinago* in Northern Ireland in 1999. *Bird Study* 49: 17–25.

Henderson, IG, Fuller, RJ, Conway, GJ & Gough, SJ 2004a. Evidence for declines in populations of grassland-associated birds in marginal upland areas of Britain. *Bird Study* 51: 12–19.

Henderson, IG, Vickery, JA & Carter, N 2004b. The use of winter bird crops by farmland birds in lowland England. *Biological Conservation* 118: 21–32.

Henderson, IG, Holt, C & Vickery, J 2007a. National and regional patterns of habitat association with foraging Barn Swallows *Hirundo rustica* in the UK. *Bird Study* 54: 371–377.

Henderson, IG, Morris, AJ, Westbury, DB, Woodcock, BA, Potts, SG, Ramsay, A & Coombes, R 2007b. Effects of field margin management on bird distributions around cereal fields. *Aspects of Applied Biology* 81: 53–60.

Herzon, I, Avninš, A, Elts, J & Preikša, J 2008. Intensity of agricultural land-use and farmland birds in the Baltic States. *Agriculture, Ecosystems and Environment* 125: 93–100.

Hester, AJ & Sydes, C 1992. Changes in burning of Scottish heather moorland since the 1940s from aerial photographs. *Biological Conservation* 60: 25–30.

Hinsley, SA & Bellamy, PE 2000. The influence of hedge structure, management and landscape context on the value of hedgerows to birds: a review. *Journal of Environmental Management* 60: 33–49.

Hinsley, SA, Bellamy, PE, Newton, I & Sparks, TH 1995. Habitat and landscape factors influencing the presence of individual breeding bird species in woodland fragments. *Journal of Avian Biology* 26: 94–104.

Hirons, G & Bickford-Smith, P 1983. The diet and behaviour of Eurasian Woodcock wintering in Cornwall. In Kalchreuter, H (ed) *Proceedings of the 2nd European Woodcock and Snipe Workshop*, pp 11–17. Slimbridge: Wildfowl and Wetlands Trust.

Holden, J, Shotbolt, L, Bonn, A, Burt, TP, Chapman, PJ, Dougill, AJ, Fraser, EDG, Hubacek, K, Irvine, B, Kirkby, MJ, Reed, MS, Prell, C, Stagl, S, Stringer, LC, Turner, A & Worrall, F 2007. Environmental change in moorland landscapes. *Earth-Science Reviews* 82: 75–100.

Hole, DG, Whittingham, MJ, Bradbury, RB, Anderson, GQA, Lee, PLM, Wilson, JD & Krebs, JR 2002. Widespread local house sparrow extinctions. *Nature* 418: 931–932.

Hole, DG, Perkins, AJ, Wilson, JD, Alexander, IH, Grice, PV & Evans, AD 2005. Does organic farming benefit biodiversity? *Biological Conservation* 122: 113–130.

Holland, JM, Southway, S, Ewald, JA, Birkett, T, Begbie, M, Hart, J, Parrott, D & Allcock, J 2002. Invertebrate chick food for farmland birds: spatial and temporal variation in different crops. *Aspects of Applied Biology* 67: 27–34.

Holloway, S 1996. *The Historical Atlas of Breeding Birds in Britain and Ireland 1875–1900*. London: T. & A. D. Poyser.

Hoodless, AN & Hirons, GJM 2007. Habitat selection and foraging behaviour of breeding Eurasian Woodcock *Scolopax rusticola*: a comparison between contrasting landscapes. *Ibis* 149 (Suppl. 2): 234–249.

Hopkins, A (ed) 1999. *Grass: Its Production and Utilization*. Oxford: Blackwell Scientific Publications.

Hopkins, JJ & Kirby, KJ 2007. Ecological change in British broadleaved woodland since 1947. *Ibis* 149 (Suppl. 2): 29–40.

Hoskins, WG & Stamp, LD 1963. *The Common Lands of England and Wales*. London: Collins.

Houghton, JT, Ding, Y, Griggs, DJ, Noguer, M, van der Linden, PJ, Dai, X, Maskell, K & Johnson, CA (eds) (2001) *Climate Change 2001: The Scientific Basis*. Cambridge University Press.

Houston, AI, MacNamara, JM & Hutchinson, JMC 1993. General results concerning the trade-off between gaining energy and avoiding predation. *Philosophical Transactions of the Royal Society of London B* 341: 375–397.

Hudson, AV, Stowe, TJ & Aspinall, SJ 1990. Status and distribution of corncrakes in Britain. *British Birds* 83: 173–186.

Hudson, PJ 1984. Some effects of sheep management on heather moorlands in northern England. In Jenkins, D (ed) *Agriculture and the Environment*, pp 143–149. Cambridge: Institute of Terrestrial Ecology.

Hudson, PJ 1992. *Grouse in Space and Time*. Fordingbridge: Game Conservancy Trust.

Hughes, J, Baines, D, Grant, M, Robert, J, Williams, I & Bayes, K 1998. Black Grouse: the challenge of reversing the decline. *RSPB Conservation Review* 12: 18–28.

Huntley, B, Green, RE, Collingham, YC & Willis, SG 2007. *A Climatic Atlas of European Breeding Birds*. Barcelona: Lynx Editions.

Hustings, F. 1997. The decline of the Corn Bunting in The Netherlands. In Donald, PF & Aebischer, NJ (eds) *The Ecology and Conservation of Corn Buntings* Miliaria calandra, UK Nature Conservation No. 13, pp 42–51. Peterborough: Joint Nature Conservation Committee.

Hutson, AM, Mickleburgh, SP & Racey, PA 2001. *Global Status Survey and Conservation Action Plan: Microchiropteran Bats*. Gland, Switzerland and Cambridge, UK: International Union for the Conservation of Nature.

IGER/BTO/CAER 2007. *Potential for Enhancing Biodiversity on Intensive Livestock Farms*, Report of Project BD1444 to Defra, Institute of Grassland and Environmental Research, British Trust for Ornithology and Centre for Agri-Environmental Research. (www.defra.gov.uk)

Inglis, IR, Thearle, RJP & Isaacson, AJ 1989. Woodpigeon (*Columba palumbus*) damage to oilseed rape. *Crop Protection* 8: 299–309.

Inglis, IR, Isaacson, AJ, Thearle, RJP & Westwood, NJ 1990. The effects of changing agricultural practice upon Woodpigeon *Columba palumbus* numbers. *Ibis* 132: 262–272.

Inglis, IR, Isaacson, AJ & Thearle, RJP 1994. Long term changes in the breeding biology of the woodpigeon *Columba palumbus* in eastern England. *Ecography* 17: 182–188.

Inglis, IR, Isaacson, AJ, Smith, GC, Haynes, PJ & Thearle, RJP 1997. The effect on the woodpigeon (*Columba palumbus*) of the introduction of oilseed rape into Britain. *Agriculture, Ecosystems and Environment* 61: 113–121.

Irvine, J 1977. Breeding birds in the New Forest broad-leaved woodland. *Bird Study* 24: 105–111.

Jackson, DB 2001. Experimental removal of introduced hedgehogs improves wader nest success in the Western Isles, Scotland. *Journal of Applied Ecology* 38: 802–812.

Jackson, DB & Green, RE 2000. The importance of the introduced hedgehog (*Erinaceus europaeus*) as a predator of the eggs of waders (Charadrii) on machair in South Uist, Scotland. *Biological Conservation* 93: 333–348.

Jackson, DB, Fuller, RJ & Campbell, ST 2004. Long-term population changes among breeding shorebirds in the Outer Hebrides, Scotland, in relation to introduced hedgehogs (*Erinaceus europaeus*). *Biological Conservation* 117: 151–166.

Jarry, G 1994. Turtle Dove *Streptopelia turtur*. In Tucker, GM & Heath, MF (eds) *Birds in Europe: Their Conservation Status*, BirdLife Conservation Series No. 3, pp 320–321. Cambridge: BirdLife International.

Jarry, G & Baillon, F 1991. *Hivernage de la Tourtelle des Bois* (Streptopelia turtur) *au Sénégal: Etude d'une Population dans la Région de Nianing*. Paris: CRBPO/ ORSTOM.

Jenkins, M 2003. Prospects for biodiversity. *Science* 302: 1175–1177.

Jenny, M 1990a. Terriorialität und Brutbiologie der Feldlerche *Alauda arvensis* in einer intensiv genutzten Agrarlandschaft. *Journal für Ornithologie* 131: 241–265.

Jenny, M 1990b. Nahrungsökologie der Feldlerche *Alauda arvensis* in einer intensiv genutzten Agrarlandschaft des schweizerischen Mittellandes. *Ornithologische Beobachter* 87: 31–53.

Jenny, M 1990c. Populationsdynamic der Feldlerche *Alauda arvensis* in einter intensiv genutzten Agrarlandschaft des schweizerischen Mittellandes. *Ornithologische Beobachter* 87: 153–163.

Johansson, OC & Blomqvist, D 1996. Habitat selection and diet of Lapwing *Vanellus vanellus* chicks on coastal farmland in S.W. Sweden. *Journal of Applied Ecology* 33: 1030–1040.

Johnstone, I, Whitehead, S & Lamacraft, D 2002. The importance of grazed habitat for foraging choughs *Pyrrhocorax pyrrhocorax*, and its implications for agri-environment schemes. *Aspects of Applied Biology* 67: 59–66.

Johnstone, I, Thorpe, R, Moore, A & Finney, S 2007. Breeding status of Choughs *Pyrrhocorax pyrrhocorax* in the UK and Isle of Man in 2002. *Bird Study* 54: 23–34.

Jonsson, PE 1992. The Corn Bunting *Miliaria calandra* in Scania, S. Sweden, 1990–91: a report from the conservation project. *Anser* 31: 101–108. (in Swedish)

Kennedy, CEJ & Southwood, TRE 1984. The number of British insects associated with British trees: a re-analysis. *Journal of Animal Ecology* 53: 455–478.

Kirby, KJ 2001. The impact of deer on the ground flora of British broadleaved woodland. *Forestry* 74: 219–230.

Kirby, KJ, Thomas, RC, Key, RS, McLean, IPG & Hodgetts, N 1995. Pasture woodland and its conservation in Britain. *Biological Journal of the Linnean Society* 56 (Suppl.): 135–153.

Kirkham, FW, Mountford, JO & Wilkins, RJ 1996. The effects of nitrogen, potassium and phosphorus addition on the vegetation of a Somerset peat moor under cutting management. *Journal of Applied Ecology* 33: 1013–1029.

Kleijn, D & Sutherland, WJ 2003. How effective are European agri-environment schemes in conserving and promoting biodiversity? *Journal of Applied Ecology* 40: 947–969.

Kleijn, D & van Zuijlen, GJC 2004. The conservation effects of meadow bird agreements on farmland in Zeeland, The Netherlands, in the period 1989–1995. *Biological Conservation* 117: 443–451.

Kleijn, D, Berendse, F, Smit, R & Gilissen, N 2001. Agri-environment schemes do not effectively protect biodiversity in Dutch agricultural landscapes. *Nature* 413: 723–725.

Kleijn, D, Berendse, F, Smit, R, Gilissen, N, Smit, J, Brak, B & Groeneveld, R 2004. Ecological effectiveness of agri-environment schemes in different agricultural landscapes in the Netherlands. *Conservation Biology* 18: 775–786.

Kleijn, D, Baquero, RA, Clough, Y, Diaz, M, De Esteban, J, Fernandez, F, Gabriel, D, Herzog, F, Holzschuh, A, Jöhl, R, Knop, E, Kruess, A, Marshall, EJP, Steffan-Dewenter, I, Tscharntke, T, Verhulst, J, West, TM & Yela, JL 2006. Mixed biodiversity benefits of agri-environment schemes in five European countries. *Ecology Letters* 9: 243–254.

Knop, E, Kleijn, D, Herzog, F & Schmid, B 2006. Effectiveness of the Swiss agri-environment scheme in promoting biodiversity. *Journal of Applied Ecology* 43: 120–127.

Kohler, F, Verhulst, J, Knop, E, Herzog, F & Kleijn, D 2007. Indirect effects of extensification schemes on pollinators in two contrasting European countries. *Biological Conservation* 135: 302–307.

Kragten, S & de Snoo, GR 2007. Nest success of Lapwings *Vanellus vanellus* on organic and conventional arable farms in the Netherlands. *Ibis* 149: 742–749.

Kragten, S, Trimbos, KB & de Snoo, GR 2008. Breeding skylarks (*Alauda arvensis*) on organic and conventional arable farms in the Netherlands. *Agriculture, Ecosystems and Environment* 126: 163–167.

Krebs, JR 1971. Territory and breeding density in the Great Tit, *Parus major* L. *Ecology* 52: 1–22.

Krebs, JR 1977. Song and territory in the Great Tit. In Stonehouse, B & Perrins, CM (eds) *Evolutionary Ecology*, pp 47–62. London: Macmillan.

Krebs, JR, Wilson, JD, Bradbury, RB & Siriwardena, GM 1999. The second Silent Spring? *Nature* 400: 611–612.

Kremen, C 2005. Managing ecosystem services: what do we need to know about their ecology? *Ecology Letters* 8: 468–479.

Kruk, M, Noordervliet, MAW & ter Keurs, WJ 1996. Hatching dates of waders and mowing dates in intensively exploited grassland areas in different years. *Biological Conservation* 77: 213–218.

Kurki, S & Lindén, H 1995. Forest fragmentation due to agriculture affects the reproductive success of the ground nesting black grouse. *Ecography* 18: 109–113.

Kurki, S, Helle, P, Lindén, H & Nikula, P 1997. Breeding success of black grouse and capercaillie in relation to mammalian predator densities on two spatial scales. *Oikos* 79: 301–310.

Kurki, S, Nikula, A, Helle, P & Lindén, H 1998. Abundance of fox and pine marten in relation to composition of the boreal forest. *Journal of Animal Ecology* 67: 874–886.

Kyrkos, A 1997. Behavioural and demographic responses of Yellowhammers to variation in agricultural practices. Unpublished DPhil thesis, University of Oxford.

Kyrkos, A, Wilson, JD & Fuller, RJ 1998. Farmland habitat change and abundance of yellowhammers (*Emberiza citrinella*): an analysis of Common Birds Census data. *Bird Study* 45: 232–246.

Laaksonen, T 2006. Climate change, migratory connectivity and change in laying date and clutch size of the pied flycatcher. *Oikos* 114: 277–290.

Lack, PC 1986. *The Atlas of Wintering Birds in Britain and Ireland*. Calton: T. & A. D. Poyser.

Lack, P 1992. *Birds on Lowland Farms*. London: Her Majesty's Stationery Office.

Lal, R 2007. Soil science and the carbon civilization. *Soil Science Society of America Journal* 71: 1425–1437.

Lampkin, N 2002. *Organic Farming*. Ipswich: Old Pond.

Lane, AB 1984. An enquiry into the responses of growers to attacks by pests on oilseed rape (*Brassica napus*), a relatively new crop in the United Kingdom. *Protection Ecology* 7: 73–78.

Langston, RHW, Smith, T, Brown, AF & Gregory, RD 2006. Status of breeding Twite *Carduelis flavirostris* in the UK. *Bird Study* 53: 64–72.

Langton, S 2008. *Change in the Area and Distribution of Set-Aside in England: January 2008 Update*, Defra Agricultural Change and Environment Observatory Research Report No. 10. London: Defra.

Leech, SM & Leonard, ML 1997. Begging and the risk of predation in nestling birds. *Behavioural Ecology* 8: 644–646.

Liley, D & Clarke, RT 2003. The impact of urban development and human disturbance on the numbers of nightjar *Caprimulgus europaeus* on heathlands in Dorset, England. *Biological Conservation* 114: 219–230.

Lima, SL 1986. Predation risk and unpredictable feeding conditions: determinants of body mass in birds. *Ecology* 67: 377–385.

Lindley, P & Smith, M 2002. Seeing black grouse through the trees. *Natur Cymru* 5: 18–21.

Lokemoen, JT & Beiser, JA 1997. Bird use and nesting in conventional, minimum tillage and organic cropland. *Journal of Wildlife Management* 61: 644–655.

Love, RA, Webbon, C, Glue, DE & Harris, S 2000. Changes in the food of British Barn Owls *Tyto alba* between 1947 and 1997. *Mammal Review* 30: 107–129.

Lovegrove, R, Williams, G & Williams, I 1994. *Birds in Wales*. London: T. & A. D. Poyser.

Ludwig, GX, Alatalo, RV, Helle, P, Linden, H, Lindstrom, J & Siitari, H 2006. Short- and long-term population dynamical consequences of asymmetric climate change in black grouse. *Proceedings of the Royal Society of London B* 273: 2009–2016.

Ludwig, GX, Alatalo, RV, Helle, P, Nissinen, K & Siitari, H 2008. Large scale drainage and breeding success in boreal forest grouse. *Journal of Applied Ecology* 45: 325–333.

Lumaret, JP, Galante, E, Lumbreras, C, Mena, J, Bertrand, M, Bernal, JL, Cooper, JF, Kadiri, N & Crowe, D 1993. Field effects of ivermectin residues on dung beetles. *Journal of Applied Ecology* 30: 428–436.

Lutz, W, Sanderson, W & Scherbov, S 2001. The end of world population growth. *Nature* 412: 543–545.

Mabey, R 1980. *The Common Ground: A Place for Nature in Britain's Future*. London: Hutchinson and Nature Conservancy Council.

Macdonald, DW & Johnson, PJ 1995. The relationship between bird distribution and the botanical and structural characteristics of hedges. *Journal of Applied Ecology* 32: 492–505.

Macdonald, MA & Bolton, M 2008a. Predation on wader nests in Europe. *Ibis* 150 (Suppl. 1): 54–73.

Macdonald, MA & Bolton, M 2008b. Predation of lapwing *Vanellus vanellus* nests on lowland wet grassland in England and Wales: effects of nest density, habitat and predator abundance. *Journal of Ornithology* 149: 555–563.

Maclean, IMD & Austin, GE 2006. *Wetland Bird Survey Alerts 2004/2005: Changes in Numbers of Wintering Waterbirds in the Constituent Countries of the United Kingdom, Special Protection Areas (SPAs) and Sites of Special Scientific Interest (SSSIs)*, BTO Research Report No. 458. Thetford: British Trust for Ornithology.

MacLeod, A, Wratten, SD & Thomas, MB 2004. 'Beetle banks' as refuges for beneficial arthropods in farmland: long-term changes in predator communities and habitat. *Agricultural and Forest Entomology* 6: 147–154.

Macleod, R, Barnett, P, Clark, JA & Cresswell, W 2005. Body mass change strategies in blackbirds *Turdus merula*: the starvation–predation risk trade-off. *Journal of Animal Ecology* 74: 292–302.

Macleod, R, Barnett, P, Clark, J & Cresswell, W 2006. Mass-dependent predation risk as a mechanism for house sparrow declines? *Biology Letters* 2: 43–46.

Macleod, R, Lind, J, Clark, J & Cresswell, W 2007. Mass regulation in response to predation risk can indicate population declines. *Ecology Letters* 10: 945–955.

Macleod, RA, Clark, J & Cresswell, W 2008. The starvation–predation risk trade-off: body mass and population status in the common starling *Sturnus vulgois*. *Ibis* 150 (Suppl. 1): 199–208.

MacMillan, D, Hanley, N & Daw, M 2004. Costs and benefits of wild goose conservation in Scotland. *Biological Conservation* 119: 475–485.

Madders, M 2000. Habitat selection and foraging success of Hen Harriers *Circus cyaneus* in west Scotland. *Bird Study* 47: 32–40.

Madders, M 2003. Hen Harrier *Circus cyaneus* foraging activity in relation to habitat and prey. *Bird Study* 50: 55–60.

Madsen, M, Overgaard Nielsen, B, Holter, P, Pedersen, OC, Brøchner Jespersen, J, Vagn-Jensen, K-M, Nansen, P & Grønvold, J 1990. Treating cattle with ivermectin: effects on the fauna and decomposition of dung pats. *Journal of Applied Ecology* 27: 1–15.

Maes, J, Musters, CJM & De Snoo, GR 2008. The effect of agri-environment schemes on amphibian diversity and abundance. *Biological Conservation* 141: 635–645.

Mal'chevskiy, AS & Pukinskiy, YB 1983. *Birds of the Leningrad Region and Adjacent Territories*. Leningrad University Press.

Mallord, JW, Dolman, PM, Brown, AF & Sutherland, WJ 2007. Linking recreational disturbance to population size in a ground-nesting passerine. *Journal of Applied Ecology* 44: 185–195.

Marchant, JH, Hudson, R, Carter, SP & Whittington, PA 1990. *Population Trends in British Breeding Birds*. Tring: British Trust for Ornithology.

Marquiss, M 2007. Seasonal pattern in hawk predation on Common Bullfinches *Pyrrhula pyrrhula*: evidence of an interaction with habitat affecting food availability. *Bird Study* 54: 1–11.

Marren, P 2002. *Nature Conservation*. London: HarperCollins.

Marrs, RH 1993a. An assessment of change in *Calluna* heathlands in Breckland, eastern England, between 1983 and 1991. *Biological Conservation* 63: 133–139.

Marrs, RH 1993b. Soil fertility and nature conservation in Europe: theoretical considerations and practical management solutions. *Advances in Ecological Research* 24: 241–300.

Marshall, EJP & Hopkins, A 1990. Plant species composition and dispersal in agricultural land. In Bunce, RGH & Howard, DC (eds) *Species Dispersal in Agricultural Habitats*, pp 98–116. London: Belhaven Press.

Marshall, EJP, Wade, PM & Clare, P 1978. Land drainage channels in England and Wales. *Geographical Journal* 144: 254–263.

Marshall, EJP, West, TM & Kleijn, D 2006. Impacts of an agri-environment field margin prescription on the flora and fauna of arable farmland in different landscapes. *Agriculture, Ecosystems and Environment* 113: 36–44.

Mason, CF 1998. Habitats of the song thrush *Turdus philomelos* in a largely arable landscape. *Journal of Zoology* 244: 89–93.
Mason, CF 2000. Thrushes now largely restricted to the built environment in eastern England. *Diversity and Distributions* 6: 189–194.
Mason, CF & Lyczynski, F 1980. Breeding biology of the Pied and Yellow Wagtails. *Bird Study* 27: 1–10.
Mason, CF & Macdonald, SM 1999a. Winter bird numbers and land-use preferences in an arable landscape in eastern England. *Bird Conservation International* 9: 119–127.
Mason, CF & Macdonald, SM 1999b. Habitat use by Lapwings and Golden Plovers in a largely arable landscape. *Bird Study* 46: 89–99.
Mason, CF & Macdonald, SM 2000. Corn Bunting *Miliaria calandra* populations, landscape and land-use in an arable district of eastern England. *Bird Conservation International* 10: 169–186.
Matson, PA & Vitousek, PM 2006. Agricultural intensification: will land spared from farming be land spared for nature? *Conservation Biology* 20: 709–710.
Mattison, EHA & Norris, K 2005. Bridging the gaps between agricultural policy, land-use and biodiversity. *Trends in Ecology and Evolution* 20: 610–616.
Maudsley, MJ 2000. A review of the ecology and conservation of hedgerow invertebrates in Britain. *Journal of Environmental Management* 60: 65–76.
Maurer, BA & Brown, JH 1989. Distributional consequences of spatial variation in local demographic processes. *Annales Zoologici Fennici* 26: 121–131.
McCanch, N 2000. The relationship between Red-billed Chough *Pyrrhocorax pyrrhocorax* (L) breeding populations and grazing pressure on the Calf of Man. *Bird Study* 47: 295–303.
McCracken, DI 1993. The potential for avermectins to affect wildlife. *Veterinary Parasitology* 48: 273–280.
McCracken, DI & Foster, GN 1993. The effect of ivermectin on the invertebrate fauna associated with cow dung. *Environmental Toxicology and Chemistry* 12: 73–84.
McCracken, DI & Tallowin, JR 2004. Swards and structure: the interactions between farming practices and bird food resources in lowland grasslands. *Ibis* 146 (Suppl. 2): 108–114.
McCracken, DI, Foster, GN, Bignal, EM & Bignal, S 1992. An assessment of Chough *Pyrrhocorax pyrrhocorax* diet using multivariate analysis techniques. *Avocetta* 16: 19–29.
McCracken, DI, Foster, GN & Kelly, A 1995. Factors affecting the size of leatherjacket (Diptera: Tipulidae) populations in pastures in the west of Scotland. *Applied Soil Ecology* 2: 203–213.
McKay, HV, Bishop, JD, Feare, CJ & Stevens, MC 1993. Feeding by brent geese can reduce yield of oilseed rape. *Crop Protection* 12: 101–105.
McKay, HV, Milsom, TP, Feare, CJ, Ennis, DC, O'Connell, DP & Haskell, DJ 2001. Selection of forage species and the creation of alternative feeding areas for dark-bellied brent geese *Branta bernicla bernicla* in southern UK coastal areas. *Agriculture, Ecosystems and Environment* 84: 99–113.
McKeever, CU 2003. Linking grassland management, invertebrates and Northern Lapwing productivity. Unpublished PhD thesis, University of Stirling.
Mellanby, K 1981. *Farming and Wildlife*. London: Collins.
Meyer, RM 1990. Observations on two Red-billed Choughs *Pyrrhocorax pyrrhocorax* in Cornwall: habitat use and food intake. *Bird Study* 37: 190–209.
Millenbah, KF, Winterstein, SR, Campa, H, Furrow, LT & Minnis, RB 1996. Effects of conservation reserve program field age on avian relative abundance, diversity and productivity. *Wilson Bulletin* 108: 760–770.
Milsom, TP 2005. Decline of Northern Lapwing *Vanellus vanellus* breeding on arable farmland in relation to spring tillage. *Bird Study* 52: 297–306.
Milsom, TP, Holditch, RS & Rochard, JBA 1985. Diurnal use of an airfield and adjacent agricultural habitats by lapwings *Vanellus vanellus*. *Journal of Applied Ecology* 22: 313–326.

Milsom, TP, Rochard, JBA & Poole, SJ 1990. Activity patterns of Lapwings *Vanellus vanellus* in relation to the lunar cycle. *Ornis Scandinavica* 21: 147–156.

Milsom, TP, Ennis, DC, Haskell, DJ, Langton, SD & McKay, HV 1998. Design of grassland feeding areas for waders during winter: the relative importance of swards, landscape factors and human disturbance. *Biological Conservation* 84: 119–129.

Milsom, TP, Langton, SD, Parkin, WK, Peel, S, Bishop, JD, Hart, JD & Moore, NP 2000. Habitat models of bird species' distribution: an aid to the management of coastal grazing marshes. *Journal of Applied Ecology* 37: 706–727.

Milsom, TP, Langton, SD, Parkin, WK, Allen, DS, Bishop, JD & Hart, JD 2001. Coastal grazing marshes as a breeding habitat for skylarks *Alauda arvensis*. In Donald, PF & Vickery, JA (eds) *The Ecology and Conservation of Skylarks* Alauda arvensis, pp 41–51. Sandy: Royal Society for the Protection of Birds.

Milsom, TP, Hart, JD, Parkin, WK & Peel, S 2002. Management of coastal grazing marshes for breeding waders: the importance of surface topography and wetness. *Biological Conservation* 103: 199–207.

Møller, AP 2001. The effect of dairy farming on barn swallow *Hirundo rustica* abundance, distribution and reproduction. *Journal of Applied Ecology* 38: 378–389.

Monaghan, P, Bignal, EM, Bignal, S, Easterbee, N & McKay, CR 1989. The distribution and status of the Chough in Scotland in 1986. *Scottish Birds* 15: 114–118.

Moorcroft, D 2000. The causes of decline in the linnet *Carduelis cannabina* within the agricultural landscape. Unpublished DPhil thesis, University of Oxford.

Moorcroft, D & Wilson, JD 2000. The ecology of Linnets *Carduelis cannabina* on lowland farmland. In Aebsicher, NJ, Evans, AD, Grice, PV & Vickery, JA (eds) *Ecology and Conservation of Lowland Farmland Birds*, pp 173–181. Tring: British Ornithologists' Union.

Moorcroft, D, Whittingham, MJ, Bradbury, RB & Wilson, JD 2002. The selection of stubble fields by wintering granivorous birds reflects vegetation cover and food abundance. *Journal of Applied Ecology* 39: 535–547.

Moorcroft, D, Wilson, JD & Bradbury, RB 2006. The diet of nestling Linnets *Carduelis cannabina* on lowland farmland before and after agricultural intensification. *Bird Study* 53: 156–162.

Moore, I 1966. *Grass and Grasslands*. London: Collins.

Moreby, SJ & Aebischer, NJ 1992. Invertebrate abundance on cereal fields and set-aside land: implications for wild gamebird chicks. In Clarke, J (ed) *Set-Aside*, British Crop Protection Council Monograph No. 58, pp 181–187. Farnham: British Crop Protection Council.

Moreby, SJ & Southway, S 2002. Cropping and year effects on the availability of invertebrate groups important in the diet of farmland birds. *Aspects of Applied Biology* 67: 107–112.

Morris, AJ & Gilroy, JJ 2008. Close to the edge: predation risks for two declining farmland passerines. *Ibis* 150 (Suppl. 1): 168–177.

Morris, AJ, Burges, D, Fuller, RJ, Evans, AD & Smith, KW 1994. The status and distribution of Nightjars *Caprimulgus europaeus* in Britain in 1992. *Bird Study* 41: 181–191.

Morris, AJ, Whittingham, MJ, Bradbury, RB, Wilson, JD, Kyrkos, A, Buckingham, DL & Evans, AD 2001. Foraging habitat selection by Yellowhammers (*Emberiza citrinella*) in agriculturally contrasting regions in lowland England. *Biological Conservation* 98: 197–210.

Morris, AJ, Bradbury, RB & Wilson, JD 2002a. Determinants of patch selection by yellowhammers *Emberiza citrinella* foraging in cereal crops. *Aspects of Applied Biology* 67: 43–50.

Morris, AJ, Bradbury, RB & Wilson, JD 2002b. Indirect effects of pesticides on breeding yellowhammers *Emberiza citrinella*. *British Crop Protection Council Conference – Pests & Diseases 2002*, pp 965–970. Farnham: British Crop Protection Council.

Morris, AJ, Holland, JM, Smith, B & Jones, NE 2004. Sustainable Arable Farming for an Improved Environment (SAFFIE): managing winter wheat sward structure for Skylarks *Alauda arvensis*. *Ibis* 146 (Suppl. 2), 155–162.

Morris, AJ, Wilson, JD, Bradbury, RB & Whittingham, MJ 2005. Indirect effects of pesticides on breeding yellowhammer (*Emberiza citrinella*). *Agriculture, Ecosystems and Environment* 106: 1–16.

Morris, AJ, Smith, B, Jones, NE & Cooke, SK 2007. Experiment 1.1. Manipulate within crop agronomy to increase biodiversity: crop architecture. In *The SAFFIE Project Report*. Boxworth: ADAS.

Morris, MG 2000. The effects of structure and its dynamics on the ecology and conservation of arthropods in British grasslands. *Biological Conservation* 95: 129–142.

Mortimer, SR, Westbury, DB, Dodd, S, Brook, AJ, Harris, SJ, Kessock-Philip, R, Chaney, K, Lewis, P, Buckingham, DL & Peach, WJ 2007. Cereal-based whole crop silages: potential biodiversity benefits of cereal production in pastoral landscapes. *Aspects of Applied Biology* 81: 77–86.

Muir, R & Muir, N 1987. *Hedgerows: Their History and Wildlife*. London: Michael Joseph.

Müller, M, Spaar, R, Schifferli, L & Jenni, L 2005. Effects of changes in farming of subalpine meadows on a grassland bird, the whinchat (*Saxicola rubetra*). *Journal of Ornithology* 146: 14–23.

Murray, KA, Wilcox, A & Stoate, C 2002. A simultaneous assessment of farmland habitat use by breeding skylarks and yellowhammers. *Aspects of Applied Biology* 67: 121–128.

Murton, RK 1965. *The Woodpigeon*. London: Collins.

Murton, RK 1968. Breeding, migration and survival of Turtle Doves. *British Birds* 61: 193–212.

Murton, RK & Westwood, NJ 1974. Some effects of agricultural change on the English avifauna. *British Birds* 67: 41–67.

Murton, RK, Westwood, NJ & Isaacson, AJ 1964. The feeding habits of Woodpigeon *Columba palumbus*, Stock Dove *C. oenas* and Turtle Dove *Streptopelia turtur*. *Ibis* 106: 174–188.

Murton, RK, Westwood, NJ & Isaacson, AJ 1974. A study of woodpigeon shooting: the exploitation of a natural animal population. *Journal of Applied Ecology* 11: 61–81.

Myers, N & Kent, J 2003. New consumers: the influence of affluence on the environment. *Proceedings of the National Academy of Sciences, USA* 100: 4963–4968.

Nature Conservancy Council 1984. *Nature Conservation in Great Britain*. Peterborough: Nature Conservancy Council.

New, TR 2005. *Invertebrate Conservation in Agricultural Ecosystems*. Cambridge University Press.

Newson, SE, Woodburn, RJW, Noble, DG, Baillie, SR & Gregory, RD 2005. Evaluating the Breeding Bird Survey for producing population size and density estimates. *Bird Study* 52: 42–54.

Newson, SE, Evans, KL, Noble, DG, Greenwood, JJD & Gaston, KJ 2008. Use of distance sampling to improve estimates of national population sizes for common and widespread breeding birds in the UK. *Journal of Applied Ecology* 45: 1330–1338.

Newton, I 1967. The adaptive radiation and feeding ecology of some British finches. *Ibis* 109: 33–98.

Newton, I 1979. *Population Ecology of Raptors*. Berkhamsted: T. & A. D. Poyser.

Newton, I 1986. *The Sparrowhawk*. Calton: T. & A. D. Poyser.

Newton, I 1988. A key factor analysis of a Sparrowhawk population. *Oecologia* 76: 588–596.

Newton, I 1993. Predation and the limitation of bird numbers. *Current Ornithology* 11: 143–198.

Newton, I 1998. *Population Limitation in Birds*. London: Academic Press.

Newton, I 1999. An alternative approach to the measurement of seasonal trends in bird breeding success: a case study of the bullfinch *Pyrrhula pyrrhula*. *Journal of Animal Ecology* 68: 698–707.

Newton, I 2004. The recent declines of farmland bird populations in Britain: an appraisal of causal factors and conservation actions. *Ibis* 146: 579–600.

Newton, I & Bogan, J 1978. The role of different organochlorine compounds in the breeding of British sparrowhawks. *Journal of Applied Ecology* 15: 105–116.

Newton, I & Haas, MB 1984. The return of the Sparrowhawk. *British Birds* 77: 47–70.

Newton, I & Wyllie, I 1992. Recovery of a Sparrowhawk population in relation to declining pesticide contamination. *Journal of Applied Ecology* 29: 476–484.

Newton, I, Dale, L & Rothery, P 1997. Apparent lack of impact of sparrowhawks on the breeding densities of some woodland songbirds. *Bird Study* 44: 129–135.

Norrdahl, K & Korpimäki, E 1998. Fear in farmlands: how much does predator avoidance affect bird community structure? *Journal of Avian Biology* 29: 79–85.

Norris, CA 1945. Summary of a report on the distribution and status of the corncrake (*Crex crex*). *British Birds* 38: 142–148, 162–168.

Norris, CA 1947. Report on the distribution of and status of the Corncrake. *British Birds* 40: 226–244.

Norris, K, Cook, T, O'Dowd, B & Durdin, C 1997. The density of Redshank *Tringa totanus* breeding on the saltmarshes of the Wash in relation to habitat and its grazing management. *Journal of Applied Ecology* 34: 999–1013.

Norris, K, Brindley, E, Cook, T, Babbs, S, Forster Brown, C & Yaxley, R 1998. Is the density of redshank *Tringa totanus* nesting on saltmarshes in Great Britain declining due to changes in grazing management? *Journal of Applied Ecology* 35: 621–634.

Oates, M 1994. The management of southern limestone grasslands. *British Wildlife* 5: 73–82.

O'Brien, M 2001. Factors affecting breeding wader populations on upland enclosed farmland in northern Britain. Unpublished PhD thesis, University of Edinburgh.

O'Brien, M 2002. The relationship between field occupancy rates by breeding lapwing and habitat management on upland farmland in Northern Britain. *Aspects of Applied Biology* 67: 85–92.

O'Brien, M & Smith, KW 1992. Changes in the status of waders breeding on wet lowland grasslands in England and Wales between 1982 and 1989. *Bird Study* 39: 165–176.

O'Brien, M, Tharme, A & Jackson, D 2002. Changes in breeding wader numbers on Scottish farmed land during the 1990s. *Scottish Birds* 23: 10–21.

O'Brien, M, Green, RE & Wilson, J 2006. Partial recovery of the population of Corncrake *Crex crex* in Britain 1993–2004. *Bird Study* 53: 213–224.

O'Connor, RJ & Shrubb, M 1986. *Farming and Birds*. Cambridge University Press.

Odderskær, P, Prang, A, Poulsen, JG, Elmegaard, N & Andersen, PN 1997a. Skylark (*Alauda arvensis*) utilisation of micro-habitats in spring barley fields. *Agriculture, Ecosystems and Environment* 62: 21–29.

Odderskær, P, Prang, A, Elmegaard, N & Andersen, PN 1997b. *Skylark Reproduction in Pesticide Treated and Untreated Fields*, Pesticide Research No. 32. Copenhagen: Danish Environmental Protection Agency.

Ogilvie, MA 1978. *Wild Geese*. Berkhamsted: T. & A. D. Poyser.

Olsson, O, Bruun, M & Smith, HG 2002. Starling foraging success in relation to agricultural land-use. *Ecography* 25: 363–371.

Opdam, P, Rijsdijk, G & Hustings, F 1985. Bird communities in small woods in an agricultural landscape: effects of area and isolation. *Biological Conservation* 34: 333–352.

Orr, HG, Wilby, RL, McKenzie Hedger, M & Brown, I 2008. Climate change in the uplands: a UK perspective on safeguarding regulatory ecosystem services. *Climate Research* 37: 77–98.

Orr, RJ, Parsons, AJ, Treacher, TT & Penning, PD 1988. Seasonal patterns of grass production under cutting or continuous stocking management. *Grass and Forage Science* 34: 199–207.

Osborne, P 1982. Some effects of Dutch elm disease on nesting farmland birds. *Bird Study* 29: 2–16.

Osborne, P 1984. Bird numbers and habitat characteristics in farmland hedgerows. *Journal of Applied Ecology* 21: 63–82.

Ottvall, R 2005. Nest survival among waders breeding on coastal meadows: the relative importance of predation and trampling damage by livestock. *Ornis Svecica* 15: 89–96.

Owen, M 1990. The damage–conservation interface illustrated by geese. *Ibis* 132: 238–252.

Palmer, SCF & Bacon, PJ 2001. The utilization of heather moorland by territorial Red Grouse *Lagopus lagopus scoticus*. *Ibis* 143: 222–232.

Palmer, SCF, Mitchell, RJ, Truscott, AM & Welch, D 2004. Regeneration failure in Atlantic oakwoods: The roles of ungulate grazing and invertebrates. *Forest Ecology and Management* 192: 251–265.

Paradis, E, Baillie, SR, Sutherland, WJ, Dudley, C, Crick, HQP & Gregory, RD 2000. Large-scale spatial variation in the breeding performance of song thrushes and blackbirds in Britain. *Journal of Applied Ecology* 37: 73–87.

Parish, DMB & Sotherton, NW 2004. Game crops as summer habitat for farmland songbirds in Scotland. *Agriculture, Ecosystems and Environment* 104: 429–438.

Parish, T, Lakhani, HK & Sparks, TH 1994. Modelling the relationship between bird population variables and hedgerow and other field margin attributes. 1. Species richness of winter, summer and breeding birds. *Journal of Applied Ecology* 31: 764–775.

Parish, T, Lakhani, HK & Sparks, TH 1995. Modelling the relationship between bird population variables and hedgerow and other field margin attributes. 2. Abundance of individual species and of groups of similar species. *Journal of Applied Ecology* 32: 362–371.

Park, KJ, Graham, KE, Calladine, J & Wernham, CW 2008. Impacts of birds of prey on gamebird in the UK: a review. *Ibis* 150 (Suppl. 1): 9–26.

Parr, R 1980. Population study of golden plover *Pluvialis apricaria*, using marked birds. *Ornis Scandinavica* 11: 179–189.

Parr, R 1992. The decline to extinction of a population of golden plover in north-east Scotland. *Ornis Scandinavica* 23: 152–158.

Parr, R 1993. Nest predation and numbers of golden plovers *Pluvialis apricaria* and other moorland waders. *Bird Study* 40: 223–231.

Parr, R & Watson, A 1988. Habitat preferences of black grouse on moorland-dominated ground in north-east Scotland. *Ardea* 76: 175–180.

Parslow, J 1973. *Breeding Birds of Britain and Ireland: A Historical Survey*. Berkhamsted: T. & A. D. Poyser.

Paterson, IW 1991. The status and breeding distribution of greylag geese *Anser anser* in the Uists (Scotland) and their impact upon crofting agriculture. *Ardea* 79: 243–251.

Paton, PWC 1994. The edge effect on avian nest success: how strong is the evidence? *Conservation Biology* 8: 17–26.

Patterson, IJ, Abdul Jalil, S & East, ML 1989. Damage to winter cereals by greylag and pink-footed geese in north-east Scotland. *Journal of Applied Ecology* 26: 879–895.

Peach, WJ, Thompson, PS & Coulson, JC 1994. Annual and long-term variation in the survival rates of British Lapwings *Vanellus vanellus*. *Journal of Animal Ecology* 63: 60–70.

Peach, WJ, Siriwardena, GM & Gregory, RD 1999. Long-term changes in the abundance and demography of British Reed Buntings *Emberiza schoeniclus*. *Journal of Applied Ecology* 36: 798–811.

Peach, WJ, Lovett, LJ, Wotton, SR & Jeffs, C 2001. Countryside Stewardship delivers cirl buntings in Devon, UK. *Biological Conservation* 101: 361–373

Peach, WJ, Taylor, R, Cotton, P, Gruar, D, Hill, I & Denny, M 2002. Habitat utilisation by song thrushes *Turdus philomelos* on lowland farmland during summer and winter. *Aspects of Applied Biology* 67: 11–20.

Peach, WJ, Denny, M, Cotton, PA, Hill, IF, Gruar, D, Barritt, D, Impey, A & Mallord, J 2004a. Habitat selection by song thrushes in stable and declining farmland populations. *Journal of Applied Ecology* 41: 275–293.

Peach, WJ, Robinson, RA & Murray, KA 2004b. Demographic and environmental causes of the decline of rural Song Thrushes *Turdus philomelos* in lowland Britain. *Ibis* 146 (Suppl. 2): 50–59.

Pearce-Higgins, JW & Grant, MC 2002. The effects of grazing-related variation in habitat on the distribution of moorland skylarks *Alauda arvensis* and meadow pipits *Anthus pratensis. Aspects of Applied Biology* 67: 155–163.

Pearce-Higgins, JW & Grant, MC 2006. Relationships between bird abundance and the composition and structure of moorland vegetation. *Bird Study* 53: 112–125.

Pearce-Higgins, JW & Yalden, DW 2003a. Variation in the use of pasture by breeding European Golden Plovers *Pluvialis apricaria* in relation to prey availability. *Ibis* 145: 365–381.

Pearce-Higgins, JW & Yalden, DW 2003b. Golden Plover *Pluvialis apricaria* breeding success on a moor managed for shooting Red Grouse *Lagopus lagopus. Bird Study* 50: 170–177.

Pearce-Higgins, JW & Yalden, DW 2004. Habitat selection, diet, arthropod availability and growth of a moorland wader: the ecology of European Golden Plover *Pluvialis apricaria* chicks. *Ibis* 146: 335–346.

Pearce-Higgins, JW, Yalden, DW & Whittingham, MJ 2005. Warmer springs advance the breeding phenology of golden plovers *Pluvialis apricaria* and their prey (Tipulidae). *Oecologia* 143: 470–476.

Pearce-Higgins, JW, Grant, MC, Robinson, MC & Haysom, SL 2007. The role of forest maturation in causing the decline of Black Grouse *Tetrao tetrix. Ibis* 149: 143–155.

Pearce-Higgins, JW, Grant, MC, Beale, CM, Buchanan, GM & Sim, IMW 2008. International importance and drivers of change of upland bird populations. In Bonn, A, Hubacek, K, Allott, T & Stewart, J (eds) *Drivers of Environmental Change in Uplands*, pp 209–226. Oxford: Routledge.

Pearsall, WH 1971. *Mountains and Moorlands.* London: Collins.

Percival, SM & Houston, DC 1992. The effect of winter grazing by barnacle geese on grassland yields on Islay. *Journal of Applied Ecology* 29: 35–40.

Perkins, AJ, Whittingham, MJ, Bradbury, RB, Wilson, JD, Morris, AJ & Barnett, PR 2000. Habitat characteristics affecting use of lowland agricultural grassland by birds in winter. *Biological Conservation* 95: 279–294.

Perkins, AJ, Whittingham, MJ, Morris, AJ & Bradbury, RB 2002. Use of field margins by foraging yellowhammers *Emberiza citrinella. Agriculture, Ecosystems and Environment* 93: 413–420.

Perkins, AJ, Anderson, G & Wilson, JD 2007. Seed food preferences of granivorous farmland passerines. *Bird Study* 54: 46–53.

Perkins, AJ, Maggs, HE & Wilson, JD 2008a. Winter bird use of seed-rich habitats in agri-environment schemes. *Agriculture, Ecosystems and Environment* 126: 189–194.

Perkins, AJ, Maggs, HE, Wilson, JD, Watson, A & Smout, C 2008b. Targeted management intervention reduces rate of population decline of Corn Buntings *Emberiza calandra* in eastern Scotland. *Bird Study* 55: 52–58.

Perrins, CM & Geer, TA 1980. The effect of Sparrowhawks on Tit populations. *Ardea* 68: 133–142.

Perrins, CM & Overall, R 2001. Effect of increasing numbers of deer on bird populations in Wytham Woods, Central England. *Forestry* 74: 299–309.

Peterken, GF 2000. Rebuilding networks of forest habitats in lowland England. *Landscape Research* 25: 291–303.

Petersen, S, Axelsen, JA, Tybirk, K, Aude, E & Vestergaard, P 2006. Effects of organic farming on field boundary vegetation in Denmark. *Agriculture, Ecosystems and Environment* 113: 302–306.

Petty, SJ, Patterson, IJ, Anderson, DIK, Little, B & Davison, M 1995. Numbers, breeding performance and diet of Sparrowhawk *Accipiter nisus* and Merlin *Falco columbarius* in relation to cone crops and seed-eating finches. *Forest Ecology and Management* 79: 133–146.

Picozzi, N & Hepburn, L 1986. A study of black grouse in north-east Scotland. *Proceedings of the International Grouse Symposium* 3: 462–480.

Piersma, T & Bloksma, N 1987. Large flock of Choughs *Pyrrhocorax pyrrhocorax* harvesting caterpillars in pinewood on La Palma, Canary Islands. *Bird Study* 34: 127–128.

van der Ploeg, RR, Ehlers, W & Sieker, F 1999. Floods and other possible adverse environmental effects of meadowland area decline in former West Germany. *Naturwissenschaften* 86: 313–319.

Pollard, E, Hooper, MD & Moore, NW 1974. *Hedges.* London: Collins.

Pollitt, M, Hall, C, Holloway, S, Hearn, R, Marshall, P, Musgrove, A, Robinson, J & Cranswick, P 2003. *The Wetland Bird Survey 2000–01: Wildfowl and Wader Counts.* Slimbridge: British Trust for Ornithology, Wildfowl and Wetlands Trust, Royal Society for the Protection of Birds and Joint Nature Conservation Committee.

Potts, GR 1980. The effects of modern agriculture, nest predation and game management on the population ecology of partridges (*Perdix perdix* and *Alectoris rufa*). *Advances in Ecological Research* 11: 1–79.

Potts, GR 1986. *The Partridge: Pesticides, Predation and Conservation.* London: Collins.

Potts, GR 2002. Grey Partridge *Perdix perdix.* In Wernham, CV, Toms, MP, Marchant, JH, Clark, JA, Siriwardena, GM & Baillie SR (eds) *The Migration Atlas: Movements of the Birds of Britain and Ireland*, pp 259–260. London: T. & A. D. Poyser.

Potts, GR & Aebischer, NJ 1995. Population dynamics of the Grey Partridge *Perdix perdix* 1793–1993: monitoring, modelling and management. *Ibis* 137 (Suppl.): 29–37.

Poulsen, JG & Sotherton, NW 1992. Nest predation in recently cut set-aside land. *British Birds* 85: 674–675.

Poulsen, JG, Sotherton, NW & Aebischer, NJ 1998. Comparative nesting and feeding ecology of skylarks *Alauda arvensis* on arable farmland in southern England with special reference to set-aside. *Journal of Applied Ecology* 35: 131–147.

Prentice, IC, Cramer, W, Harrison, SP, Leemans, R, Monserud, RA & Solomon, AM 1992. A global biome model based on plant physiology and dominance, soil properties and climate. *Journal of Biogeography* 19: 117–34.

Preston, CD, Pearman, DA & Dines, TD 2002. *The New Atlas of the British and Irish Flora.* Oxford University Press.

Prestt, I 1965. An enquiry into the recent breeding status of some of the smaller birds of prey and crows in Britain. *Bird Study* 12: 196–221.

Proffitt, FM 2002. Causes of population decline of the bullfinch *Pyrrhula pyrrhula* in agricultural environments. Unpublished DPhil thesis, University of Oxford.

Proffitt, FM, Newton, I, Wilson, JD & Siriwardena, GM 2004. Bullfinch *Pyrrhula pyrrhula pileata* breeding ecology in lowland farmland and woodland: comparisons across time and habitat. *Ibis* 146 (Suppl. 2): 78–86.

Pulliam, HR 1988. Sources, sinks and population regulation. *American Naturalist* 132: 652–661.

Pywell, RF, Warman, EA, Hulmes, L, Hulmes, S, Nuttall, P, Sparks, TH, Critchley, CNR & Sherwood, A 2006. Effectiveness of new agri-environment schemes in providing foraging resources for bumblebees in intensively farmed landscapes. *Biological Conservation* 129: 192–206.

Rackham, O 1986. *The History of the Countryside.* London: J. M. Dent.

Raine, AF 2006. The breeding ecology of the Twite *Carduelis flavirostris* and the effect of upland agricultural intensification. Unpublished PhD thesis, University of East Anglia.

Ramsden, DJ 1998. Effect of barn conversions on local populations of Barn Owl *Tyto alba. Bird Study* 45: 68–76.

Rands, MRW 1985. Pesticide use on cereals and the survival of grey partridge chicks: a field experiment. *Journal of Applied Ecology* 22: 49–54.

Rands, MRW 1986. Effect of hedgerow characteristics on partridge breeding densities. *Journal of Applied Ecology* 23: 479–487.

Rands, MRW 1987. Hedgerow management for the conservation of partridges *Perdix perdix* and *Alectoris rufa*. *Biological Conservation* 40: 127–139.

Ratcliffe, DA 1970. Changes attributable to pesticides in egg breakage frequency and eggshell thickness in some British birds. *Journal of Applied Ecology* 7: 67–107.

Ratcliffe, DA 1976. Observations on the breeding of the golden plover in Great Britain. *Bird Study* 23: 63–116.

Ratcliffe, DA 1980. *The Peregrine Falcon*. Calton: T. & A. D. Poyser.

Ratcliffe, DA 1990. *Bird Life of Mountain and Upland*. Cambridge University Press.

Ratcliffe, DA 2007. *Galloway and the Borders*. London: Collins.

Ratcliffe, DA & Thompson, DBA 1988. The British uplands: their ecological character and international significance. In Usher, MB & Thompson, DBA (eds.) *Ecological Change in the Uplands*, pp 9–36. Oxford: Blackwell Scientific Publications.

Ratcliffe, N, Schmitt, S & Whiffin, M 2005. Sink or swim? Viability of a black-tailed godwit population in relation to flooding. *Journal of Applied Ecology* 42: 834–843.

Raven, MJ, Noble, DG & Baillie, SR 2007. *The Breeding Bird Survey 2006*, BTO Research Report No. 471. Thetford: British Trust for Ornithology.

Ravenscroft, NOM 1989. The status and habitat of the Nightjar *Caprimulgus europaeus* in coastal Suffolk. *Bird Study* 36: 161–169.

Rebecca, GW 2006. The breeding ecology of the Merlin (*Falco columbarius aesalon*), with particular reference to north-east Scotland and land-use change. Unpublished PhD thesis, Open University.

Redpath, SM & Thirgood, SJ 1997. *Birds of Prey and Red Grouse*. London: Stationery Office.

Redpath, SM, Thirgood, SJ & Clarke, R 2002. Field vole *Microtus agrestis* abundance and hen harrier *Circus cyaneus* diet and breeding in Scotland. *Ibis* 144: E33–E38.

Reed, S 1996. Factors limiting the distribution and population size of Twite *Carduelis flavirostris* in the Pennines. *Naturalist* 120: 93–102.

Rehfisch, MM, Austin, GE, Armitage, MJS, Atkinson, PW, Holloway, SJ, Musgrove, AJ & Pollitt, MS 2003. Numbers of wintering waterbirds in Great Britain and the Isle of Man (1994/1995–1998/1999). 2. Coastal waders (Charadrii). *Biological Conservation* 112: 329–341.

Reid, JM, Bignal, EM, Bignal, S, McCracken, DI & Monaghan, P 2003. Environmental variability, life-history covariation and cohort effects in red-billed choughs *Pyrrhocorax pyrrhocorax*. *Journal of Animal Ecology* 72: 36–46.

Reid, JM, Bignal, EM, Bignal, S, McCracken, DI & Monaghan, P 2004. Identifying the demographic determinants of population growth rate: a case study of red-billed choughs *Pyrrhocorax pyrrhocorax*. *Journal of Animal Ecology* 73: 777–788.

Relton, J 1972. Breeding biology of Moorhens on Huntingdonshire farm ponds. *British Birds* 65: 248–256.

Rich, TCG & Woodruff, ER 1996. Changes in the vascular plant floras of England and Scotland between 1930–60 and 1987–88: the BSBI monitoring scheme. *Biological Conservation* 75: 217–229.

Risely, K, Noble, DG & Baillie, SR 2008. *The Breeding Bird Survey 2007*, BTO Research Report 508. Thetford: British Trust for Ornithology.

Roberts, P 1982. Foods of the Chough on Bardsey Island, Wales. *Bird Study* 29: 155–161.

Roberts, PD & Pullin, AS 2007. *The Effectiveness of Land-Based Schemes at Conserving Farmland Bird Densities within the UK*, Systematic Review No. 11. Birmingham: Centre for Evidence-Based Conservation.

Robertson, GP & Swinton, SM 2005. Reconciling agricultural productivity and environmental integrity: a grand challenge for agriculture. *Frontiers of Ecology and Environment* 3: 38–46.

Robins, M & Bibby, CJ 1985. Dartford Warblers in Britain. *British Birds* 78: 269–280.

Robinson, M & Armstrong, AC 1988. The extent of agricultural field drainage in England and Wales, 1971–1980. *Transactions of the Institute of British Geographers* 13: 19–28.

Robinson, RA 2001. Feeding ecology of skylarks *Alauda arvensis* in winter: a possible mechanism for population decline? In Donald, PF & Vickery, JA (eds.) *The Ecology and Conservation of Skylarks* Alauda arvensis, pp 129–138. Sandy: Royal Society for the Protection of Birds.

Robinson, RA & Sutherland, WJ 1999. The winter distribution of seed-eating birds: habitat structure, seed density and seasonal depletion. *Ecography* 22: 447–454.

Robinson, RA & Sutherland, WJ 2002. Post-war changes in arable farming and biodiversity in Great Britain. *Journal of Applied Ecology* 39: 157–176.

Robinson, RA, Wilson, JD & Crick, HQP 2001. The importance of arable habitat for farmland birds in grassland landscapes. *Journal of Applied Ecology* 38: 1059–1069.

Robinson, RA, Crick, HQP & Peach, WJ 2003. Population trends of Swallows *Hirundo rustica* breeding in Britain 1964–1998. *Bird Study* 50: 1–7.

Robinson, RA, Green, RE, Baillie, SR, Peach, WJ & Thomson, DL 2004. Demographic mechanisms of the population decline of the song thrush *Turdus philomelos* in Britain. *Journal of Animal Ecology* 73: 670–682.

Robinson, RA, Siriwardena, GM & Crick, HQP 2005a. Size and trends of the House Sparrow *Passer domesticus* population in Great Britain. *Ibis* 147: 552–562.

Robinson, RA, Siriwardena, GM & Crick, HQP 2005b. Status and population trends of Starling *Sturnus vulgaris* in Great Britain. *Bird Study* 52: 252–260.

Robson, G 1998. The breeding ecology of curlew *Numenius arquata* on north Pennine moorland. DPhil thesis, University of Sunderland.

Rodwell, J 1992. *British Plant Communities*, vol. 3, *Grassland and Montane Communities*. Cambridge University Press.

Rolfe, R 1966. The status of the Chough in the British Isles. *Bird Study* 13: 221–226.

Roschewitz, I, Gabriel, D, Tscharntke, T & Thies, C 2006. The effects of landscape complexity on arable weed species diversity in organic and conventional farming. *Journal of Applied Ecology* 42: 873–882.

Rose, RJ, Webb, NR, Clarke, RT & Traynor, CH 2000. Changes on the heathlands in Dorset, England, between 1987 and 1996. *Biological Conservation* 93: 117–125.

Rounsevell, MDA, Ewert, F, Reginster, I, Leemans, R & Carter, TR 2005. Future scenarios of European agricultural land use. 2. Projecting changes in cropland and grassland. *Agriculture, Ecosystems and Environment* 107: 117–135.

RSPB 2001. *Futurescapes: Large-Scale Habitat Restoration for Wildlife and People*. Sandy: Royal Society for the Protection of Birds.

RSPB 2007. *The Uplands: Time to Change*? Sandy: Royal Society for the Protection of Birds.

Rundlöf, M & Smith, HG 2006. The effect of organic farming on butterfly diversity depends on landscape context. *Journal of Applied Ecology* 43: 1121–1137.

Rundlöf, M, Bengtsson, J & Smith, HG 2008a. Local and landscape effects of organic farming on butterfly species richness and abundance. *Journal of Applied Ecology* 45: 813–820.

Rundlöf, M, Nilsson, H & Smith, HG 2008b. Interacting effects of farming practice and landscape context on bumble bees. *Biological Conservation* 141: 417–426.

Rushton, SP, Lurz, WW, Fuller, RM & Grason, PJ 1997. Modelling the distribution of red and grey squirrel at the landscape scale: a combined GIS and population dynamics approach. *Journal of Applied Ecology* 34: 1137–1154.

Ryves, BH & Ryves, BH 1934. The breeding-habits of the corn bunting as observed in North Cornwall, with reference to its polygamous habit. *British Birds* 28: 2–26.

Sæther, B-E, Ringsby, TH & Røskaft, E 1996. Life history variation, population processes and priorities in species conservation: towards a reunion of research paradigms. *Oikos* 77: 217–226.

Sage, RB 1998. Short rotation coppice for energy: towards ecological guidelines. *Biomass and Bioenergy* 15: 39–47.

Sage, RB & Robertson, PA 1996. Factors affecting songbird communities using new short rotation coppice habitats in spring. *Bird Study* 43: 201–213.

Sage, RB, Hollins, K, Gregory, CL, Woodburn, MIA & Carroll, JP 2004. Impact of roe deer *Capreolus capreolus* browsing on understorey vegetation in small farm woodlands. *Wildlife Biology* 10: 115–120.

Sage, RB, Ludolf, C & Robertson, PA 2005. The ground flora of ancient semi-natural woodlands in pheasant release pens in England. *Biological Conservation* 122: 243–252.

Schekkerman, H, Teunissen, W & Oosterveld, E 2008. The effect of 'mosaic management' on the demography of black-tailed godwit *Limosa limosa* on farmland. *Journal of Applied Ecology* 45: 1067–1075.

Schläpfer, A 1988. Populationsökologie der Feldlerche *Alauda arvensis* in der intensiv genutzten Agrarlandschaft. *Der Ornithologische Beobachter* 85: 305–371.

Schläpfer, A 2001. A conceptual model of skylark *Alauda arvensis* territory distribution in different landscape types. In Donald, PF & Vickery, JA (eds) *The Ecology and Conservation of Skylarks* Alauda arvensis, pp 3–9. Sandy: Royal Society for the Protection of Birds.

Schön, M 1999. On the significance of micro-structures in arable land: does the Skylark (*Alauda arvensis*) show a preference for places with stunted growth? *Journal für Ornithologie* 140: 87–91.

Scott, GW, Jardine, DC, Hills, G & Sweeney, B 1998. Changes in Nightjar *Caprimulgus europaeus* populations in upland forests in Yorkshire. *Bird Study* 45: 219–225.

Sears, J & Hunt, A 1991. Lead poisoning in Mute Swans, *Cygnus olor*, in England. In Sears, J & Bacon, PJ (eds) *Proceedings of the 3rd IWRB International Swan Symposium*, Wildfowl Supplement No. 1, pp 383–388.

Sekercioğlu, CH, Daily, GC & Ehrlich, PR 2004. Ecosystem consequences of bird declines. *Proceedings of the National Academy of Sciences, USA* 101: 18 042–18 047.

Selas, V 1993. Selection of avian prey by breeding Sparrowhawks *Accipiter nisus* in Southern Norway: the importance of size and foraging behaviour of prey. *Ornis Fennica* 70: 144–154.

Semere, T & Slater, FM 2007. Ground flora, small mammal and bird species diversity in miscanthus (*Miscanthus × giganteus*) and reed canary-grass (*Phalaris arundinacea*) fields. *Biomass and Bioenergy* 31: 20–29.

Seymour, AS, Harris, S, Ralston, C & White, PCL 2003. Factors influencing the nesting success of Lapwings *Vanellus vanellus* and behaviour of Red Fox *Vulpes vulpes* in Lapwing nest sites. *Bird Study* 50: 39–46.

Sharrock, JTR 1974. The changing status of breeding birds in Britain and Ireland. In Hawksworth, DL (ed) *The Changing Flora and Fauna of Britain*, pp 203–220. London: Academic Press.

Sharrock, JTR 1976. *The Atlas of Breeding Birds in Britain and Ireland*. Tring: British Trust for Ornithology.

Sheldon, R 2002a. The breeding success and chick survival of Lapwings *Vanellus vanellus* in arable landscapes: with reference to the Arable Stewardship Pilot Scheme. Unpublished PhD thesis, Harper Adams University College.

Sheldon, R 2002b. Lapwings in Britain: a new approach to their conservation. *British Wildlife* 13: 109–116.

Sheldon, RD, Chaney, K & Tyler, GA 2007. Factors affecting nest survival of Northern Lapwings *Vanellus vanellus* in arable farmland: an agri-environment scheme prescription can enhance nest survival. *Bird Study* 54: 168–175.

Shkedy, Y & Safriel, UN 1992. Nest predation and nestling growth-rate of two lark species in the Negev Desert. *Ibis* 134: 268–272.

Shoard, M 1980. *The Theft of the Countryside*. London: Temple Smith.

Shore, RF, Walker, LA, Turk, CL, Wienburg, J, Wright, J, Murk, A & Wanless, S 2006. *Wildlife and Pollution: 2003/04 Annual Report*, JNCC Report No. 391. Peterborough: Joint Nature Conservation Committee.

Shrubb, M 1988. The influence of crop rotation and field size on a wintering Lapwing *Vanellus vanellus* population in an area of mixed farmland in West Sussex. *Bird Study* 35: 123–131.

Shrubb, M 1990. Effects of agricultural change on nesting Lapwings *Vanellus vanellus* on farmland. *Bird Study* 37: 115–128.

Shrubb, M 1997. Historical trends in British and Irish Corn Bunting *Miliaria calandra* populations: evidence for the effects of agricultural change. In Donald, PF & Aebischer, NJ (eds) *The Ecology and Conservation of Corn Buntings* Miliaria calandra, UK Nature Conservation No. 13, pp 27–41. Peterborough: Joint Nature Conservation Committee.

Shrubb, M 2003. *Birds, Scythes and Combines: A History of Birds and Agricultural Change*. Cambridge University Press.

Shrubb, M & Lack, PC 1991. The numbers and distribution of Lapwings *V. vanellus* nesting in England and Wales in 1987. *Bird Study* 38: 20–37.

Sim, IMW, Gregory, RD, Hancock, MH & Brown, AF 2005. Recent changes in the abundance of British upland breeding birds. *Bird Study* 52: 261–275.

Sim, IMW, Burfield, IJ, Grant, MC, Pearce-Higgins, JW & Brooke, M de L 2007a. The role of habitat composition in determining breeding site occupancy in a declining Ring Ouzel *Turdus torquatus* population. *Ibis* 149: 378–385.

Sim, IMW, Dillon, IA, Eaton, MA, Etheridge, B, Lindley, P, Riley, H, Saunders, R, Sharpe, C & Tickner, M 2007b. Status of the Hen Harrier *Circus cyaneus* in the UK and Isle of Man in 2004, and a comparison with the 1988/89 and 1998 surveys. *Bird Study* 54: 256–267.

Sim, IMW, Eaton, MA, Setchfield, R, Warren, P & Lindley, P 2008. Abundance of male Black Grouse *Tetrao tetrix* in Britain in 2005, and change since 1995. *Bird Study*, in press.

Simms, E 1971. *Woodland Birds*. London: Collins.

Simms, E 1978. *British Thrushes*. London: Collins.

Siriwardena, GM, Baillie, SR, Buckland, ST, Fewster, RM, Marchant, JH & Wilson, JD 1998a. Trends in the abundance of farmland birds: a quantitative comparison of smoothed Common Birds Census indices. *Journal of Applied Ecology* 35: 24–43.

Siriwardena, GM, Baillie, SR & Wilson, JD 1998b. Variation in survival rates of British farmland passerines with respect to their population trends. *Bird Study* 45: 276–292.

Siriwardena, GM, Baillie, SR & Wilson, JD 1999. Temporal variation in the annual survival rates of six granivorous passerines with contrasting population trends. *Ibis* 141: 621–636.

Siriwardena, GM, Baillie, SR, Crick, HQP & Wilson, JD 2000a. The importance of variation in the breeding performance of seed-eating birds in determining their population trends on farmland. *Journal of Applied Ecology* 37: 128–148.

Siriwardena, GM, Baillie, SR, Crick, HQP, Wilson, JD & Gates, S 2000b. The demography of lowland farmland birds. In Aebsicher, NJ, Evans, AD, Grice, PV & Vickery, JA (eds) *Ecology and Conservation of Lowland Farmland Birds*, pp 117–133. Tring: British Ornithologists' Union.

Siriwardena, GM, Crick, HQP, Baillie, SR & Wilson, JD 2000c. Agricultural land-use and the spatial distribution of granivorous lowland farmland birds. *Ecography* 23: 702–719.

Siriwardena, GM, Crick, HQP, Baillie, SR & Wilson, JD 2000d. Agricultural habitat-type and the breeding performance of granivorous farmland birds in Britain. *Bird Study* 47: 66–81.

Siriwardena, GM, Baillie, SR, Crick, HQP & Wilson, JD 2001a. Changes in agricultural land-use and breeding performance of some granivorous farmland passerines in Britain. *Agriculture, Ecosystems and Environment* 84: 191–206.

Siriwardena, GM, Freeman, SN and Crick, HQP 2001b. The decline of the Bullfinch *Pyrrhula pyrrhula* in Britain: is the mechanism known? *Acta Ornithologica* 36: 143–152.

Siriwardena, GM, Calbrade, NA, Vickery, JA & Sutherland, WJ 2006. The effect of the spatial distribution of winter seed food resources on their use by farmland birds. *Journal of Applied Ecology* 43: 628–639.

Siriwardena, GM, Stevens, DK, Anderson, GQA, Vickery, JA, Calbrade, NA & Dodd, S 2007. The effect of supplementary winter seed food on breeding populations of farmland birds: evidence from two large-scale experiments. *Journal of Applied Ecology* 44: 920–932.

Siriwardena, GM, Calbrade, NA & Vickery, JA 2008. Farmland birds and late winter seed supply: does seed supply fail to meet demand? *Ibis* 150: 585–595.

Small, CJ 2002. Waders, habitats and landscape in the Pennine Dales. Unpublished PhD thesis, University of Lancaster.

Smart, J, Gill, JA, Sutherland, WJ & Watkinson, AR 2006. Grassland-breeding waders: identifying key habitat requirements for management. *Journal of Applied Ecology* 43: 454–463.

Smart, SM, Firbank, LG, Bunce, RGH & Watkins, JW 2000. Quantifying changes in abundance of food plants for butterfly larvae and farmland birds. *Journal of Applied Ecology* 37: 398–414.

Smart, SM, Bunce, RGH, Marrs, R, LeDuc, M, Firbank, LG, Maskell, LC, Scott, WA, Thompson, K & Walker, KJ 2005. Large-scale changes in the abundance of common higher plant species across Britain between 1987, 1990 and 1998 as a consequence of human activity: tests of hypothesised changes in trait representation. *Biological Conservation* 124: 355–371.

Smith, A, Redpath, S & Campbell, S 2000. *The Influence of Moorland Management on Grouse and their Predators*. London: The Stationery Office.

Smith, AA, Redpath, SM, Campbell, ST & Thirgood, SJ 2001. Meadow pipits, red grouse and the habitat characteristics of managed grouse moors. *Journal of Applied Ecology* 38: 390–400.

Smith, HG & Bruun, M 2002. The effect of pasture on starling (*Sturnus vulgaris*) breeding success and population density in a heterogeneous agricultural landscape in southern Sweden. *Agriculture, Ecosystems and Environment* 92: 107–114.

Smith, KW 1983. The status and distribution of waders breeding on lowland wet grassland in England and Wales. *Bird Study* 30: 177–192.

Smith, P 2004. Carbon sequestration in croplands: the potential in Europe and the global context. *European Journal of Agronomy* 20: 229–236.

Smith, P, Andren, O, Karlsson, T, Perala, P, Regina, K, Rounsevell, M & vanWesemael, B 2005. Carbon sequestration potential in European croplands has been over-estimated. *Global Change Biology* 11: 2153–2163.

Smith, RK, Jennings, NV, Robinson, A & Harris, S 2004. Conservation of European hares *Lepus europaeus* in Britain: is increasing habitat heterogeneity in farmland the answer? *Journal of Applied Ecology* 41: 1092–1102.

Smith, RS & Jones, L 1991. The phenology of mesotrophic grassland in the Pennine Dales, northern England: historic hay cutting dates, vegetation variation and plant species phenologies. *Journal of Applied Ecology* 28: 42–59.

Smout, TC 2000. *Nature Contested: Environmental History in Scotland and Northern England since 1600*. Edinburgh University Press.

Sotherton, NW 1982a. Observations on the biology and ecology of the chrysomelid beetle *Gastrophysa polygoni* in cereals. *Ecological Entomology* 7: 197–206.

Sotherton, NW 1982b. Effects of herbicides on the chrysomelid beetle *Gastrophysa polygoni* (L.) in laboratory and field. *Zeitschrift für Angewandte Entomologie* 94: 446–451.

Sotherton, NW 1991. Conservation headlands: a practical combination of intensive cereal farming and conservation. In Firbank, LG, Carter, N, Darbyshire, JF & Potts, GR (eds) *The Ecology of Temperate Cereal Fields*, pp 373–397. Oxford: Blackwell Scientific Publications.

Sotherton, NW 1992. The environmental benefits of conservation headlands in cereal fields. *Outlook on Agriculture* 21: 219–224.

Sotherton, NW 1998. Land use changes and the decline of farmland wildlife: an appraisal of the set-aside approach. *Biological Conservation* 83: 259–268.

Sotherton, NW, Boatman, ND & Rands, MRW 1989. The 'Conservation Headland' experiment in cereal ecosystems. *Entomologist* 108: 135–143.

Southwood, TRE & Cross, DJ 1969. The ecology of the Partridge. 3. Breeding success and the abundance of insects in natural habitats. *Journal of Animal Ecology* 38: 497–509.

Spaepen, JF 1995. A study of the migration of the skylark *Alauda arvensis* based on European ringing data. *Le Gerfaut* 85: 63–89.

Sparks, TH & Martin, T 1999. Yields of hawthorn *Crataegus monogyna* berries under different hedgerow management. *Agriculture, Ecosystems and Environment* 72: 107–110.

Sparks, TH, Parish, T & Hinsley, SA 1996. Breeding birds in field boundaries in an agricultural landscape. *Agriculture, Ecosystems and Environment* 60: 1–8.

Spurgeon, J 1998. The socio-economic costs and benefits of coastal habitat rehabilitation and creation. *Marine Pollution Bulletin* 37: 373–382.

Squire, GR, Brooks, DR, Bohan, DA, Champion, GT, Daniels, RE, Haughton, AJ, Hawes, C, Heard, MS, Hill, MO, May, MJ, Osborne, JL, Perry, JN, Roy, DB, Woiwod, IP & Firbank, LG 2003. On the rationale and interpretation of the Farm Scale Evaluations of genetically modified herbicide-tolerant crops. *Philosophical Transactions of the Royal Society of London B* 358: 1779–1799.

Staines, BW, Balharry, R & Welch, D 1995. The impact of red deer and their management on the natural heritage in the uplands. In Thompson, DBA, Hester, AJ & Usher, MB (eds) *Heaths and Moorland: Cultural Landscapes*, pp 294–308. Edinburgh: Her Majesty's Stationery Office.

Stamp, D 1969. *Nature Conservation in Britain*. London: Collins.

Stanbury, A 2002. Bird communities on chalk grassland: a case study of Salisbury Plain Training Area. *British Wildlife* 13: 344–350.

Starling-Westerberg, A 2001. The habitat use and diet of Black Grouse *Tetrao tetrix* in the Pennine hills of northern England. *Bird Study* 48: 76–89.

Stephens, PA, Freckleton, RP, Watkinson, AR & Sutherland, WJ 2003. Predicting the response of farmland bird populations to changing food supplies. *Journal of Applied Ecology* 40: 970–983.

Stevens, CJ, Dise, NB, Mountford, JO & Gowing, DJ 2004. Impact of nitrogen deposition on the species richness of grasslands. *Nature* 303: 1876–1879.

Stevens, DK & Bradbury, RB 2006. Effects of the Arable Stewardship Pilot Scheme on breeding birds at field and farm-scales. *Agriculture, Ecosystems and Environment* 112: 283–290.

Stevens, DK, Donald, PF, Evans, AD, Buckingham, DL & Evans, J 2002. Territory distribution and foraging patterns of Cirl Buntings (*Emberiza cirlus*) breeding in the UK. *Biological Conservation* 107: 307–313.

Still, E, Monaghan, P & Bignal, E 1987. Social structuring at a communal roost of Choughs *Pyrrhocorax pyrrhocorax*. *Ibis* 129: 398–403.

Stillman, RA & Brown, AF 1994. Population size and habitat associations of upland breeding birds in the south Pennines, England. *Biological Conservation* 69: 307–314.

Stillman, RA, MacDonald, MA, Bolton, MR, le V dit Durell, SEA, Caldow, RG & West, AD 2006. *Management of Wet Grassland Habitat to Reduce the Impact of Predation on Breeding Waders: Phase 1*. Unpublished CEH/RSPB report to Defra.

Stiven, R & Holl, K 2004. *Wood-pasture*. Edinburgh: Scottish Natural Heritage.

Stoate, C 1995. The changing face of lowland farming and wildlife. 1. 1845–1945. *British Wildlife* 6: 341–350.

Stoate, C 1996. The changing face of lowland farming and wildlife. 2. 1945–1995. *British Wildlife* 7: 162–172.

Stoate, C & Szczur, J. 2001. Whitethroat *Sylvia communis* and yellowhammer *Emberiza citrinella* nesting success and breeding distribution in relation to field boundary vegetation. *Bird Study* 48: 229–235.

Stoate, C, Moreby, SJ & Szczur, J 1998. Breeding ecology of farmland yellowhammers *Emberiza citrinella*. *Bird Study* 45: 109–121.

Stoate, C, Szczur, J & Aebischer, NJ 2003. Winter use of wild bird cover crops by passerines on farmland in northeast England. *Bird Study* 50: 15–21.

Stoate, C, Henderson, IG & Parish, DMB 2004. Development of an agri-environment scheme option: seed-bearing crops for farmland birds. *Ibis* 146 (Suppl. 2): 203–209.

Stokstad, E 2005. What's wrong with the Endangered Species Act? *Science* 309: 2150–2152.

Stowe, TJ, Newton, AV, Green, RE & Mayes, E 1993. The decline of the corncrake *Crex crex* in Britain and Ireland in relation to habitat. *Journal of Applied Ecology* 30: 53–62.

Strong, L & Wall, R 1994. Effects of ivermectin and moxidectin on the insects of cattle dung. *Bulletin of Entomological Research* 84: 403–409.

Stroud, DA 2002. The UK network of Special Protection Areas for Birds. *British Wildlife* 14: 7–14.

Stroud, DA, Reed, TM, Pienkowski, MW & Lindsay, RA 1987. *Birds, Bogs and Forestry*. Peterborough: Nature Conservancy Council.

Stroud, DA, Reed, TM & Harding, NJ 1990. Do moorland breeding waders avoid plantation edges? *Bird Study* 37: 177–186.

Stroud, DA, Chambers, D, Cook, S, Buxton, N, Fraser, B, Clement, P, Lewis, P, McLean, I, Baker, H & Whitehead, S (eds) 2001. *The UK SPA Network: Its Scope and Content*. Peterborough: Joint Nature Conservation Committee.

Suhonen, J, Norrdahl, K & Korpimäki, E 1994. Avian predation risk modifies breeding bird community on a farmland area. *Ecology* 75: 1626–1634.

Summers, RW 1990. The effect of grazing on winter wheat by brent geese *Branta b. bernicla*. *Journal of Applied Ecology* 27: 821–833.

Summers, RW, Green, RE, Proctor, R, Dugan, D, Lambie, D, Moncrieff, R, Moss, R & Baines, D 2004. An experimental study of the effects of predation on the breeding productivity of capercaillie and black grouse. *Journal of Applied Ecology* 41: 513–525.

Sutcliffe, OL & Kay, QON 2000. Changes in the arable flora of central southern England since the 1960s. *Biological Conservation* 93: 1–8.

Sutherland, WJ 2002. Restoring a sustainable countryside. *Trends in Ecology and Evolution* 17: 148–150.

Sutherland, WJ 2004. A blueprint for the countryside. *Ibis* 146 (Suppl. 2): 230–238.

Sutherland, WJ, Armstrong-Brown, S, Armsworth, PR, Brereton, T, Brickland, J, Campbell, CD, Chamberlain, DE, Cooke, AI, Dulvy, NK, Dusic, NR, Fitton, M, Freckleton, RP, Godfray, CJ, Grout, N, Harvey, HJ, Hedley, C, Hopkins, JJ, Kift, NB, Kirby, J, Kunin, WE, Macdonald, DW, Marker, B, Naura, M, Neale, AR, Oliver, T, Osborn, D, Pullin, AS, Shardlow, MEA, Showler, DA, Smith, PL, Smithers, RJ, Solandt, J-L, Spencer, J, Spray, CJ, Thomas, CD, Thompson, J, Webb, SE, Yalden, DW & Watkinson, AR 2006. The identification of 100 ecological questions of high policy relevance in the UK. *Journal of Applied Ecology* 43: 617–627.

Sutherland, WJ, Bailey, MJ, Bainbridge, IP, Brereton, T, Dick, JTA, Drewitt, J, Dulvy, NK, Dusic, NR, Freckleton, RP, Gaston, KJ, Gilder, PM, Green, RE, Heathwaite, AL, Johnson, SM, Macdonald, DW, Mitchell, R, Osborn, D, Owen, RP, Pretty, J, Prior, SV, Prosser, H, Pullin, AS, Rose, P, Stott, A, Tew, T, Thomas, CD, Thompson, DBA, Vickery, JA, Walker, M, Walmsley, C, Warrington, S, Watkinson, AR, Williams, RJ, Woodroffe, R & Woodroof,

HJ 2008. Future novel threats and opportunities facing UK biodiversity identified by horizon scanning. *Journal of Applied Ecology* 45: 821–833.

Swash, A, Grice, PV & Smallshire, D 2000. The contribution of the UK Biodiversity Action Plan and agri-environment schemes to the conservation of farmland birds in England. In Aebsicher, NJ, Evans, AD, Grice, PV & Vickery, JA (eds) *Ecology and Conservation of Lowland Farmland Birds*, pp 36–42. Tring: British Ornithologists' Union.

Swetnam, RD, Mountford, JO, Manchester, SJ & Broughton, RK 2004. Agri-environmental schemes: their role in reversing floral decline in the Brue floodplain, Somerset, UK. *Journal of Environmental Management* 71: 79–93.

Swetnam, RD, Wilson, JD, Whittingham, MJ & Grice, PV 2005. Designing lowland landscapes for farmland birds: scenario testing with GIS. *Computers, Environment and Urban Systems* 29: 275–296.

Sydes, C & Miller, GR 1988. Range management and nature conservation in the British uplands. In Usher, MB & Thompson, DBA (eds) *Ecological Change in the Uplands*, pp 323–338. Oxford: Blackwell Scientific Publications.

Symon, JA 1959. *Scottish Farming*. Edinburgh: Oliver & Boyd.

Tallowin, JRB, Williams, JHH & Kirkham, FW 1989. Some consequences of imposing different continuous grazing pressures in the spring on tiller demography and leaf growth. *Journal of Agricultural Science* 112: 115–122.

Tallowin, JRB, Smith, REN, Goodyear, J & Vickery, JA 2005. Spatial and structural uniformity of lowland agricultural grassland in England: a context for low biodiversity. *Grass and Forage Science* 60: 225–236.

Tapper, SC 1992. *Game Heritage: An Ecological Review from Shooting and Game-Keeping Records*. Fordingbridge: Game Conservancy Ltd.

Tapper, S (ed) 1999. *A Question of Balance: Game Animals and Their Role in the British Countryside*. Fordingbridge: Game Conservancy Trust.

Tapper, SC 2001. *Conserving the Grey Partridge*. Fordingbridge: Game Conservancy Trust.

Tapper, SC, Potts, GR & Brockless, MH 1996. The effect of an experimental reduction in predation pressure on the breeding success and population density of Grey Partridges (*Perdix perdix*). *Journal of Applied Ecology* 33: 965–978.

Taylor, IR & Grant, MC 2004. Long-term trends in the abundance of breeding Lapwing *Vanellus vanellus* in relation to land-use change on upland farmland in southern Scotland. *Bird Study* 51: 133–142.

Tharme, AP, Green, RE, Baines, D, Bainbridge, IP & O'Brien, M 2001. The effect of management for red grouse shooting on the population density of breeding birds on heather dominated moorland. *Journal of Applied Ecology* 38: 439–457.

Thies, C, Roschewitz, I & Tscharntke, T 2005. Landscape context of cereal aphid–parasitoid interactions. *Proceedings of the Royal Society of London B* 272: 203–210.

Thirgood, SJ, Redpath, S, Haydon, DT, Rothery, P, Newton, I & Hudson, P 2000a. Habitat loss and raptor predation: disentangling long and short term causes of red grouse declines. *Proceedings of the Royal Society of London B* 267: 651–656.

Thirgood, SJ, Redpath, S, Newton, I & Hudson, P 2000b. Raptors and red grouse: conservation conflicts and management solutions. *Conservation Biology* 14: 95–104.

Thomas, CD, Cameron, A, Green, RE, Bakkenes, M, Neaumont, LJ, Collingham, YC, Erasmus, BFN, de Siqueira, MF, Grainger, A, Hannah, L, Hughes, L, Huntley, B, van Jaarsveld, AS, Midgley, GF, Miles, L, Ortega-Huerta, MA, Peterson, AT, Phillips, OL & Williams, SE 2004. Extinction risk from climate change. *Nature* 427: 145–148.

Thomas, JA, Telfer, MG, Roy, DB, Preston, CD, Greenwood, JJD, Asher, J, Fox, R, Clarke, RT & Lawton, JH 2004. Comparative losses of British butterflies, birds, and plants and the global extinction crisis. *Science* 303: 1879–1881.

Thomas, MB, Wratten, SD & Sotherton, NW 1991. Creation of 'island' habitats in farmland to manipulate populations of beneficial arthropods: predator densities and emigration. *Journal of Applied Ecology* 28: 906–917.

Thomas, SR, Goulson, D & Holland, JM 2001. Resource provision for farmland gamebirds: the value of beetle banks. *Annals of Applied Biology* 139: 111–118.

Thompson, DBA & Brown, A 1992. Biodiversity in montane Britain: habitat variation, vegetation diversity and some objectives for nature conservation. *Biodiversity and Conservation* 1: 179–208.

Thompson, DBA & Gribbin, S 1986. Ecology of Corn Buntings (*Miliaria calandra*) in NW England. *Bulletin of the British Ecological Society* 17: 69–75.

Thompson, DBA, Stroud, DA & Pienkowski, MW 1988. Afforestation and upland birds: consequences for population ecology. In Usher, MB & Thompson, DBA (eds) *Ecological Change in the Uplands*, pp 237–259. Oxford: Blackwell Scientific Publications.

Thompson, DBA, Macdonald, AJ, Marsden, JH & Galbraith, CA 1995. Upland heather moorland in Great Britain: a review of international importance, vegetation change and some objectives for nature conservation. *Biological Conservation* 71: 163–178.

Thompson, DBA, Gillings, SD, Galbraith, CA, Redpath, SM & Drewitt, J 1997. The contribution of game management to biodiversity: a review of the importance of grouse moors for upland birds. In Fleming, V, Newton, AC, Vickery, JA & Usher, MB (eds) *Biodiversity in Scotland: Status, Trends and Initiatives*, pp 198–212. Edinburgh: The Stationery Office.

Thompson, FR, Donovan, TM, DeGraaf, RM, Faaborg, J & Robinson, SK 2002. A multi-scale perspective of the effects of forest fragmentation on birds in eastern forests. *Studies in Avian Biology* 25: 8–19.

Thompson, PS & Hale, WG 1993. Adult survival and numbers in a coastal breeding population of Redshank *Tringa totanus* in northwest England. *Ibis* 135: 61–69.

Thompson, PS, Baines, D, Coulson, JC & Longrigg, G 1994. Age at first breeding, philopatry and breeding site fidelity in the Lapwing *Vanellus vanellus*. *Ibis* 136: 474–484.

Thomson, DL 2002. Song Thrush *Turdus philomelos*. In Wernham, CV, Toms, MP, Marchant, JH, Clark, JA, Siriwardena, GM & Baillie, SR (eds) *The Migration Atlas: Movements of the Birds of Britain and Ireland*, pp 530–533. London: T. & A. D. Poyser.

Thomson, DL & Cotton, PA 2000. Understanding the decline of the British population of Song Thrushes *Turdus philomelos*. In Aebsicher, NJ, Evans, AD, Grice, PV & Vickery, JA (eds) *Ecology and Conservation of Lowland Farmland Birds*, pp 151–155. Tring: British Ornithologists' Union.

Thomson, DL, Baillie, SR & Peach, WJ 1997. The demography and age-specific annual survival of British Song Thrushes *Turdus philomelos* during periods of population stability and decline. *Journal of Animal Ecology* 66: 414–424.

Thomson, DL, Green, RE, Gregory, RD & Baillie, SR 1998. The widespread declines of songbirds in rural Britain do not correlate with the spread of their avian predators. *Proceedings of the Royal Society of London B* 265: 2057–2062.

Thomson, DL, Baillie, SR & Peach, WJ 1999. A method for studying post-fledging survival rates using data from ringing recoveries. *Bird Study* 46: S104–S112.

Thorup, O (comp.) 2006. *Breeding Waders in Europe 2000*, International Wader Studies No. 14. UK: International Wader Study Group.

Tilman, D, Fargione, J, Wolff, B, D'Antonio, C, Dobson, A, Howarth, R, Schindler, D, Schlesinger, WH, Simberloff, D & Swackhamer, D 2001. Forecasting agriculturally driven global environmental change. *Science* 292: 281–284.

Tilman, D, Cassman, KG, Matson, PA, Naylor, P & Polarsky, S 2002. Agricultural sustainability and intensive production practices. *Nature* 418: 671–677.

Toms, MP, Crick, HQP & Shawyer, CR 2001. The status of breeding Barn Owls *Tyto alba* in the United Kingdom 1995–1997. *Bird Study* 48: 23–37.

Tracy, M 1989. *Government and Regulation in Western Europe, 1880–1988*. New York: New York University Press.

Tscharntke, T, Klein, AM, Kruess, A, Steffan-Dewenter, I & Thies, C 2005. Landscape perspectives on agricultural intensification and biodiversity: ecosystem service management. *Ecology Letters* 8: 857–874.

Tubbs, CR 1991. Grazing the lowland heaths. *British Wildlife* 2: 276–289.

Tucker, GM 1992. Effects of agricultural practices on field use by invertebrate-feeding birds in winter. *Journal of Applied Ecology* 29: 779–790.

Tucker, GM & Evans, MI 1997. *Habitats for Birds in Europe: A Conservation Strategy for the Wider Environment*, BirdLife Conservation Series No. 6. Cambridge: BirdLife International.

Tucker, GM & Heath, MF 1994. *Birds in Europe: Their Conservation Status*, BirdLife Conservation Series No. 3. Cambridge: BirdLife International.

Tyler, GA & Green, RE 1996. The incidence of nocturnal song by male Corncrakes *Crex crex* is reduced during pairing. *Bird Study* 43: 214–219.

Tyler, GA, Green, RE & Casey, C 1998. Survival and behaviour of Corncrake *Crex crex* chicks during the mowing of agricultural grassland. *Bird Study* 45: 35–50.

Underhill-Day, JC 1998. Breeding Marsh Harriers in the United Kingdom, 1983–95. *British Birds* 91: 210–218.

van der Waal, R, Pearce, I, Brooker, R, Scott, D, Welch, D & Woodin, S 2003. Interplay between nitrogen deposition and grazing causes habitat degradation. *Ecology Letters* 6: 141–146.

Vanhinsbergh, D & Evans, A 2002. Habitat associations of the Red-backed Shrike (*Lanius collurio*) in Carinthia, Austria. *Journal of Ornithology* 143: 405–415.

Vera, FWM 2000. *Grazing Ecology and Forest History*. Wallingford: CABI Publishing.

Verhulst, J, Kleijn, D & Berendse, F 2007. Direct and indirect effects of the most widely implemented Dutch agri-environment schemes on breeding waders. *Journal of Applied Ecology* 44: 70–80.

Vickery, JA & Buckingham, DL 2001. The value of set-aside for skylarks *Alauda arvensis* in Britain. In Donald, PF & Vickery, JA (eds) *The Ecology and Conservation of Skylarks* Alauda arvensis, pp 161–175. Sandy: Royal Society for the Protection of Birds.

Vickery, JA & Gill, JA 1999. Managing grassland for wild geese in Britain: a review. *Biological Conservation* 89: 93–106.

Vickery, JA, Sutherland, WJ & Lane, SJ 1994a. The management of grass pastures for brent geese. *Journal of Applied Ecology* 31: 282–290.

Vickery, JA, Watkinson, AR & Sutherland, WJ 1994b. The solutions to the brent goose problem: an economic analysis. *Journal of Applied Ecology* 31: 371–382.

Vickery, JA, Tallowin, JR, Feber, RE, Asteraki, EJ, Atkinson, PW, Fuller, RJ & Brown, VK 2001. The management of lowland neutral grasslands in Britain: effects of agricultural practices on birds and their food sources. *Journal of Applied Ecology* 38: 647–664.

Vickery, JA, Carter, N & Fuller, RJ 2002. The potential value of managed cereal field margins as foraging habitats for farmland birds in the UK. *Agriculture, Ecosystems and Environment* 89: 41–52.

Vickery, JA, Bradbury, RB, Henderson, IG, Eaton, MA & Grice, PV 2004. The role of agri-environment schemes and farm management practices in reversing the decline of farmland birds in England. *Biological Conservation* 119: 19–39.

Vickery, J, Chamberlain, D, Evans, A, Ewing, S, Boatman, N, Pietravalle, S, Norris, K & Butler, S 2007. *Predicting the Impact of Future Agricultural Change and the Uptake of Entry Level Stewardship on Farmland Birds*, BTO Research Report No. 485. Thetford: British Trust for Ornithology.

Visser, ME, van Noordwijk, AJ, Tinbergen, JM & Lessells, CM 1998. Warmer springs lead to mistimed reproduction in great tits (*Parus major*). *Proceedings of the Royal Society of London B* 265: 1867–1870.

Wakeham-Dawson, A & Aebsicher, NJ 1998. Factors determining winter densities of birds on Environmentally Sensitive Area arable reversion grassland in southern England, with special reference to skylarks (*Alauda arvensis*). *Agriculture, Ecosystems and Environment* 70: 189–201.

Wakeham-Dawson, A & Aebischer, NJ 2001. Management of grassland for skylarks *Alauda arvensis* in downland Environmentally Sensitive Areas in southern England. In Donald, PF & Vickery, JA (eds) *The Ecology and Conservation of Skylarks* Alauda arvensis, pp 189–201. Sandy: Royal Society for the Protection of Birds.

Wakeham-Dawson, A, Szoskiewicz, K, Stern, K & Aebischer, NJ 1998. Breeding skylarks *Alauda arvensis* on Environmentally Sensitive Area arable reversion grass in southern England: survey-based and experimental determination of density. *Journal of Applied Ecology* 35: 635–648.

Walker, KJ, Critchley, CNR, Sherwood, AJ, Large, R, Nuttall, P, Hulmes, S, Rose, R & Mountford, JO 2007. The conservation of arable plants on cereal field margins: an assessment of new agri-environment scheme options in England, UK. *Biological Conservation* 136: 260–270.

Walker, MP, Dover, JW, Hinsley, SA & Sparks, TH 2005. Birds and green lanes: breeding season bird abundance, territories and species richness. *Biological Conservation* 126: 540–547.

Walpole-Bond, J 1938. *A History of Sussex Birds*. London: Witherby.

Walter, C, Merot, P, Layer, B & Dutin, G 2003. The effect of hedgerows on soil organic carbon storage in hillslopes. *Soil Use and Management* 19: 201–207.

Warren, MS, Hill, JK, Thomas, JA, Asher, J, Fox, R, Huntley, B, Roy, DB, Telfer, MG, Jeffcoate, S, Harding, P, Jeffcoate, G, Willis, SG, Greatorex-Davies, JN, Moss, D & Thomas, CD 2001. Rapid responses of British butterflies to opposing forces of climate and habitat change. *Nature* 414: 65–69.

Warren, P & Baines, D 2008. Current status and distribution of black grouse *Tetrao tetrix* in northern England. *Bird Study* 55: 94–99.

Watkinson, AR, Freckleton, RP, Robinson, RA & Sutherland, WJ 2000. Predictions of biodiversity response to genetically modified herbicide-tolerant crops. *Science* 289: 1554–1557.

Watson, A 1992. Unripe grain, a major food for young finch-like birds. *Scottish Birds* 16: 287.

Watson, A & Miller, GR 1976. *Grouse Management*, The Game Conservancy Booklet No. 12. Banchory: Institute of Terrestrial Ecology.

Watson, A & Rae, R 1997. Some effects of set-aside on breeding birds in northeast Scotland. *Bird Study* 44: 245–251.

Watson, A & Rae, R 1998. Use by birds of rape fields in east Scotland. *British Birds* 91: 144–145.

Watson, J, Rae, SR & Stillman, R 1992. Nesting density and breeding success of golden eagles in relation to food supply in Scotland. *Journal of Animal Ecology* 61: 543–550.

Watson, M, Aebischer, NJ, Potts, GR & Ewald, JA 2007. The relative effects of raptor predation and shooting on overwinter mortality of grey partridges in the United Kingdom. *Journal of Applied Ecology* 44: 972–982.

Watt, TA & Buckley, GP 1994. *Hedgerow Management and Nature Conservation*. Ashford: Wye College Press.

Webb, L, Beaumont, DJ, Nager, RG & McCracken, DI 2007. Effects of avermectin residues in cattle dung on yellow dung fly *Scathophaga stercoraria* (Diptera: Scathophagidae) populations in grazed pastures. *Bulletin of Entomological Research* 97: 129–138.

Webb, N 1986. *Heathlands*. London: Collins.

Webber, MI 1975. Some aspects of the non-breeding population dynamics of the Great Tit (*Parus major*). Unpublished DPhil thesis, University of Oxford.

Weibel, U 1998. Habitat use of foraging skylarks (*Alauda arvensis* L.) in an arable landscape with wild flower strips. *Bulletin of the Geobotanical Institute, ETH* 64: 37–45.

Welch, D 1998. Response of bilberry *Vaccinium myrtillus* L. stands in the Derbyshire Peak District to sheep grazing, and implications for moorland conservation. *Biological Conservation* 83: 155–164.

Westerhoff, DV & Tubbs, CR 1991. Dartford Warblers *Sylvia undata*, their habitat and conservation in the New Forest, Hampshire, England, in 1988. *Biological Conservation* 56: 89–100.

Wheeler, P 2008. Effects of sheep grazing on abundance and predators of field vole (*Microtus agrestis*) in upland Britain. *Agriculture, Ecosystems and Environment* 123: 49–55.

White, G 1789. *The Natural History of Selborne* (1936 abridged edn). London: Methuen.

Whitehead, SC, Wright, J & Cotton, PA 1995. Winter field use by the European Starling *Sturnus vulgaris*: habitat preferences and the availability of prey. *Journal of Avian Biology* 26: 193–202.

Whitehead, S, Johnstone, I & Wilson, JD 2005. Choughs *Pyrrhocorax pyrrhocorax* breeding in Wales select foraging habitat at different spatial scales. *Bird Study* 52: 193–203.

Whitfield, DP, McLeod, DRA, Fielding, AH, Broad, RA, Evans, RJ & Haworth, PF 2001. The effects of forestry on golden eagles on the island of Mull, western Scotland. *Journal of Applied Ecology* 38: 1208–1220.

Whitfield, DP, McLeod, DRA, Watson, J, Fielding, AH & Haworth, PF 2003. The association of grouse moor with the illegal use of poisons to control predators. *Biological Conservation* 114: 157–163.

Whitfield, DP, Fielding, AH, McLeod, DRA & Haworth, PF 2004a. The effects of illegal persecution on age of breeding and territory occupation in golden eagles in Scotland. *Biological Conservation* 118: 249–259.

Whitfield, DP, Fielding, AH, McLeod, DRA & Haworth, PF 2004b. Modelling the effects of persecution on the population dynamics of golden eagles in Scotland. *Biological Conservation* 119: 319–333.

Whitlock, RE, Aebischer, NJ & Reynolds, JC 2003. *The National Gamebag Census as a Tool for Monitoring Mammalian Abundance in the UK*. Fordingbridge: Game Conservancy Trust.

Whittingham, MJ 1996. The habitat requirements of breeding Golden Plover. Unpublished PhD thesis, University of Sunderland.

Whittingham, MJ 2007. Will agri-environment schemes deliver substantial biodiversity gain, and if not, why not? *Journal of Applied Ecology* 44: 1–5.

Whittingham, MJ & Evans, K 2004. The effects of habitat structure on predation risk of birds in agricultural landscapes. *Ibis* 146 (Suppl. 2): 210–220.

Whittingham, MJ, Percival, SM & Brown, AF 1999. Evaluation of radio telemetry methods in measuring habitat choice by young Golden Plover *Pluvialis apricaria* chicks. *Bird Study* 46: 363–368.

Whittingham, MJ, Percival, SM & Brown, AF 2000. Time budgets and foraging of breeding golden plover *Pluvialis apricaria*. *Journal of Applied Ecology* 37: 632–646.

Whittingham, MJ, Bradbury, RB, Wilson, JD, Morris, AJ, Perkins, AJ & Siriwardena, GM 2001a. Chaffinch *Fringilla coelebs* foraging patterns, nestling survival and territory density on lowland farmland. *Bird Study* 48: 257–270.

Whittingham, MJ, Percival, SM & Brown, AF 2001b. Habitat selection by golden plover *Pluvialis apricaria* chicks. *Basic and Applied Ecology* 2: 177–191.

Whittingham, MJ, Percival, SM & Brown, AF 2002. Nest-site selection by golden plover: why do shorebirds avoid nesting on slopes? *Journal of Avian Biology* 33: 184–190.

Whittingham, MJ, Butler, SJ, Quinn, JL & Cresswell, W 2004. The effect of limited visibility on vigilance behaviour and speed of predator detection: implications for the conservation of granivorous passerines. *Oikos* 106: 377–385.

Whittingham, MJ, Devereux, CL, Evans, AD & Bradbury, RB 2006. Altering perceived predation risk and food availability: management prescriptions to benefit farmland birds on stubble fields. *Journal of Applied Ecology* 43: 640–650.

Whittingham, MJ, Krebs, JR, Swetnam, RD, Vickery, JA, Wilson, JD & Freckleton, RP 2007. Should conservation strategies consider spatial generality? Farmland birds show regional not national patterns of habitat association. *Ecology Letters* 10: 25–35.

Whittingham, MJ, Krebs, JR, Swetnam, RD, Thewlis, RM, Wilson, JD & Freckleton, RP 2008. Habitat associations of British breeding farmland birds. *Bird Study*, in press.

Wickramsinghe, LP, Harris, S, Jones, G & Vaughan, N 2003. Bat activity and species richness on organic and conventional farms: impact of agricultural intensification. *Journal of Applied Ecology* 40: 984–993.

Wickramsinghe, LP, Harris, S, Jones, G & Jennings, NV 2004. Abundance and species richness of nocturnal insects on organic and conventional farms: effects of agricultural intensification on bat foraging. *Conservation Biology* 18: 1283–1292.

Wiens, JA 1989. Spatial scaling in ecology. *Functional Ecology* 3: 385–397.

Williams, G, Green, RE, Casey, C, Deceuninck, B & Stowe, TJ 1997. Halting declines in globally threatened species: the case of the Corncrake. *RSPB Conservation Review* 11: 22–31.

Williams, P, Whitfield, M, Biggs, J, Bray, S, Fox, G, Nicolet, P & Sear, D 2003. Comparative biodiversity of rivers, streams, ditches and ponds in an agricultural landscape in southern England. *Biological Conservation* 115: 329–341.

Williamson, K 1969. Habitat preferences of the Wren on English farmland. *Bird Study* 16: 53–59.

Williamson, K 1972. Breeding birds of the Ariundle Oakwood Forest Nature Reserve. *Quarterly Journal of Forestry* 66: 243–255.

Wilson, AM, Vickery, JA & Browne, SJ 2001. Numbers and distribution of Northern Lapwings *Vanellus vanellus* breeding in England and Wales in 1998. *Bird Study* 48: 2–17.

Wilson, AM, Ausden, M & Milsom, TP 2004. Changes in breeding wader populations on lowland wet grassland in England and Wales: causes and potential solutions. *Ibis* 146 (Suppl. 2): 32–40.

Wilson, AM, Vickery, JA, Brown, A, Langston, RHW, Smallshire, D, Wotton, S & Vanhinsbergh, D 2005. Changes in the numbers of breeding waders on lowland wet grasslands in England and Wales between 1982 and 2002. *Bird Study* 52: 55–69.

Wilson, A, Vickery, J & Pendlebury, C 2007. Agri-environment schemes as a tool for reversing declining populations of grassland waders: mixed benefits from Environmentally Sensitive Areas. *Biological Conservation* 136: 128–135.

Wilson, G, Harris, S & McLaren, R 1997. *Changes in the British Badger Population 1988–1997*. London: People's Trust for Endangered Species.

Wilson, JD 2001. Foraging habitat selection by skylarks *Alauda arvensis* on lowland farmland during the nestling period. In Donald, PF & Vickery, JA (eds) *The Ecology and Conservation of Skylarks* Alauda arvensis, pp 91–101. Sandy: Royal Society for the Protection of Birds.

Wilson, JD 2002. Linnet *Carduelis cannabina*. In Wernham, CV, Toms, MP, Marchant, JH, Clark, JA, Siriwardena, GM & Baillie, SR (eds) *The Migration Atlas: Movements of the Birds of Britain and Ireland*, pp 654–656. London: T. & A. D. Poyser.

Wilson, JD & Fuller, RJ 1992. Set-aside: potential and management for wildlife conservation. *Ecos* 13: 24–29.

Wilson, JD, Evans, AD, Poulsen, JG & Evans, J 1995. Wasteland or oasis? The use of set-aside by breeding and wintering birds. *British Wildlife* 6: 214–223.

Wilson, JD, Arroyo, BE & Clark, SC 1996a. *The Diet of Bird Species of Lowland Farmland: A Review*, unpublished report to Joint Nature Conservation Committee and Department of the Environment. Sandy: Royal Society for the Protection of Birds.

Wilson, JD, Taylor, R & Muirhead, LB 1996b. Field use by farmland birds in winter: an analysis of field type preferences using resampling methods. *Bird Study* 43: 320–332.

Wilson, JD, Evans, J, Browne, SJ & King, JR 1997. Territory distribution and breeding success of skylarks *Alauda arvensis* on organic and intensive farmland in southern England. *Journal of Applied Ecology* 34: 1462–1478.

Wilson, JD, Morris, AJ, Arroyo, BE, Clark, SC & Bradbury, RB 1999. A review of the abundance and diversity of invertebrate and plant foods of granivorous birds in northern Europe in relation to agricultural change. *Agriculture, Ecosystems and Environment* 75: 13–30.

Wilson, JD, Whittingham, MJ & Bradbury, RB 2005. Managing crop structure: a general approach to reversing the impact of agricultural intensification on birds? *Ibis* 147: 453–463.

Wilson, J, Anderson, G, Perkins, A, Wilkinson, N & Maggs, H 2007a. Adapting agri-environment management to multiple drivers of decline of corn buntings *Emberiza calandra* across their UK range. *Aspects of Applied Biology* 81: 191–198.

Wilson, JD, Boyle, J, Jackson, DB, Lowe, B & Wilkinson, NI 2007b. Effect of cereal harvesting on a recent population decline of Corn Buntings *Emberiza calandra* on the Western Isles of Scotland. *Bird Study* 54: 362–370.

Wilson, PJ 1992. Britain's arable weeds. *British Wildlife* 3: 149–161.

van Wingerden, WKRE, van Kreveld, AR & Bongers, W 1992. Analysis of species composition and abundance of grasshoppers (Orth, Acrididae) in natural and fertilized grasslands. *Journal of Applied Entomology* 113: 138–152.

Witter, MS & Cuthill, IC 1993. The ecological costs of avian fat storage. *Philosophical Transactions of the Royal Society of London B* 340: 73–92.

Woiwod, IP & Harrington, R 1994. Flying in the face of change: the Rothamsted Insect Survey. In Leigh, RA & Johnston, AE (eds) *Long-Term Experiments in Agricultural and Ecological Sciences*, pp 321–342. Wallingford: CABI Publishing.

Wolfenden, IH & Peach, WJ 2001. Temporal survival rates of skylarks *Alauda arvensis* breeding in duneland in northwest England. In Donald, PF & Vickery, JA (eds) *The Ecology and Conservation of Skylarks* Alauda arvensis, pp 79–89. Sandy: Royal Society for the Protection of Birds.

Woodcock, BA, Potts, SG, Pilgrim, E, Ramsay, AJ, Tscheulin, T, Parkinson, A, Smith, REN, Gundrey, AL, Brown, VK, Tallowin, JR 2007. The potential of grass field margin management for enhancing beetle diversity in intensive livestock farms. *Journal of Applied Ecology* 44: 60–69.

Woodhouse, SP, Good, JEG, Lovett, AA, Fuller, RJ & Dolman, PM 2005. Effects of land-use and agricultural management on birds of marginal farmland: a case study in the Llŷn peninsula, Wales. *Agriculture, Ecosystems and Environment* 107: 331–340.

Worrall, F, Armstrong, A & Adamson, JK 2007. The effects of burning and sheep-grazing on water table depth and soil water quality in an upland peat. *Journal of Hydrology* 339: 1–14.

Wotton, SR & Gillings, S 2000. The status of breeding Woodlarks *Lullula arborea* in Britain in 1997. *Bird Study* 47: 212–224.

Wotton, SR & Peach, WJ 2008. *Population Changes and Summer Habitat Associations of Breeding Cirl Buntings* Emberiza cirlus *and Other Farmland Birds in Relation to Measures Provided through the Countryside Stewardship Scheme in Deron, England*, RSPB Research Report No. 30. Sandy: Royal Society for the Protection of Birds.

Wotton, SR, Langston, RHW, Gibbons, DW and Pierce, AJ 2000. The status of the Cirl Bunting in the United Kingdom and the Channel Islands in 1998. *Bird Study* 47: 138–146.

Wotton, SR, Langston, RHW & Gregory, RD 2002a. The breeding status of the Ring Ouzel *Turdus torquatus* in the UK in 1999. *Bird Study* 49: 26–34.

Wotton, SR, Carter, I, Cross, AV, Etheridge, B, Snell, N, Duffy, K, Thorpe, R & Gregory, RD 2002b. Breeding status of the Red Kite *Milvus milvus* in Britain in 2000. *Bird Study* 49: 278–286.

Wotton, S, Rylands, K, Grice, P, Smallshire, D & Gregory, R 2004. The status of the Cirl Bunting in Britain and the Channel Islands in 2003. *British Birds* 97: 376–384.

Wretenberg, J, Lindström, Å, Svensson, S, Thierfelder, T & Pärt, T 2006. Population trends of farmland birds in Sweden and England: similar trends but different patterns of agricultural intensification. *Journal of Applied Ecology* 43: 1110–1120.

Wretenberg, J, Lindström, Å, Svensson, S & Pärt, T 2007. Linking agricultural policies to population trends of Swedish farmland birds in different agricultural regions. *Journal of Applied Ecology* 44: 933–941.

Wright, LJ, Hoblyn, RA, Sutherland, WJ & Dolman, PM 2007. Reproductive success of Woodlarks *Lullula arborea* in traditional and recently colonized habitats. *Bird Study* 54: 315–323.

Wyllie, I 1995. The return of the Sparrowhawk in Rockingham Forest. *Northamptonshire Bird Report* 1995: 66–70.

Yalden, DW & Pearce-Higgins, JW 1997. Density-dependence and winter weather as factors affecting the size of a population of Golden Plovers *Pluvialis apricaria*. *Bird Study* 44: 227–234.

Yallop, AR, Thacker, JI, Thomas, G, Stephens, M, Clutterbuck, B, Brewer, T & Sannier, CAD 2006. The extent and intensity of management burning in the English uplands. *Journal of Applied Ecology* 43: 1138–1148.

Yom-Tov, Y 1992. Clutch size and laying dates of three species of Buntings *Emberiza* in England. *Bird Study* 39: 111–114.

Index

Printed in the United States
By Bookmasters